II International Scientific Conference on Advances in Material Science and Technology (CAMSTech-II 2021)

IOP Conference Series: Materials Science and Engineering
Volume 1181

Krasnoyarsk, Russia
29 – 31 July 2021

Editors:

Igor V. Kovalev
Alena A. Stupina
Anna A. Voroshilova

ISBN: 978-1-7138-4641-3
ISSN: 1757-8981

Printed from e-media with permission by:

Curran Associates, Inc.
57 Morehouse Lane
Red Hook, NY 12571

Some format issues inherent in the e-media version may also appear in this print version.

Institute of Physics
Dirac House, Temple Back
Bristol BS1 6BE UK

Phone: 44 1 17 929 7481
Fax: 44 1 17 920 0979

techtracking@iop.org

Additional copies of this publication are available from:

Curran Associates, Inc.
57 Morehouse Lane
Red Hook, NY 12571 USA
Phone: 845-758-0400
Fax: 845-758-2633
Email: curran@proceedings.com
Web: www.proceedings.com

TABLE OF CONTENTS

Overview of II International Conference on Advances in Material Science and Technology - CAMSTech-2021

I V Kovalev[1,2,3,4], N A Testoyedov[4,5] and A A Voroshilova[2]

[1] Siberian Federal University, 79, Svobodny pr., Krasnoyarsk, 660041, Russia
[2] Krasnoyarsk Science and Technology City Hall, 61, Uritskogo Str., Krasnoyarsk, 660049, Russia
[3] China Aviation Industry General Aircraft Zhejiang Institute Co., Ltd, China
[4] Reshetnev Siberian State University of Science and Technology, 31, Krasnoyarsky Rabochy Av., Krasnoyarsk, 660037, Russia
[5] JSC "Academician M F Reshetnev Information satellite systems", 52 Lenin Street, Zheleznogorsk, Krasnoyarsk region, 662972, Russia

E-mail: krasnio@bk.ru

Abstract. The overview describes the main directions and results of the II International Conference "CAMSTech-2020" held in Krasnoyarsk in July 29-31, 2021. It gives the details about the participants and the proceedings. The purpose of the conference is to share the experience of leading experts in the application of innovative technologies, information management systems in industrial production, in the fields of aerospace, energy and chemical engineering, materials science and design of new materials, as well as engineering and automation of technological processes.

We are pleased to present this issue of IOP Conference Series: Materials Science and Engineering (MSE) for the selected revised papers presented in the II International Conference on Advances in Material Science and Technology - CAMSTech-2021, held on July 29-31, 2021, at the Krasnoyarsk Science and Technology City Hall (Russia). The purpose of the Conference is to share the experience of leading experts in the application of innovative technologies, mathematical methods and information management systems in industrial production, in the fields of aerospace, energy and chemical engineering, materials science and design of new materials, as well as engineering and automation of technological processes, etc.

CAMSTech-2021 Conference was jointly organised by the International and Russian Union of Scientific and Engineering Associations (Moscow, Russia), Krasnoyarsk Regional Union of Scientific and Engineering Associations and Krasnoyarsk Science and Technology City Hall.

The main partner organisations in preparing and organising the Conference in Krasnoyarsk as well as of the previous scientific events [1-13] were: Siberian Federal University, JSC "Academician M F Reshetnev Information Satellite Systems" and Krasnoyarsk Scientific Centre of the Siberian Branch of the Russian Academy of Sciences.

Tashkent Institute of Irrigation and Agricultural Mechanization Engineers (Tashkent, Uzbekistan) and Bukhara Engineering Technological Institute (Bukhara, Uzbekistan) are international partner universities of the Conference.

The papers of this volume are related to the following themes:

- Material science and innovative technology;
- Mechanical engineering and automation of technological processes for Industry 4.0;
- Cybernetics, economics and organization of mechanical engineering production;
- Reliability and data protection in automated technological systems;
- Energy, chemical technologies and ecological engineering.

The Program Committee and Editorial Board included prominent professors and scientists from the Russian Academy of Sciences, Amity University (India), Republican Centre for Space Communications and Electromagnetic Compatibility of Radio-Electronic Devices (Kazakhstan), Bukhara Engineering Technological Institute and Tashkent Institute of Irrigation and Agricultural Mechanization Engineers (Uzbekistan), Siberian Federal University (Russia), Voronezh State Technical University (Russia), Bialystok University of Technology and West Pomeranian University of Technology (Poland) as well as specialists from Russian and foreign leading industrial enterprises (Appendix 1).

The event has offered a platform for bringing together students, postdocs, innovative academics and industrial experts to exchange their ideas and contribute new engineering approaches to research technological and automation processes in various production systems. Great interest was aroused by the section about material science and innovative technology and the section about energy, chemical technologies and ecological engineering.

The participants from Russia represented universities, scientific institutes and industrial enterprises from more than 20 regions:

- Bauman Moscow State Technical University;
- Blagonravov Institute of Mechanical Engineering, Russian Academy of Sciences, Moscow;
- Kant Baltic Federal University, Kaliningrad;
- Institute of Technical Chemistry, Branch of the Perm Federal Research Center, Ural Branch, Russian Academy of Sciences, Perm;
- Kaliningrad State Technical University;
- Baltic State Fishing Fleet Academy, Kaliningrad;
- Kazan National Research Technological University;
- Krasnoyarsk State Agrarian University;
- Kuzbass State Technical University named after T.F. Gorbachev in Novokuznetsk;
- Mechanical Engineering Research Institute of the Russian Academy of Sciences, Moscow;
- Reshetnev Siberian State University of Science and Technology, Krasnoyarsk;
- Samara National Research University;
- Samara Scientific Center of the Russian Academy of Sciences;
- Sankt-Petersburg State University of Aerospace Instrumentation;
- Siberian Federal University, Krasnoyarsk;
- Togliatti State University;
- Tomsk State University of Control Systems and Radioelectronics;
- Ufa State Petroleum Technological University, Branch in Sterlitamak;
- Ufa State Petroleum Technological University;
- Ural engineering center, Chelyabinsk;
- Voronezh State Technical University;
- Voronezh State University of Forestry and Technologies named after G.F. Morozov;
- GAZ METALLOKOMPLEKT LLC, Taganrog;
- Royal Carbon LLC, Kemerovo.

Foreign participants represented universities and scientific organizations from Ukraine - Innovative Initiative LLC, Alchevsk; Uzbekistan - Physical-Technical Institute RPA "Physics-Sun" AS RUz,

Tashkent, Gulistan State University, Gulistan city; Kazakhstan - Serpin Consult Group, Almaty and Republican Centre for Space Communications and Electromagnetic Compatibility of Radio-Electronic Devices, Nur-Sultan; China - China Aviation Industry General Aircraft Zhejiang Institute Co., Ltd.

The Conference provided the premier interdisciplinary and multidisciplinary forum for researchers, practitioners and educators to present and discuss the most recent innovations, mathematical methods, practical challenges encountered and the solutions adopted in the fields of materials sciences, mechanical, chemical, ecological and power engineering, aerospace technologies, cybernetics, reliability and data protection in automation systems, etc.

Although the schedule of the Conference was very tight, there were very vivid discussions among the participants. The program of the Conference consisted of keynote and plenary talks and numerous presentations.

All participants were invited to present their papers in this Volume and all submitted manuscripts went through the independent peer review process. We are very grateful to all reviewers from Russia, Poland, India, Kazakhstan and Uzbekistan for their time and highly professional comments. We deeply believe that their reviews gave opportunity to improve the scientific quality of the presented papers which may be useful for academic, scientific and industrial partners.

Each paper has been peer reviewed by the reviewers specialized in the related field. All papers that have been selected and accepted for publication reflect modern engineering approaches in all Conference directions, broaden the researches of the previous CAMSTech Conference [9]. We hope that you enjoy the reading of the papers.

We chose the IOP Conference Series: Materials Science and Engineering to provide all contributors with the opportunity to publish their papers in an international, peer-reviewed journal. This is understood and appreciated by all the participants of our Conference, and therefore this Volume provides an excellent overview of the main topics of our Conference.

Acknowledgements

The JSC "Academician M F Reshetnev Information Satellite Systems" and Siberian Scientific Centre DNIT sponsored the Conference CAMSTech-2021.

CAMSTech Conference Committee highly appreciate the contribution of Professor Nikolay Testoyedov, corresponding member of the Russian Academy of Sciences, Director General of Information Satellite Systems – Reshetnev Company (Zheleznogorsk, Krasnoyarsk region, Russia) for cooperation in organization of CAMSTech-2021 and preparation of the papers for this Volume.

We are grateful to scientists and professors of Siberian Federal University, Reshetnev Siberian State University of Science and Technology, Institute of Computational Modeling of Siberian Branch of Russian Academy of Sciences, Ufa State Petroleum Technological University, Sankt-Petersburg State University of Aerospace Instrumentation, Samara National Research University and Kazan National Research Technological University for active participation in the Conference.

On behalf of the Conference Committee and organizers, we would like to thank all the authors who contributed to this Volume as well as to the reviewers, speakers and all the conference participants for their support to CAMSTech-2021.

We express gratitude to IOP Publishing for an opportunity to publish the Proceedings of the Conference to provide open access and to make them available for worldwide recognition.

References

[1] 2018 *IOP Conf. Ser.: Mater. Sci. Eng.* **450** 011001
[2] Kovalev I V, *et al.* 2019 *IOP Conf. Ser.: Earth and Environ. Sci.* **315** 011001
[3] Kovalev I V, *et al.* 2019 *IOP Conf. Ser.: Mater. Sci. Eng.* **537** 011001
[4] Kovalev I V, *et al.* 2019 *J. Phys.: Conf. Ser.* **1399** 011001
[5] Kovalev I V and Voroshilova A A 2020 *IOP Conf. Ser.: Mater. Sci. Eng.* **734** 011001
[6] Kovalev I V, Pyzhikova N V and Voroshilova A A 2020 *IOP Conf. Ser.: Earth and Environ. Sci.* **421** 011001
[7] Antokhina Yu A, *et al.* 2020 *J. Phys.: Conf. Ser.* **1515** 011001

[8] Kovalev I V and Voroshilova A A 2020 *IOP Conf. Ser.: Mater. Sci. Eng.* **862** 011001
[9] Kovalev I V, *et al*. 2020 *IOP Conf. Ser.: 8,9Mater. Sci. Eng.* **919** 011001
[10] Kovalev Igor V, *et al* 2020 *J. Phys.: Conf. Ser.* **1679** 011001
[11] Kovalev Igor V, *et al* 2021 *IOP Conf. Ser.: Mater. Sci. Eng.* **1047** 011001
[12] Antokhina Yu A, *et al* 2021 *J. Phys.: Conf. Ser.* **1889** 011001
[13] Kovalev Igor V, *et al* 2021 *IOP Conf. Ser.: Mater. Sci. Eng.* **1155** 011001

Appendix 1

INTERNATIONAL PROGRAMME COMMITTEE

Yuri V. GULYAEV	Academician of the Russian Academy of Sciences, Doctor of Physical and Mathematical Sciences, Professor, President of the International and Russian Union of Scientific and Engineering Associations	Moscow, Russia
Igor V. KOVALEV	Doctor of Technical Sciences, Professor, President of the Krasnoyarsk Regional Union of Scientific and Engineering Associations, Director of Krasnoyarsk Regional Science and Technology City Hall, Professor of Siberian Federal University	Krasnoyarsk, Russia
Uktam P. UMURZAKOV	Dr. Sc., Professor, Rector of the Tashkent Institute of Irrigation and Agricultural Mechanization Engineers	Tashkent, Uzbekistan
Nusratillo R. BARAKAEV	Doctor of Technical Sciences, Professor, Rector of Bukhara Engineering Technological Institute	Bukhara, Uzbekistan
Vasiliy F. SHABANOV	Academician of the Russian Academy of Science, Doctor of Physics and Mathematics, Professor, Scientific Supervisor of the Krasnoyarsk Science Centre of the Siberian Branch of the Russian Academy of Science	Krasnoyarsk, Russia
Sergey N. VARNAKOV	Doctor of Physical and Mathematical Sciences, Professor, Vice-Director of Krasnoyarsk Science Center of the Siberian Branch of the Russian Academy of Sciences	Krasnoyarsk, Russia
Viktor LEFTER	Doctor of Technical Sciences, Professor, Scientific Advisor of the joint-stock company "Republican center for space communications and electromagnetic compatibility of radio-electronic devices»	Nur-Sultan, Kazakhstan
Takhir Z. SULTANOV	Doctor of Technical Sciences, Professor, Vice-Rector for Research and Innovation of the Tashkent Institute of Irrigation and Agricultural Mechanization Engineers	Tashkent, Uzbekistan
Mirzo Z. SHARIPOV	Doctor of Physics and Mathematics, Professor, Vice-rector for research and innovation of Bukhara Engineering Technological Institute	Bukhara, Uzbekistan
Sergey P. DRUKARENKO	Vice-president, First Secretary the International and Russian Union of Scientific and Engineering Associations	Moscow, Russia

IOP Conf. Series: Materials Science and Engineering 1181 (2021) 011001 doi:10.1088/1757-899X/1181/1/011001

Alena A. STUPINA	Doctor of Technical Sciences, Professor of the Siberian Federal University	Krasnoyarsk, Russia
Dr Krystyna ARASZKIEWICZ	West Pomeranian University of Technology	Szczecin, Poland
Krishna Mohan AGARWAL	Assistant Professor, Grade III, Mechanical Engineering Department, Amity University Uttar Pradesh	Noida, India
Bakhytbek ZHANKUBAEV	Serpin Consult Group	Almaty, Republic of Kazakhstan
Prof (Dr) Satya N MANDAL	RICS School of Built Environment, Amity University	Noida, India
Prof (Dr) Amit Kumar PANDEY	Amity University	Noida, India
Prof (Dr) Adam MUSIUK	Bialystok University of Technology	Poland
Bahadir S. MIRZAEV	Doctor of Technical Sciences, Professor, First Vice-Rector of the Tashkent Institute of Irrigation and Agricultural Mechanization Engineers	Tashkent, Uzbekistan
Kadyrzhan A. SHOVAZOV	Candidate of Technical Sciences, Associate Professor, Dean of the Faculty of Agricultural Mechanization of the Tashkent Institute of Irrigation and Agricultural Mechanization Engineers	Tashkent, Uzbekistan
Khamro B. DUSTOV	Editor-in-chief of the journal "Development of Science and Technology", Doctor of Chemistry, Professor of the Department of Oil and Gas Engineering, Bukhara Engineering Technological Institute	Bukhara, Uzbekistan
Ismoil B. ISABAEV	Doctor of Technical Sciences, Professor of the Department of Food Technology, Bukhara Engineering Technological Institute	Bukhara, Uzbekistan
Sadriddin F. FOZILOV	Doctor of Technical Sciences, Professor, Head of the Department "Technology of Chemical Gas Processing", Bukhara Engineering Technological Institute	Bukhara, Uzbekistan
Komil D. ASTANAKULOV	Doctor of Technical Sciences, Senior Researcher, Head of the Department of	Tashkent, Uzbekistan

EDITORIAL BOARD

IOP Conf. Series: Materials Science and Engineering 1181 (2021) 011002 doi:10.1088/1757-899X/1181/1/011002

Peer review declaration

II International Conference on Advances in Material Science and Technology (CAMSTech-2021), 29-31 July 2021, Krasnoyarsk, Russian Federation

All papers published in this volume of IOP Conference Series: Materials Science and Engineering have been peer reviewed through processes administered by the Editors. Reviews were conducted by expert referees to the professional and scientific standards expected of a proceedings journal published by IOP Publishing.

- **Type of peer review: Double-blind**

All papers came through the basic review which included an initial technical criteria check (paper field, structure of submission, adherence to the submission instructions, English language usage and a check for the similarity rate). Any papers out of the scope or containing plagiarism, including self-plagiarism, were rejected.

We used a double-blind system for peer review; both reviewers' and authors' identities remained anonymous. The submitted papers were reviewed by at least two experts: one editorial staff member as well as at least one external reviewer. The third/fourth reviewer is involved in case two reviewers have disputes about the content of the papers or the authors do not agree with the review result. The review process took from 7 to 15 days as a rule. The reviews were conducted to the professional and scientific standards.

The decision to accept or reject the paper was based on the suggestions of reviewers. Acceptance/rejecting notifications were sent to the corresponding author(s).

- **Conference submission management system:**

Conference submission is carried out through the conference management system which was developed by the Organizer. It easy to use, and has features to make it suitable for simultaneous conference calls.

More than 20,500 registration forms have been already processed by the system which showed it stability and flexibility.

The conference management system is integrated with the conference website of the Organizer which helps the Organizer to cope with the complexity of the submission, reviewing, refereeing and accepting processes. The current version supports:

1. management of the access of reviewers and editors to papers;
2. automatic registration for conferences and paper submission;
3. list of the current, forthcoming and held conferences with all necessary information;
4. list of all conference materials with online presentations and/or video reports of participants, list of published proceedings with direct links to publications and journals;
5. sending emails to programme committee members, reviewers and authors;
6. preparation and publishing of conference programmes;

7. all authors submitting the papers are assumed to accept the terms of Publication Licence;
8. all requirements for papers, templates and sample papers are accessible for the authors.

- **Number of submissions received:** 100
- **Number of submissions sent for review:** 100
- **Number of submissions accepted:** 33
- **Acceptance Rate (Number of Submissions Accepted / Number of Submissions Received X 100):** 33
- **Average number of reviews per paper:** 2
- **Total number of reviewers involved:** 25
- **Any additional info on review process:**

The Checklist for the reviewers included the following criteria: relevance to the scope of the conference topics, the scientific quality, assessing the errors, organisation of the paper, the title, the references, the level of the English language, the quality of the figures and tables, etc.

The reviewers could give the following recommendation: to publish the paper as it is, to publish it after mandatory/optional minor/major revision, to reject the paper.

- **Contact person for queries:**

Anna Voroshilova, Krasnoyarsk Science and Technology City Hall, 61, Uritskogo Street, Krasnoyarsk, 660049, Russia, E-mail: krasnio@bk.ru. Tel + 7 913 538 6990

Study of used marine oils by the photon correlation spectroscopy method

N Y Sinyavsky[1], I P Korneva[2], A M Ivanov[1] and N A Kostrikova[1]

[1] Kaliningrad State Technical University, Sovetskiy prospect, 1, Kaliningrad, 236022, Russia

[2] Baltic State Fishing Fleet Academy, Molodezhnaya, 6, Kaliningrad, 236029, Russia

E-mail: n_sinyavsky@mail.ru

Abstract. In this paper, we investigate the possibility of using the method of photon correlation spectroscopy for diagnostics of marine engines based on used engine oil in order to improve efficiency and safety of operation of sea vessels. The study used the dynamic light scattering (DLS) method to measure correlation functions and then to obtain distributions of correlation times, diffusion coefficients and particle sizes. To invert the Laplace transform, a method based on standard deviation minimization and regularization (RILT) was used, modified to find the distribution of correlation times. It has been found that the distributions of the hydrodynamic radii of the particles for fresh and used oils differ significantly. By the nature of the distribution of large particles, it has been found that the level of aggregation or micelle formation of the polymer of the viscosity modifier decreases in waste oils. It is shown that the dynamic light scattering method makes it possible to obtain parameters suitable for analyzing the state of engine oil and for obtaining diagnostic data on the state of the engine for used oil.

1. Introduction

The overwhelming majority of ship traffic and marine power engineering is based on turbocharged piston diesel engines. Marine engines are very complex technical objects with many important functional systems that affect the current technical condition of a marine engine, including fuel consumption and failures that threaten the safety of the ship. There is a direct link between reliability of these engines, safety of navigation and operating costs. Improving diagnostic methods can improve and increase reliability of ship engines, maintenance, economic operation, as well as affect safe operation of ships.

Marine diesel engine oil can be used as a medium for diagnosing component malfunctions since it contains particles resulting from wear. Knowledge of the materials of mechanism construction based on the results of the analysis of used oil makes it possible to identify a unit that needs repair or replacement. Changes in the properties of the oil and its composition, depending on the duration of engine operation, make it possible to use them for troubleshooting. At the same time, delivery of equipment to the place of repair, dismantling of units and assemblies is not required. The presence of lead, tin and iron in the waste oil indicates wear on the plain bearings and the crankshaft. Engine overheating, fuel entering the oil, and the presence of abrasive particles in it cause wear on pistons and cylinders. Piston ring wear is detected by the presence of chromium in the engine oil, etc.

The work (Zhu et al. 2013) is devoted to a review of methods and systems for monitoring the condition of lubricating oils, diagnostics and forecasting. The considered methods are analyzed and divided into four categories: electrical (magnetic), physical, chemical and optical methods. The ultimate goal of all sensor systems under development is to achieve online monitoring of the condition of the lubricating oil on board and predict the remaining service life. The development of online analysis of marine lubricating oil and the development of intelligent sensor systems in diagnostics is discussed in the publication (Knowles and Baglee 2012). The article describes a developed system that includes real-time condition monitoring sensors that monitor the properties of the lubricating oil and special monitoring software. A wide range of in-service oil analysis methods are presented in the handbook (Zhao 2014). In (Berestova 2002), a technique for monitoring the condition of a diesel engine by the parameters of engine oil measured by infrared spectroscopy and ferrography is proposed.

The work (Witkowski 2017) is devoted to improving the operational safety of ships by improving diagnostic methods for a marine diesel engine. In this work, to diagnose diesel injection systems, the author used the heat release characteristics of ship engines. The work (Sinyavsky et al. 2018) investigated possibilities of using NMR relaxometry with the Laplace transform for quality control and identification of marine fuels and oils. The specificity of the distributions of the relaxation times of the studied petroleum products is shown; recommendations are formulated for the creation of promising control and measuring equipment.

Photon correlation spectroscopy is an indirect method for measuring the physical parameters of a medium. In order to determine the quantities of interest, for example, the particle size distribution in a liquid or the diffusion coefficient distribution, the measured autocorrelation of the scattered light intensity fluctuations must be subjected to further mathematical processing.

Dynamic Light Scattering (DLS) (User's manual 2020) is a widely used method for determining particle size in colloidal systems. The resolution of the simple one-angle method is limited by the low information content of the initial data and the ambiguous nature of their analysis.

A multi-angle approach proposed in (Bryant and Thomas 1995) provides significantly more reliable particle size determination, allowing better resolution of closely spaced components in multimodal distributions. Multi-angle dynamic light scattering (MDLS) can provide more accurate particle size distribution information than single angle dynamic light scattering. In MDLS data analysis, the intensities of the autocorrelation functions of light for each scattering angle are combined with the appropriate weighting factors.

Used engine oil is a polydisperse system, for which the photon correlation spectroscopy method allows one to obtain distribution of the effective particle radii of each component and contribution of each component of the system to the scattering signal (Sinyavsky et al. 2020). By the nature of the decay of the autocorrelation function for radiation scattered by impurity particles in oil, one can determine the self-diffusion coefficients of the particles and calculate their hydrodynamic radii. Since the used oil contains a suspension of particles with different effective radii, a special mathematical algorithm can be used to determine the particle size distribution curve. This algorithm must find a solution to the ill-posed inverse problem.

The DynaLS program (Goldin 2002) uses the SVD (Singular Value Decomposition) method as the main analysis tool. Singular value expansion is an effective tool for solving unstable systems of linear equations (ill-conditioned problems). To limit the solution to non-negative values, it is also very important to use the non-negative least squares (NNLS) algorithm.

The most famous program using the Tikhonov regularization method for processing experimental data in the dynamic light scattering method is a program using the CONTIN algorithm (Provencher 1982). To determine the distribution of effective particle radii from the experimental correlogram, Danovich and Serdyuk (1983) also proposes Tikhonov's regularization in the SIPP (Solution Ill-Posed Problems) program.

The purpose of this work was to study some fresh and used oils by the method of dynamic light scattering to determine the changes occurring with oil during the operation of ship equipment, and to clarify the relationship of these changes with malfunctions of the units. There are no publications on such studies in the literature.

2. Experimental research procedure

The scattered light of a single-mode helium-neon laser was recorded in real time using a photomultiplier tube. A *Photocor-FC* correlator was used to measure the scattered light autocorrelation function. *Photocor* software (User's manual 2020) was used to control the measurement process and processing of measurement results. The correlation function for light scattering at an angle θ will have the following form for a polydisperse medium:

$$G(t) = \int_0^\infty P(T_d)\exp\left(-\frac{t}{T_d}\right)dT_d, \quad (1)$$

where T_d – decay time. The particle diffusion coefficient was found as:

$$D = \frac{1}{T_d q^2}, \quad (2)$$

where the scattering vector

$$q = \frac{4\pi n}{\lambda}sin\frac{\theta}{2}, \quad (3)$$

Θ – scattering angle, n – refraction index, λ – emission wave length длина. The effective radius of a particle is determined by a well-known formula, which depends on temperature, diffusion coefficients, and fluid viscosity:

$$R_h = \frac{kT}{6\pi\eta D}, \quad (4)$$

where η - dynamic viscosity. Integral equation (1) actually describes the relationship between the measured data obtained during the experiment and the physical properties of the investigated liquid. If the distribution of decay times T_d is known, then the distribution of the diffusion coefficient or the particle size distribution can be easily calculated using formulas (2) and (4).
Formula (1) can be represented as:

$$G(t,\theta) = \int_0^\infty f(R_h)exp\left[-\frac{\Omega(\theta)t}{R_h}\right]dR_h, \quad (5)$$

where $\Omega(\theta) = \frac{16\pi kT}{3\eta}\left(\frac{n}{\lambda}\right)^2 sin^2\frac{\theta}{2}$. Formula (5) is written without taking into account the constant component and the fraction of light scattered at an angle θ according to the Mie scattering theory (Mie 1908).

The intensity of the light scattered by the sample depends on the size and refractive index of the particles, as well as on the angle at which the scattering is detected. This means that not all particles of different sizes in the same sample are detected with the same sensitivity. Traditional single angle DLS measurement can give different particle size distributions for a mixture depending on the angle at which the measurement is made. In MADLS measurements, correlation data for multiple detection angles are combined using Mie theory to obtain a higher resolution particle size distribution.

To calculate the particle size distribution in used engine oil from the measured correlation function, we used the RILT program (Marino 2004) based on the CONTIN algorithm. The program has been modified to determine the correlation functions.

The samples for the study were M-20G2SD marine oil, Mobil 5w50 synthetic oil, and AUP (MG-22-B) marine hydraulic oil. AUP oil is used as a hydraulic fluid in marine engineering. The measurements were carried out without using a solvent and with a solvent. Liquid Csl.4 paraffin was used as a solvent. Waste oils are not transparent; therefore almost all measurements were carried out using a solvent. Concentrations of solutions: AUP oil (spent within 500 hours) - 6.25%, Mobil oil (spent within 100 hours) - 1.6%, marine oil M-20G2SD (spent within 300 hours) - 0.5%.

3. Experimental research results and their analysis

An important characteristic of the interaction of radiation with a medium is a scattering indicatrix - a curve that graphically displays the dependence of the scattered light intensity on the scattering angle. The form of this dependence is determined by the size and shape of the scattering particles.

In general, the scattering indicatrix is not expressed in an explicit function and is described in the table. When solving many applied problems, simplified approximations of the scattering indicatrix are used. The most common of such approximations is Henyey-Greenstein (Henyey and Greenstein 1941) function:

$$\Phi(\theta) = \frac{1-g^2}{4\pi}\frac{1}{(1+g^2-2g\cos\theta)^{3/2}}, \quad (6)$$

where θ – scattering angle, g – parameter determining the main characteristic of the indicatrix –its elongation. The parameter g is determined by the size of the scattering particles, the radiation wavelength, and the particle size distribution. When constructing the indicatrix to take into account the correction for the value of the scattering volume, the intensity was multiplied by $\sin\theta$.

In order to exclude the influence of the solvent on the measurement results, the experiment was carried out, inter alia, according to the angular and backscattering scheme. The scattering indicatrix for fresh and used AUP oil (without solvent), approximated using formula (6), is shown in Fig. 1. Due to the opacity of the used oil, there is a considerably intense backscattering of light.

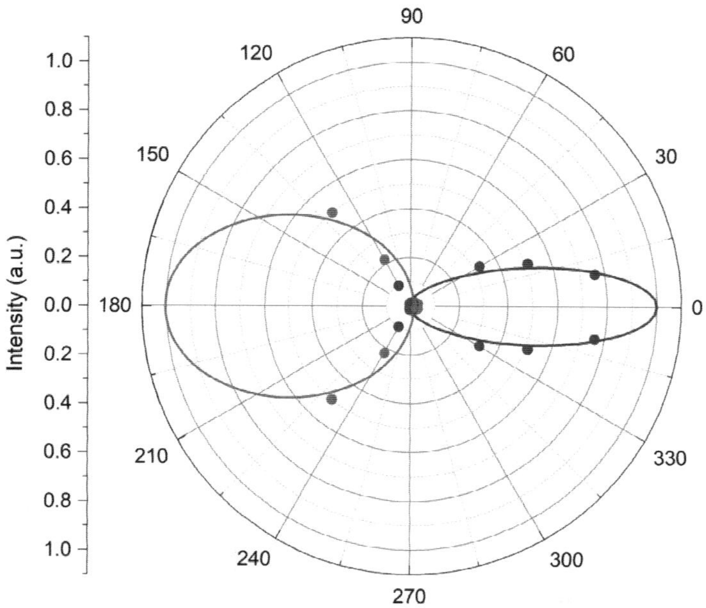

Figure 1. Scattering indicatrixes of fresh (red) and used (blue) hydraulic oil AUP (MG-22-B).

Figure 2 and Figure 3 show correlation functions for fresh and used AUP marine hydraulic oil without solvent at different scattering angles. As can be seen from the figure, the correlation times are practically independent of the scattering angle for both fresh and used oil. For used AUP oil, the correlation time is shorter than for fresh oil. The results of studying the distribution of the hydrodynamic radii of particles for different scattering angles for fresh and used AUP hydraulic oil are illustrated in Figure 4. It can be seen that distributions for used and clean oils differ from each other. For used oil, the values of hydrodynamic radii are significantly less than for fresh oil. As the scattering angle increases, the R_h distributions change from monomodal to bimodal. R_h values are determined by the ratio of the average molecular weights of different hydrocarbons, the viscosity of the oil, the presence of additives and viscosity modifiers, metal particles, soot, water, etc.

Changes in the particle size distribution in used oil are associated with a change at the molecular level (thermal polymerization, oxidation, evaporation, thermal cracking) and with changes caused by contamination (formation of soot, ingress of water and air, metal particles of rubbing parts).

Figure 5 and Figure 6 show distributions of hydrodynamic radii for different scattering angles for fresh (a) and used (b) M-20G2SD marine oil and Mobil 5w50 oil. In both cases, with an increase in the scattering angle θ, the hydrodynamic distribution of radius of oil particles increases and R_h of used oils is bigger than R_h of fresh oils.

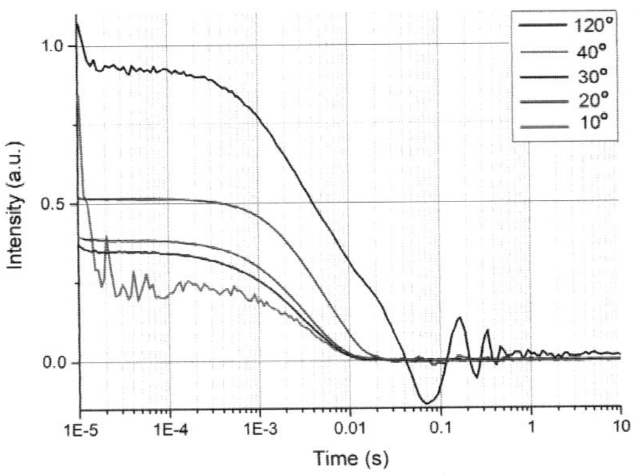

Figure 2. Correlation functions for different scattering angles for used hydraulic AUP oil.

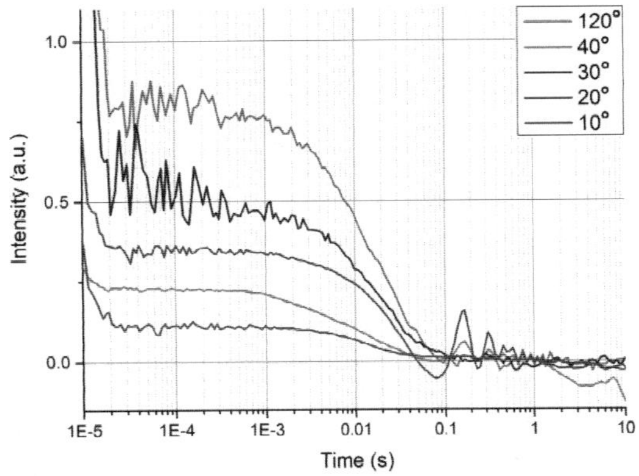

Figure 3. Correlation functions for different scattering angles for fresh hydraulic AUP oil.

Comparison of the distributions of the hydrodynamic radii of fresh and used AUP, Mobil, and M-20G2SD oils (in solvent) obtained at $\theta=\pi/2$ is shown in Figure 7. All distributions are not unimodal. The used AUP and M-20G2SD oils have an increase in the hydrodynamic radii R_h in comparison with fresh oils, in contrast to Mobil oil, where the opposite is true, the R_h values of the used oil are lower.

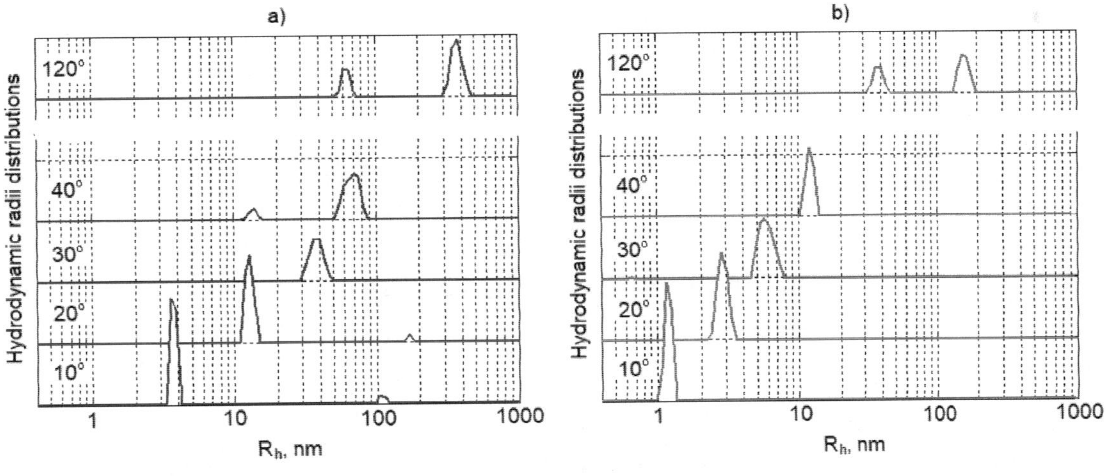

Figure 4. Distributions of hydrodynamic radii for different scattering angles for fresh (a) and used (b) AUP hydraulic oil.

When analyzing the distribution of the hydrodynamic radii of particles, it is important to take into account their dependence on the viscosity of oils. During operation, engine oil degrades and in most cases its viscosity increases. The reasons for the increase in oil viscosity are: violation of the combustion efficiency of the fuel-air mixture, thermal polymerization of the oil, oxidation, evaporation, sludge formation, water entering the oil, mixing with air. The oil viscosity can increase due to the formation of dissolved coke and oxides, contamination with soot. This leads to frictional brake, to insufficient lubricant supply to the bearings, to a drop in engine power, to a violation of the smoothness of the speed set, the appearance of a cavitation process, etc. However, the oil viscosity can also decrease, this is due to thermal cracking (thermal destruction of oil molecules), i.e. the opposite polymerization process. Polymer macromolecules can be destroyed by shear forces, and this leads to a loss of viscosity. Finally, the viscosity of the engine oil drops due to fuel contamination. Low viscosity results in a too thin oil film, which, as a result, causes severe wear at the friction points, excessive heat generation, loss of oil cooling efficiency and an increase in the oxidation process.

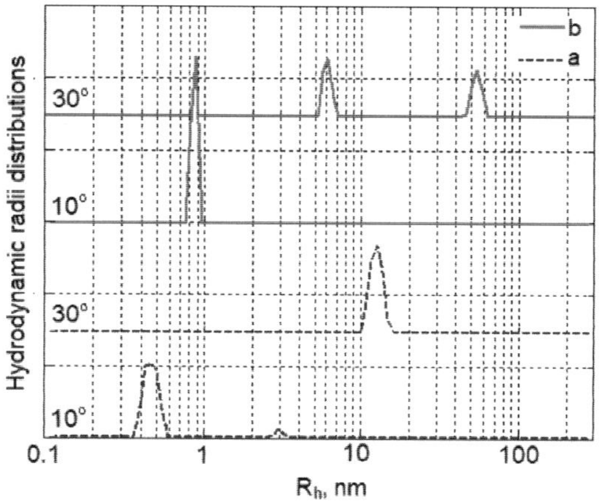

Figure 5. Distributions of hydrodynamic radii for different scattering angles for fresh (a) and used (b) M-20G2SD marine oil.

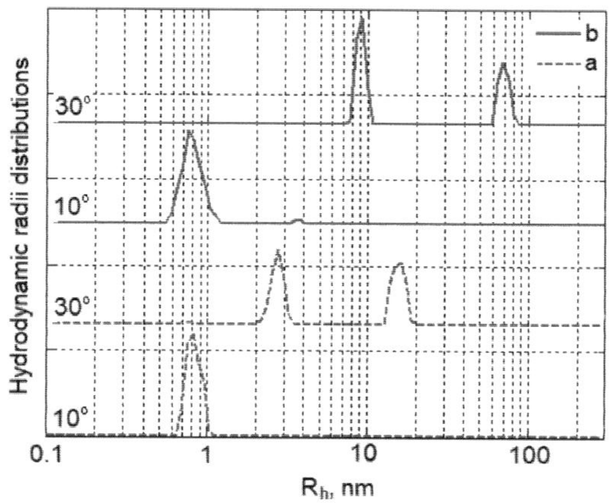

Figure 6. Distributions of hydrodynamic radii for different scattering angles for fresh (a) and used (b) Mobil 5w50 oil.

As can be seen from Table 1, the average hydrodynamic radii of particles of used oils M-20G2SD and AUP in the solvent Csl.4 (liquid paraffin) are significantly larger than for fresh oils. Due to the bimodality (multimodality) of the distributions of the correlation times, all values in Table 1 have two (three) values.

But for Mobil oil, the opposite is true: in used oil, the average hydrodynamic radius of particles is smaller than that of fresh oil. The particle diffusion coefficient depends on the correlation time and the scattering vector. The smallest diffusion coefficients were found for fresh Mobil oil and used M-20G2SD oil.

Table 1. Results of studying oils at t=20 °C, $\theta=\pi/2$.

Substance	Correlation time τ, ms	Diffusion coefficient D, $cm^2/s \cdot 10^{-10}$	Hydrodynamic radius R_h, nm
Marine engine oil M-20G2SD (fresh)	12; 250	19; 0.9	25; 562
M-20G2SD oil (used)	40; 425	5.6; 0.5	79; 794
Mobil 5w50 oil (fresh)	40; 520	5.6; 0.4	63; 1259
Mobil 5w50 oil (used)	10; 60; 620	22; 3.7; 0.4	25; 158; 1585
Marine hydraulic AUP oil (fresh)	10; 900	22; 0.2	32; 1995
AUP oil (used)	25; 200	9; 1.1	63; 398
Liquid paraffin Csl.4 (solvent)	5; 540	45; 0.4	20; 1200

Even fresh engine oil is a complex system of many substances: base oil containing a mixture of different hydrocarbons, a viscosity modifier, additives (detergents, anti-corrosion, etc.). Large particles scatter light better, and the intensities of the peaks associated with them in the distributions are bigger. As a result, even a small amount of large particles in the oil can mask the detection of small particles, even if they are large. Large particles in oil are usually micelles of the viscosity modifier polymer and dust.

As seen from Figure 7, the size distributions for fresh and used oils have different polydispersities. The nature of the distributions of the hydrodynamic radii of particles in the studied samples is predominantly bimodal. The distribution peaks are located in two main domains: from 10 nm to 100 nm and above 100 nm, respectively. The presence of viscous-thickening additives in fresh oils leads to the appearance of large particles (micelles) in the oils, and on the distributions of the hydrodynamic radii of the peaks in the second domain. In used oils, these peaks either decrease in magnitude (Mobil, AUP oils), or disappear altogether, which corresponds to a state of low aggregation or micelle formation of the viscosity modifier polymer. The level of aggregation of the viscous-thickening polymer (viscosity modifier) changes in used oils.

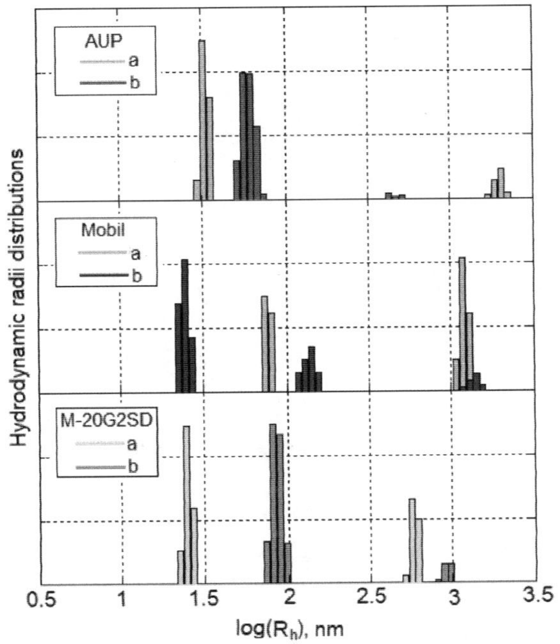

Figure 7. Distributions of hydrodynamic radii for the studied oils in solvent: a – fresh, b – used.

4. Conclusion

The In this work, some fresh and used marine oils have been investigated by the method of photon correlation spectroscopy, which made it possible to obtain the distributions of correlation times, diffusion coefficients and hydrodynamic radii of particles of these substances. The reasons leading to changes in these distributions in used oils have been analyzed. The effect on particle size distribution of viscosity is shown, which, depending on the reasons (engine malfunctions), can change both upward and downward. By the nature of the distribution of large particles, it has been established that the level of aggregation or micelle formation of the polymer of the viscosity modifier can decrease in used oils. Compared to fresh oil, used oil has an increase in the content of aromatic hydrocarbons; in addition, due to the oxidation of naphthenic and alkane hydrocarbons, as a rule, the average molecular weight of the oil changes, which manifests itself in the distributions of diffusion coefficients and hydrodynamic radii.

The results of the study performed show that the method of dynamic light scattering makes it possible to obtain parameters suitable for analysing the condition of engine oil and for obtaining diagnostic data on the condition of the engine for used oil. The development of new diagnostic methods for ship engines is the basis for creating a method for detecting defects at an early stage of their formation, thereby improving the efficiency of preventing malfunctions and ensuring navigation safety.

Acknowledgements
The research has been carried out within the framework of the state assignment of FSEI HE "KSTU" (topic No. AAA-A17-117041810027-8).

References

[1] Berestova G 2002 *Assessment of the state of engine oils in marine diesel engines by the complex method of ferrography and infrared spectroscopy* (Murmansk) p 178
[2] Bryant G and Thomas J 1995 *Langmuir* **11** 2480
[3] Danovich G and Serdyuk I 1983 *Photon Correlation Techniques in Fluid Mechanics Springer Series in Optical Sciences* vol 38 ed E O Schulz-DuBois (Berlin Heidelberg: Springer) p 315
[4] Goldin A 2002 *Software for Particle Size Distribution Analysis in Photon Correlation Spectroscopy* Retrieved from: http://www.softscientific.com/science/WhitePapers/dynals1/dynals100.htm
[5] Henyey L and Greenstein J 1941 *Astrophysical Journal* **93** 70
[6] Knowles M and Baglee D 2012 *J Phys Conf Ser* **364** 012007
[7] Marino I 2007 *Regularized Inverse Laplace Transform* Retrieved from: http://www.mathworks.com/matlabcentral/fileexchange/6523-rilt/content/rilt.m
[8] Mie G 1908 *Ann Phys* **330** 377
[9] Provencher S 1982 *Comput Phys Commun* **27(3)** 213
[10] Sinyavsky N, Korneva I, Ivanov A and Kostrikova N 2020 *Marine intellectual technologies* **4(1)** 103
[11] Sinyavsky N, Mershiev I and Kupriyanova G 2018 *Marine intellectual technologies* **42(4)** 82
[12] *Particle size analyzers of the Photocor series User's manual* 2020 Retrieved from: https://bio.pnpi.nrcki.ru/wp-content/uploads/2020/01/Photocor-Compact-Z_Manual.pdf
[13] Witkowski K 2017 *TransNav* **11(2)** 317
[14] Zhao Y 2014 *Oil Analysis Handbook for Predictive Equipment Maintenance* Retrieved from: https://www.spectrosci.com/default/assets/File/SpectroSci_OilAnalysisHandbook_FINAL_2014-08.pdf
[15] Zhu J, He D and Bechhoefer E 2013 *Int. j. chem. sci. technol.* **2(3)** 100

Growing the solid solution of molekular substitution $(GaAs)_{1-z}(ZnSe)_z$

Amin S Saidov[1], Dadajon V Saparov [1], Shukrullo N Usmonov[1,3], Kurban G Gaimnazarov[2] and Ilhom I Maripov[2]

[1] Physical-Technical Institute SPA "Physics-Sun" AS RUz, st. Bodomzor yuli 2B, Tashkent, 100084, Uzbekistan
[2] Gulistan State University, 4th microdistrict, Gulistan city, Sirdarin region, 120100, Uzbekistan
[3] Chirchiq State Pedagogical University, st. Amir Temur 104, Chirchiq city, Tashkent region, 111700, Uzbekistan

E-mail: kgaymnazarov@mail.ru

Abstract: Epitaxial layers of semiconductor solid solutions of molecular substitution $(GaAs)_{1-z}(ZnSe)_z$ ($0 < z < 0.80$) with a smoothly (but not monotonously) varying composition were obtained from the liquid phase on monocrystal GaAs (100) substrates. The solubility of gallium arsenide and zinc selenide in tin and the phase diagrams of binary systems Ga-Sn, As-Sn, Zn-Sn and Sn-Se indicate that binary compounds in the tin solvent at temperatures of 650–750°C, below the melting point of the corresponding materials, are not in the form of individual atoms of gallium, arsenic, zinc and selenium, but in the form of molecules of gallium arsenide and zinc selenide. Moreover, in the temperature range 650-750°C, the solubility of GaAs in tin varies from ~ 2.6 to ~ 5.5 mol%, and ZnSe - from ~ 0.2 to ~ 0.35 mol%, the solubility of gallium and zinc is unlimited, and the solubility of arsenic is more than 70 at. %, selenium - more than 5 at.%. It is shown that due to the difference in the valence of gallium and arsenic atoms, the forces of chemical bonds between neighboring Ga and As atoms located in the tetrahedral lattice are not the same. In the tetrahedral lattice, each Ga atom forms 4 bonds with four neighboring As atoms, three of which are formed by the sharing of electrons of neighboring Ga and As atoms, and the fourth one only due to the two valence electrons of the As atom. Apparently, the ionic fraction of the fourth chemical bond is stronger than the other three.

1. Introduction

The development of modern semiconductor instrumentation is uninterruptedly linked with the success in improving the technology for producing traditional semiconductor materials, the development and research of new promising semiconductors. The great possibilities of semiconductor microelectronics are realized only with the development of the production of multifunctional semiconductor materials with various physical properties. In this aspect, semiconductor solid solutions of binary compounds III-V and II-VI have great opportunities [1-7]. The crystalline perfection of solid solutions is mainly determined by the crystal structures of the solution-forming components, as well as the geometric dimensions and charge states of the molecules of these components. Binary compounds GaAs and ZnSe have the same crystal structure (like zinc blende), and the lattice constants are $a_{GaAs} = 5.6532$ Å and a_{ZnSe}

= 5.6676 Å, respectively. The sums of the valence of gallium (z_{Ga}) and arsenic (z_{As}) atoms, as well as zinc (z_{Zn}) and selenium (z_{Se}), are equal: $z_{Ga} + z_{As} = z_{Zn} + z_{Se} = 8$. Therefore, the binary compounds GaAs and ZnSe are suitable pairs for growing the semiconductor solid solutions of molecular substitution such as high quality $(GaAs)_{1-z}(ZnSe)_z$.

2. Objects and research methods

This paper is devoted to the technological features of liquid-phase epitaxy of the monocrystal solid solution of molecular substitution $(GaAs)_{1-z}(ZnSe)_z$.

The technological process of growing the semiconductor solid solution of molecular substitution $(GaAs)_{1-z}(ZnSe)_z$ is given in [8]. Monocrystals GaAs(100) with a thickness of ~ 400 μm were used as substrates. For the growth of the epitaxial layer, the solution-melt (Sn-GaAs-ZnSe) was poured into the space bounded between two horizontally located substrates (Fig. 1). The substrates were fastened inside a graphite cassette. On the sides of the cassette there were slots for pouring the solution-melt. The technological growth process was carried out in the atmosphere of purified hydrogen.

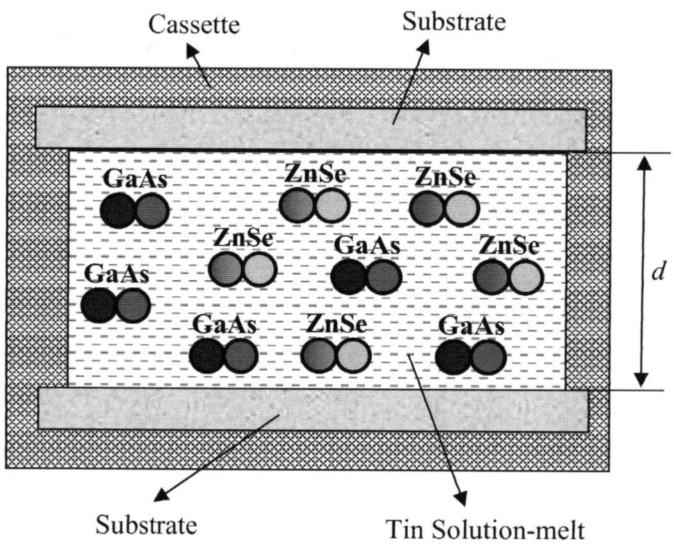

Figure 1. Schematic of the cassette with the Sn-GaAs-ZnSe solution-melt (the distance between the substrates is d = 1 mm).

The solution-melt consisted of Sn-GaAs-ZnSe components. The process of growth of the epitaxial layer was carried out at 650-750°C, that is, at temperatures higher than the melting point of tin (231°C), but less than the melting temperature of gallium arsenide (1238°C) and zinc selenide (1525°C). At these temperatures, tin is in a liquid state. For the crystallization of the $(GaAs)_{1-z}(ZnSe)_z$ solid solution, it is necessary that at these temperatures the tin solution be saturated with both gallium arsenide and zinc selenide. To ensure this condition, the solubilities of GaAs and ZnSe in Sn were studied. The state diagrams of binary systems Ga-Sn, As-Sn, Zn-Sn and Sn-Se were also studied.

3. Results and its discussion

The temperature dependence of the solubility of GaAs in tin is shown in Fig. 2a. We used the results of Vikram Kumar's work on the solubility of ZnSe in tin [9] at different temperatures (Fig. 2 b).

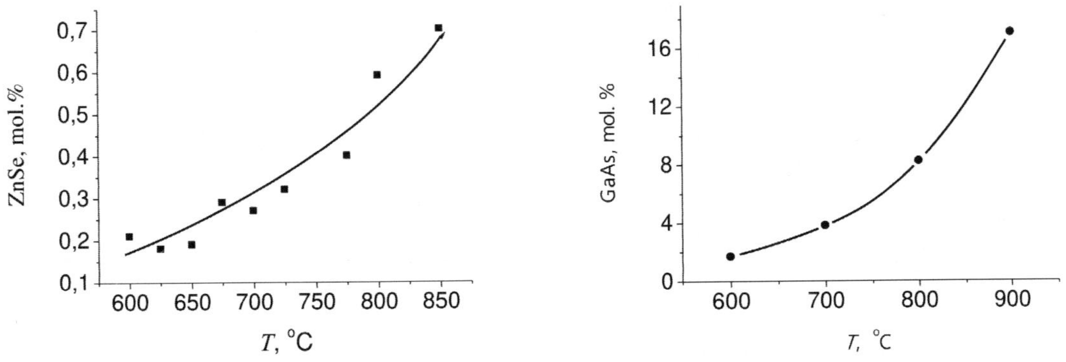

a) b)

Figure 2. Temperature dependence of the solubility of gallium arsenide (a) and zinc selenide (b) in tin.

In fig. 3 the diagrams of states of binary systems Ga-Sn (a), As-Sn (b), Zn-Sn (c) and Sn-Se (d) are presented. These results were taken from the work of M. Hansen and K. Anderko [10].
Analysis of the solubility of gallium arsenide and zinc selenide in tin at temperatures of 750 – 650°C (Fig. 2) as well as the phase diagrams of binary systems Ga-Sn, As-Sn, Zn-Sn and Sn-Se (Fig. 3) shows that these compounds at selected temperatures in tin are not in the form of individual Ga, As, Zn, and Se atoms, but in the form of GaAs and ZnSe molecules. If, upon dissolution of GaAs in tin, the substance decomposes into gallium and arsenic atoms, then this would be equivalent to the simultaneous dissolution of Ga and As in tin (Fig. 4 b).

As can be seen from Fig. 3 a, at temperatures above 231°C Ga and Sn are in a molten state, at temperatures 750–650°C the solubility of Ga in Sn is unlimited. At temperatures of 750–650°C, the solubility of As in Sn varies from 85 to 70 at.% (See Fig. 3 b). Consequently, if GaAs molecules decompose into separate Ga and As atoms upon dissolution of GaAs in Sn, then the solubility of GaAs in Sn should be at least 70 mol.% At 750–650°C. However, the solubility of gallium arsenide in tin at temperatures from 650 to 750°C varies from ~ 2.6 to ~ 5.5 mol.% (Fig. 2, a). Consequently, gallium arsenide in tin does not decompose into individual Ga and As atoms, but are in the form of GaAs molecules. This is clearly seen in Fig. 4.

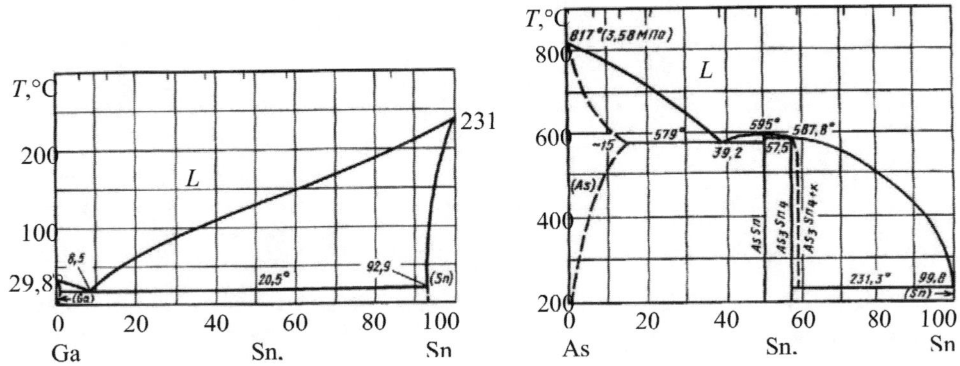

a) b)

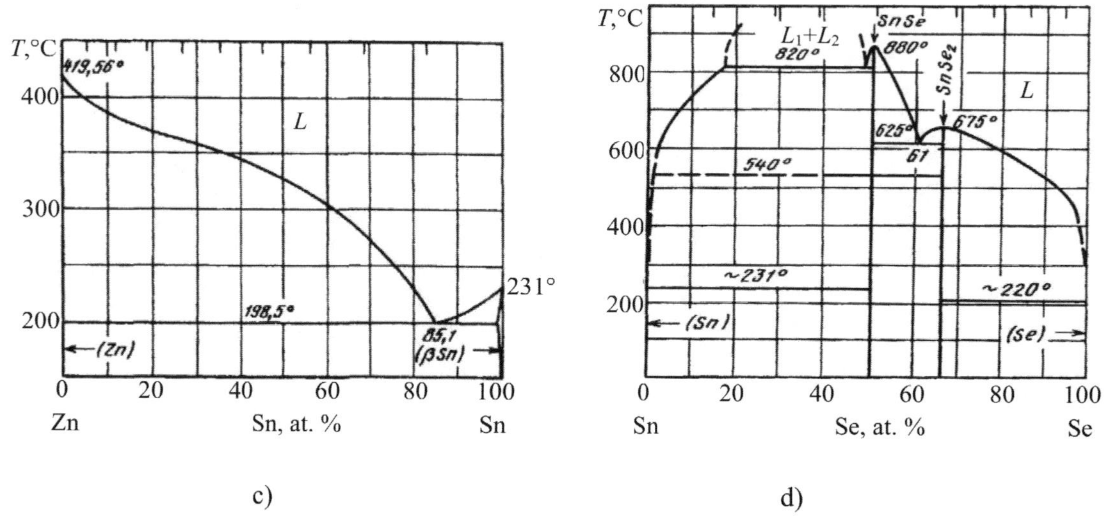

Figure 3. State diagrams of the binary system Ga-Sn (a), As-Sn (b), Zn-Sn (c) and Sn-Se (d).

Similarly from Fig. 3 s it can be seen that at temperatures above 420°C Zn and Sn are in a molten state, the solubility of Zn in Sn at 650-750°C is unlimited. It can be seen from the Sn-Se phase diagram that when Se dissolves in Sn, the Se solubility at 650-750°C varies within the range from ~ 5 to ~ 12 at.% (Fig. 3d, tin angle of the phase diagram). Whereas, the solubility of ZnSe in Sn at 650-750°C varies from ~ 0.2 to 0.35 mol.% (See Fig. 2). Consequently, as in the case of gallium arsenide, zinc selenide in tin at 650–750°C does not decompose into individual Zn and Se atoms, but is in the form of ZnSe molecules, as shown in Fig. 5.

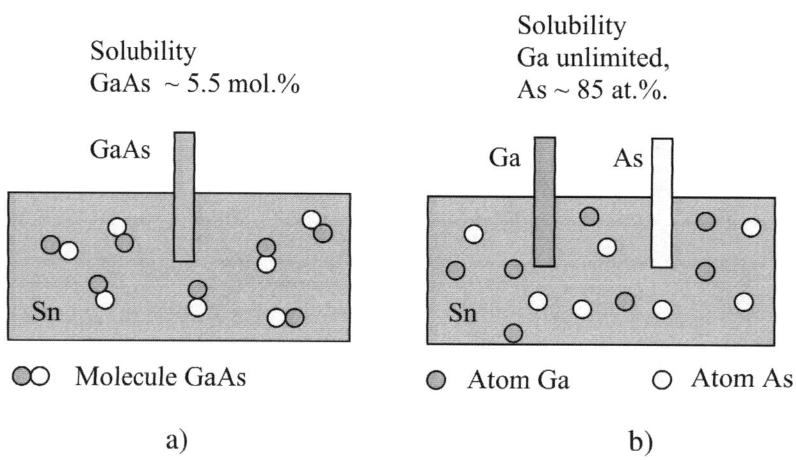

Figure 4. Dissolution of the binary compound gallium arsenide in the form of molecules (a), as well as gallium and arsenic in the form of individual atoms (b) in tin at the temperature of 750°C.

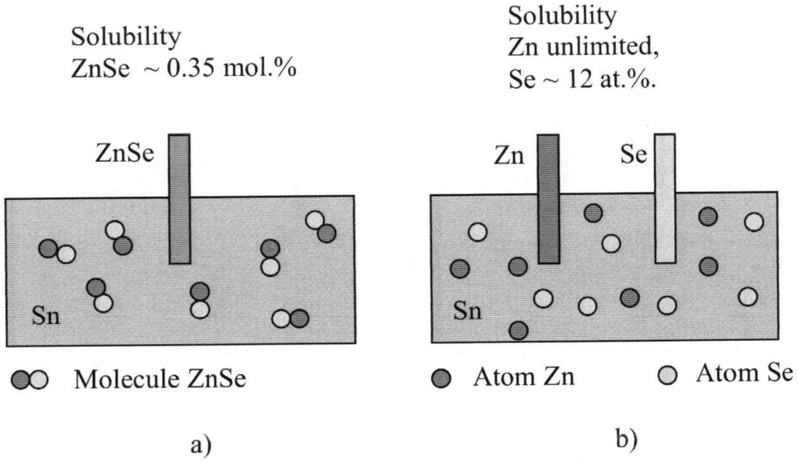

Figure 5. Dissolution of the binary compound zinc selenide in the form of molecules (a), as well as zinc and selenium in the form of individual atoms (b) in tin at the temperature of 750°C.

One of the necessary conditions for the crystallization of a solid solution from the liquid phase is the need to saturate the tin solution-melt at a selected temperature with the components of the solid solution, that is, gallium arsenide and zinc selenide. The $(GaAs)_{1-z}(ZnSe)_z$ solid solution from the tin solution-melt (Sn-GaAs-ZnSe) at the component content of 4.6 mol.% GaAs and 0.3 mol.% ZnSe at the crystallization initiation temperature of 720°C was grown. But if binary compounds GaAs and ZnSe decompose into individual Ga, As, Zn and Se atoms, then according to the phase diagram of binary alloys (Fig. 3) at the specified content of components at 720°C the solution-melt will not be saturated not with gallium, not arsenic, not zinc and not selenium. Then the crystallization of the $(GaAs)_{1-z}(ZnSe)_z$ epitaxial layer should not have occurred. Crystallization of the solid solution $(GaAs)_{1-z}(ZnSe)_z$ under the chosen conditions indicates that the solution-melt is saturated with the molecules of the components, and there is no decomposition of the GaAs and ZnSe molecules into individual Ga, As, Zn, and Se atoms.

In fig. 6 shows the tetrahedral lattice of the GaAs binary compound. Atom Ga* (cationic lattice site) and As* (anionic lattice site) are in the center of the corresponding tetrahedra. The Ga* atom has 3 valence electrons, and all of them will participate in the formation of covalent bonds with three neighboring As atoms - bonds (1), (2) and (3) (see Fig. 6). The As* atom has 5 valence electrons, 3 of which are involved in the formation of covalent bonds with three neighboring Ga atoms - bonds (5), (6) and (7) (see Fig. 6). The remaining two valence electrons participate in the formation of the bond with the Ga* atom - bond (4) (see Fig. 6). Since bonds (1), (2), (3), (5), (6) and (7) are formed by the sharing of electrons of neighboring Ga and As atoms, and bond (4) is only due to two electrons of the As* atom, then, apparently, the ionic fraction of the chemical bond (4) is stronger than the others. Consequently, relationship (4) is stronger than the others.

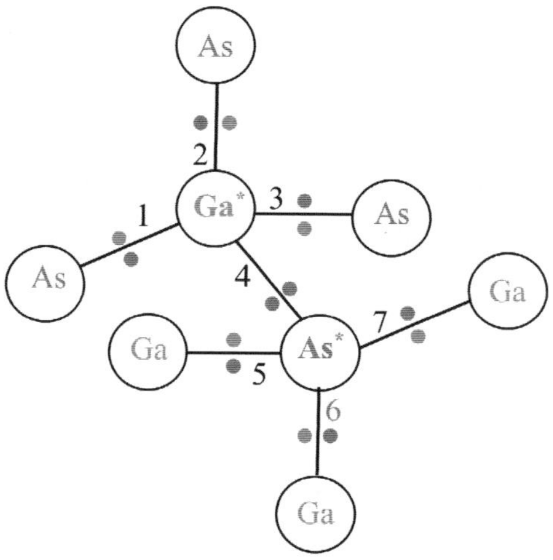

Figure 6. GaAs tetrahedral lattice. The dark circles represent electrons forming covalent bonds 1–7.

When GaAs is dissolved in the tin solvent at 650-750°C, the energy of thermal vibrations and the forces of interaction of Sn and GaAs atoms is sufficient to break covalent bonds (1), (2), (3), (5), (6) and (7), however, not enough to break the connection (4). Due to this, dissolved GaAs molecules will exist in the solution-melt, and not separate Ga and As atoms. A similar situation occurs when ZnSe dissolves in Sn at 650-750°C. Apparently, in binary compounds, there are asymmetries of the forces of interaction between the atoms of cations and anions.

The solid solution $(GaAs)_{1-z}(ZnSe)_z$ was grown on monocrystal GaAs substrates. The solution-melt was in close contact with the substrates. At the initial moment of crystallization, the solution-melt was supersaturated with gallium arsenide. Therefore, upon cooling, the epitaxial sublayer of gallium arsenide crystallizes. While cooling, a moment comes when the solution-melt is oversaturated with zinc selenide. From this moment on, the epitaxial layers of the $(GaAs)_{1-z}(ZnSe)_z$ solid solution with a variable composition crystallize. The optimization of the technological process of the growth of epitaxial layers was carried out by varying parameters such as the thickness of the solution-melt (d), the growth temperature interval (T), and the cooling rate of the solution-melt. Crystalline perfect layers were grown at d = 1 mm, T = 720–650^0 C, cooling rate of the solution-melt - 1 deg/min. In this case, the epitaxial layers had p - type conductivity and a thickness of 4 μm. The resistivity of the layers, determined by the Hall method, at 300 K was ~ 180 mΩcm, the Hall mobility was ~ 60 $cm^2/(Vs)$, the concentration of charge carriers was ~ $3 10^{18}$ cm^{-3}. The surface density of dislocation defects at the substrate-film interface was ~ 10^7 cm^{-2}, on the film surface ~ $4 10^5$ cm^{-2}. The band gap of the near-surface layer of the solid solution $(GaAs)_{1-z}(ZnSe)_z$ (z = 0.04) is E_g =1.45 eV.

The chemical composition of epitaxial layers $(GaAs)_{1-z}(ZnSe)_z$ was determined using a Jeol JSM 5910 LV-Japan X-ray microanalyzer. Fig. 7 shows the profile of the distribution of As, Ga, Se, and Zn atoms over the thickness of the epitaxial layer. Analysis of the results of the fig. 7 shows that, upon crystallization of the epitaxial film, there is an intense substitution of gallium arsenide molecules by zinc selenide molecules at the sites of the crystal lattice of the solid solution. The maximum content of zinc selenide in the solid solution (z = 0.80) is achieved at an epitaxial layer thickness of 0.6 μm. Subsequently, as the film grows to the thickness of 1.7 μm, the content of zinc selenide molecules decreases rapidly, and then slowly to the thickness of 3.8 μm. On the surface layer, the molar content of zinc selenide is z = 0.04.

The liquid phase epitaxy process takes place in an open system and in a limited volume. The system is two-phase - liquid (solution-melt) and solid (solid solution). During the process, heat is continuously removed from the system. The solution-melt is multicomponent. In such a system, a process of self-organization certainly arises. Crystallization of a solid solution with a variable composition is the result of the self-organization process of a multicomponent system. At 720°C, the solubility of zinc selenide in tin is an order of magnitude lower than the solubility of gallium arsenide. Therefore, after the intensive introduction of zinc selenide molecules into the solid phase, the solution-melt is depleted by these molecules, which causes a further decrease in the content of zinc selenide in the solid solution. The sublayer of the solid solution enriched with zinc selenide has a higher value of the band gap. Thus, the resulting epitaxial layer is a varyband gap semiconductor solid solution $(GaAs)_{1-z}(ZnSe)_z$ ($0 \leq z \leq 0.80$) with a smoothly (but not monotonously) varying composition. A wider-gap sublayer of the $(GaAs)_{0.8}(ZnSe)_{0.2}$ solid solution is formed between the substrate and the near-surface epitaxial film.

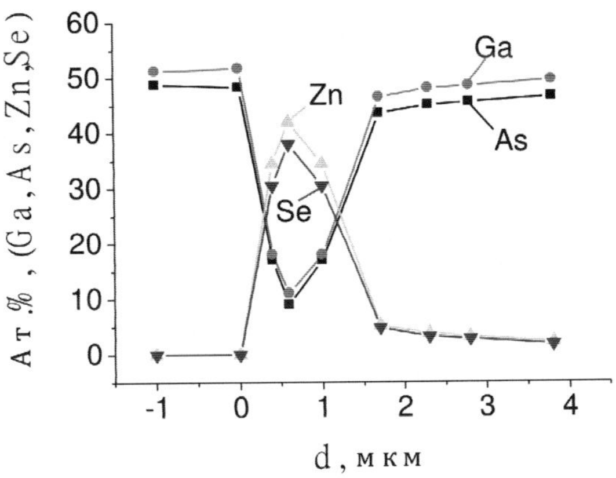

Figure 7. Distribution of As, Ga, Se, and Zn atoms over the depth of the $(GaAs)_{1-z}(ZnSe)_z$ epitaxial layer.

4. Conclusion

Thus, the technological possibilities of obtaining the monocrystal semiconductor $(GaAs)_{1-z}(ZnSe)_z$ solid solution of molecular substitution are shown. Necessary condition for obtaining such a solid solution is the existence of the components of the solid solution in the liquid solution in the form of molecules, and not in the form of individual atoms, as well as the saturation of the solution-melt with molecules of the components of the solid solution.

References

[1] Ballabio A, Bietti S, Scaccabarozzi A, Esposito L, Vichi S, Fedorov A, Vinattieri A, Mannucci C, Biccari F, Nemcsis A, Toth L, Miglio L, Gurioli M and Isella S 2019 *Sanguinetti. Sci. Rep.* **9** doi:10.1038/s41598-019-53949-x

[2] Feifel M, Lackner D, Ohlmann J, Benick J, Hermle M and Dimroth F. 2019 *Sol. RRL* **3** 1900313 doi:10.1002/solr.201900313

[3] Geldash A A, Djuplin V N, Klimin V S, Solodovnik M S, Ageev O A 2019 *J. Phys. Conf. Ser.* **1410** 012030 doi:10.1088/1742-6596/1410/1/012030

[4] Suprun S P, Sherstyakov V N, Fedosenko E V 2009 Epitaksiya ZnSe na GaAs pri ispolzovanii v kachestve istochnika soyidineniya ZnSe *Semiconductor physics and technology* **43(11)** 1570-5

[5] Bobrenko Yu N, Paveles S Yu, Paveles A M and Yaroshenko N V 2013 Fotoelektricheskie preobrazovateli s varizonnimi sloyami na osnove ZnSe *Semiconductor physics and technology*

47(10) 1381-4
[6] Farrella H H and Randall A 2004 LaViolette Cation variations at semiconductor interfaces: ZnSe(001)/GaAs(001) superlattices *Journal of Vacuum Science & Technology B* **22(4)** 2250-6
[7] Wang L G and Dilute Alex Zunger 2003 Nonisovalent (II-VI)-(III-V) Semiconductor Alloys: Monodoping, Codoping, and Cluster Doping in ZnSe-GaAs *Physical Review B* **68** 125211.1-125211.8.
[8] Saidov A S, Saidov M S, Usmonov Sh N, Leiderman A Yu, Kalanov M U, Gaimnazarov K G and Kurmantaev A N 2011 Growth of $(GaAs)_{1-x}(ZnSe)_x$ Solid Solution Films and Investigation of Their Structural and Some Photoelectric Properties *Physics of the Solid State* **53(10)** 2012-21
[9] Vikram K 2000 *International Workshop on the Physics of Semiconductor Devices* (December 14-18, 1999, Delhi) (Mumbai: Allied Publ.) **2** 1425 *Proceedings of SPIE* **3795** 2
[10] Xansen M and Anderko K 1962 *Binary alloy structures* (M.: Metallurgizdat)

Development of innovative technologies for deep mineral processing

Elvira Zabneva[1], Yulia Kuznetsova[1], Anatoly Poddubnyak[2] and Evgeny Lunev[3]

[1] Kuzbass State Technical University named after T.F. Gorbachev in Novokuznetsk, Russia
[2] Innovative Initiative LLC, Alchevsk, Ukraine
[3] Royal Carbon LLC, Kemerovo, Russia

E-mail: zabnevailvira@mail.ru, lunev21@mail.ru, acanaria2005@yandex.ru

Abstract. The rate of industrial development depends not only on the demand for products that are outdated in terms of quality, but also on the possibility of updating the quality indicators of the raw material mass, matching the demand dynamics of the raw materials quality with the indicators corresponding to the future tasks. The transition of the world economy and energy to a low-carbon development trajectory reduces the demand for Russian coal, and therefore the near future of the coal industry is associated with fundamental changes in the requirements for the quality and variety of marketable products. There is an urgent need for technological renewal in coal enrichment, restoration and further development of coal chemistry. The paper highlights the key problems and trends in the field of deep coal enrichment, reveals the methods used in the processing plants of Kuzbass. The proposed technology of deep mineral processing "Carbon-99", which received the support of the Skolkovo Innovation Center in 2021, is described. The Carbon-99 technology is a synthesis of technological solutions groups, which variety depends on the properties of the mineral groups included in a particular coal material: enrichment of rock mass cl. 2-200 mm, magnetic separation cl. 0-2 mm, enrichment of sludge masses-centrifugation cl. 0-0. 05 mm. It is shown that the proposed complex technology of deep coal enrichment will reduce the ash content in the concentrates of brown, coking and anthracite coals to 1-9%, and reduce the sulfur content by 30-70%. Reduction of moisture content in the coal mass, including dispersed materials. Also, to obtain concentrates of mineral groups extracted from the coal mass, without chemical reagents.

1. Introduction

In the complicated modern economic conditions, the search for ways to maximize the economic effect of the results of their activities becomes a relevant objective for enterprises. The solution of this problem takes on a special character for those of them whose sphere of activity is the enrichment of minerals. On the one hand, this is due to the global trend of reorientation of the raw materials model of the economy to the processing one; on the other, the specifics of mineral processing involves the implementation of a large number of sequential operations, when the effectiveness of one of them lays the foundation for the quality of the next one; on the third hand, for many enterprises, increasing the content of useful components in processed raw materials is a key direction for the intensification of technological processes.

Coal is a unique material that is used as fuel, raw materials for various industries, and a resource for extracting concentrates of rare elements. One of the areas of active discussion of modern researchers is the fact which method of enrichment ("wet" or "dry") is the most acceptable. Thus, from the point of view of A. M. Gerasimov and S. V. Dmitriev, the use of "wet" production technologies has many negative economic and environmental consequences, especially in the regions of Siberia and regions with arid climate [1]. Therefore, the authors proposed a technological scheme for mineral coal enrichment based on its thermochemical modification. B. Mandal, D. Pudasainee, V. Kurian, R. Gupta and other have a different point of view: in their view, the product quality and extraction with the "wet" method is better than with the dry method of enrichment and the former should be actively developed. But it is important to understand that wet cleaning is mainly used for metallurgical coals, whereas for energy coals there is a general tendency to use dry coal cleaning [2, 3]. It should be noted that many foreign researchers, when describing the methods of coal enrichment, place great emphasis on embedding carbon capture technology and reducing greenhouse gas emissions [4].

The currently existing methods of coal enrichment differ in a wide range of parameters, but the constant increase in the requirements of the consuming industries for the quality of coal products leads to a continuing interest in the development of new ones.These and many other features of the activities of enterprises engaged in the field of mineral processing determine the need to search for production-specific, cost-optimal, relevant to the trends and processes occurring in the industry, scientifically based and highly efficient technologies.In addition, it is important for processing enterprises that the technologies used meet the natural-geographical, climatic and geological features of the territory of their presence.

2. Identification of problems and trends in the field of deep coal processing

The discrepancy between the technological level and the quality of coal products to the requirements of the global coal market is an urgent problem of the industry. The fall in the cost of coal on the world market can be considered as a natural process associated with the imbalance between supply and demand in the competitive struggle with cleaner energy sources. The study of the coal market in Russia conducted by Alto Consulting Group company showed that 80% of all production is carried out in the Kansk-Achinsk, Kuzbass, and Pechersk basins [5], and the search for new coal formations is constantly underway. However, the increase in the production of high-ash coal will only worsen the situation on the market in the future. Considering the fact that the logistics transit is overloaded with coal with an ash content of 9-20%, a valuable mineral component containing a polymetallic mineral group and rare earth elements is sent abroad. High-quality Russian products with an ash content of less than 6% in the domestic and foreign markets are practically invisible, in contrast with foreign high-tech carbon-containing products in the form of a long range of lubricants, paints, fillers, sorbents, electrode mass, catalysts and other things [6].

Against the background of theseproblemsnew global trends in the field of deep coal enrichment are being formed:

- Strengthening of Russia's positions in the global coal market.
- Technological development of the industry and strengthening of the scientific and technical base of companies and research centers.
- Development of labor relations and corporate social responsibility of coal companies.
- Improving of professional training system of personnel for the coal industry.

3. Methods of coal enrichment at Kuzbass processing plants

Currently only in the Kemerovo region – Kuzbass there are 54 processing plants and units [7]. In addition to the Kuzbass enterprises themselves which develop and use existing coal enrichment technologies, scientific and educational institutions are working in this direction in the region (table 1).

Table 1. Key coal processing methods used at Kuzbass processing plants [8].

Company name	Methods used
JSC «SUEK-Kuzbass»	Jigging
JSC HC «SDS-Coal»	Jigging, flotation, dense medium separation
JSC OUK «Yuzhkuzbassugol»	Jigging, flotation
JSC «Sibuglemet»	Jigging, flotation, dense medium separation, spiral separation
PJSC «SouthernKuzbass»	Jigging, flotation, dense medium separation, spiral separation
JSC «Coalcompany«Kuzbassrazrezugol»	Flotation, dense medium separation, spiral separation
JSC «Stroyservice»	Jigging, flotation, dense medium separation
LLC «Processing plant «Prokopevskaya»	Jigging
JSC «Topprom»	Dense medium separation, spiral separation
LLC «Processingplant «Angersky»	Jigging, flotation
LLC «Coalcompany «Zarechnaya»	Dense medium separation, spiral separation
JSC «RaspadskayaCoalCompany»	Dense medium separation, spiral separation
JSC «Coal company «Northern Kuzbass»	Jigging, flotation, spiralseparation

The study of the disadvantages and advantages of existing technological solutions according to their main indicators has allowed to reveal the following differences (table 2).

Table 2. Quantitative indicators of key technological solutions in the field of coal processing.

Indicator name	Pneumatic vacuum separation unit	«Wet» technologies for the enrichment of mining and coal raw materials
Fractional composition of the enriched mass, mm	13-100	0.5-200
Energy costs, kW/t	1.4-2,0	2.5-4
Ash content of marketable products, %	from +5	from +5
Output of marketable products, %	30-35	45-70
Ashcontent of waste, %	+40	45-70

The presented technologies are characterized by the lack of the following features:

- Work at the same density index of coal and rock.
- Production of marketable products with reduced ash and sulfur values below the parent values.
- Extraction of mineral concentrates without chemical reagents in a separate stream using dry method.
- Extraction of toxic, mineral groups from enrichment waste using dry method.
- Deep processing of waste into construction and mineral concentrates.

4. Deep mineral processing technology “Karbon-99”

For several years the authors of the research have been working on the concept of a technology for deep enrichment of coal mass to produce a concentratewith an ash content of 1-6% without the use of chemical reagents, which in 2021 received the support of the Skolkovo Innovation Center. The approach underlying the project includes 3 technologies that are basic for achieving high-quality indicators of coal enrichment in industrial volumes (figure 1)

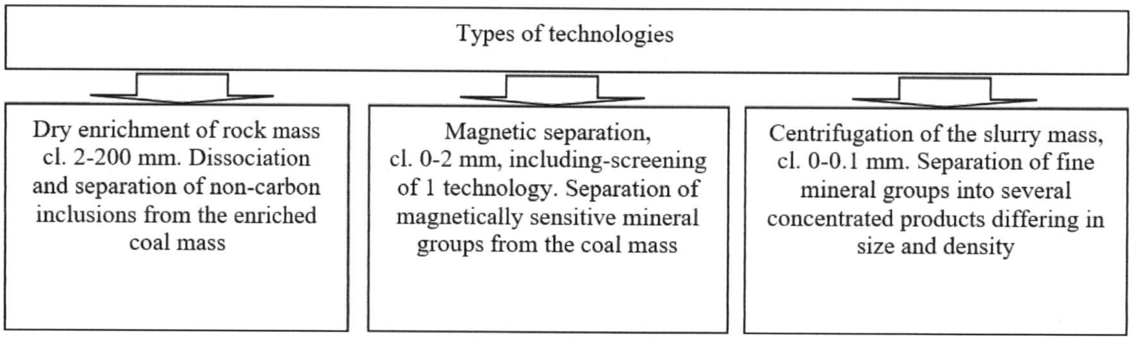

Figure 1. Types of technologies as a part of “Carbon-99”.

According to research, the existing technologies of coal enrichment use, as a rule, two main differences in the organic and mineral components (the difference in the densities of these parts and the difference in their surface properties-wettability). In addition to these differences, the proposed integrated technology for deep coal enrichment uses a significant difference in the friction properties and spectral-noise characteristics, geometric shapes of surface particles, magnetic susceptibility and specific conductivity. The proposed technology will allow the extraction of pyrite inclusions, minerals of the ferroaluminosilicate group, separation of vitrinite group, organic sulfur, alkaline earth compounds and sulfur-containing minerals from the carbon mass, as well as the separation of polymetallic mineral groups.

In terms of the first type of the technology, an automatic control system allows to control the ash content on the flow of the material going to the enrichment, including noise characteristics.There is an option that allows to control the quality of enrichment on the flow, due to the deviation of the actuator by feedback, according to the results of the analysis of noise characteristics.The automatic control system monitors the incoming and outgoing mass, humidity, volume, ash content of marketable products on the flow (according to load cells, capacitive and ultrasonic sensors, the spectrum of noise characteristics controlled on the flow is analyzed). At the exit from the processing plant, the indicators of coal, pyrite minerals, and rock varieties are similarly monitored. The operation principle of the ash content sensors is based on a two-stage analysis of the spectral noise characteristics of coal, rock, and pyrite before the enrichment process and after on the outgoing streams. The proposed technology is used for dry coal with a moisture content of no more than 7.5% for classes from +2 mm and higher.

In the second technology, the material crushed within 0.05-2mm is classified and sent for magnetic separation. The principle of the magnetic separator operation is based on the delta in the magnetic susceptibility of the minerals being enriched, followed by radial separation. For minerals that do not have magnetic susceptibility, an additional concentrating unit of magnetostatic separation is installed to separate mineral inclusions by resistivity. The magnetostatic separator can work independently with various mineral groups ranging in size from values less than a micron to 3 mm.

Settling centrifuge-classifier allows to implement several modes of operation in one unit: including counterflow and direct-flow mode. In the zone where the power is supplied through the feed pipe the counterflow mode of movement of the settled solid phase and the formed liquid phase – a centrate with non-settled solid particlesis carried out. In the opposite part of the rotor a direct-flow mode of operation of a cascade classifier centrifuge is implemented, in which the settled fine particles of the solid phase and the clarified product move along the rotor in one direction, which contributes to the controlled clarification of the centrate in several cascades discharged from the centrifuge.

Deep enrichment of coal mass will allow Russian ore enrichment specialists to provide their industry with high-quality raw materials and compete in the world market of energy sources and high-tech materials by reducing the ash content of products below the parent values in the range of 1-9%.One of the significant factors in the application of the "Carbon-99" technology is the reduction of the environmental load due to the extraction of sorption mineral groups from the coal mass, with the subsequent introduction of commercially attractive mineral concentrates into the enterprise's economy. As a result of deep processing of coal several products are being formed:

- Coal concentrate with an ash content of 1-9%.
- Mineral concentrate, which in contrast to the industrial product of existing technologies is characterized by an incomparably increased content of sorption mineral groups extracted from the coal mass (the material is of interest as a valuable ore concentrate).
- Deep enrichment tailings (of interest as a group of building materials).

Sludge coal washing discharges are the product of technological imperfections of the currently used technologies. By applying “Carbon-99” technologies to the processing of sludge discharges it is possible to change the situation in financial and environmental terms, extracting from the accumulated sludge mass for many decades in all fields of the country, carbon containing material with an ash content of 1 to 15%. With no additional mining and milling costs. To obtain concentrates of mineral groups of commercial interest in industry, to extract environmentally hazardous mineral groups. Significantly reduce the volume and financial maintenance of tailing dumps, hydraulic waste disposals, treatment of recycled and waste water. Additionally, to attract to the economy of the regions products of deep enrichment, mineral groups, both from dry waste and from pulp.

Speaking about the quantitative indicators of the “Carbon-99” technology the following values should be given (table 3).

Table 3. Quantitative indicators of the «Carbon-99» technology.

Indicator name	Deep coal processing plant with extraction of valuable mineral and organic components
Fractional composition of the enriched mass, mm	0.001-200
Energy costs, kW / t	0.1
Ash content of marketable products, %	1-6
Output of marketable products, %	75-90
Ash content of waste, %	70-90
The possibility of enriching dry rock mass	yes
The possibility of enriching wet rock mass	yes

5. Summary

The approach underlying the "Carbon-99" technology is an awareness of a set of unsolved goals, economic and environmental problems: between the quality indicators of the industrial coal market and the increasing requirements for the quality of coal products in metallurgy, coke chemistry, coal chemistry, energy in the domestic and international markets, associated with insufficient technological equipment and, as a result, with low economic profitability of coal enterprises, environmentally hazardous marketable products and toxic production waste. "Carbon-99" is a technological tool that shifts the standards of requirements from the existing quality indicators and technological capabilities to the zone of quality indicators of 1-9%.

References

[1] Gerasimov A M and DmitrievS V 2016 Combined dry coal processing technology *Ore processing* **6** 9-14
[2] Mandal B B, Bhattacharya S A, Anupam and Jain P K 2017 Vibration exposure of workmen of coal preparation plants - an exploratory study *Noise and Vibration Worldwide* **17:45(7)** 21-

31
[3] Pudasainee D, Kurian V and Gupta R 2020 2-Coal: Past, Present, and Future Sustainable Use, Future Energy (Third Edition): Improved, Sustainable and Clean Options for our Planet *Elsevier*
[4] Miller Bruce G 2017 *Clean Coal Engineering Technology* (Oxford : Butterworth-Heinemann)
[5] Coal market research *Current situation and forecast Alto Consulting Group Company* https://alto-group.ru/otchot/analiz-importa-eksporta/1179-import-i-eksport-burogo-uglya-v-rossiiv-2018-g.html
[6] Novoselov S V and Oganisyan A S 2021 Problems, risks and forecasts of the coal industry development of the Kemerovo region for the period up to 2035 *Coal* **2(1139)** 38-41
[7] Department of Coal Industry of the Kuzbass *Government Administration* https://mupk42.ru/ru/
[8] Antipenko L A and Ermakov A Yu 2012 Ways to increase the volume of coal preparation *Mining information and analytical bulletin (scientific and technical journal* **55** 59-69

Photo- and magnetic-field-induced conductivity change in the BiFe1-xCoxO3 films

V V Kretinin, O B Begisheva, F V Zelenov, P P Mashkov and A N Pavlova

Reshetnev Siberian State University of Science and Technology, 31 Krasnoyarsky Rabochy Ave., Krasnoyarsk, 660037, Russia

E-mail: kret_vas@mail.ru

Abstract. The magnetic and electrical properties of the polycrystalline BiFe1-xCoxO3 films have been investigated in the temperature range of 80–600 K and magnetic fields of up to 12 kOe. The photocurrent and the diode effect induced by the blue light laser illumination at room temperatures have been established. The ohmic conductivity has been determined from the I–V characteristics and the conductivity growth in a magnetic field has been observed. The deviation of the temperature dependence of the magnetization from the power law below room temperature has been found.

1. Introduction

Recently, multiferroics have been intensively investigated due to the presence of the magnetic and ferroelectric order in them [1, 2]. The interaction between the magnetic and electric subsystems in these materials makes it possible to control the electric polarization by a magnetic field and govern the magnetic characteristics by an electric field [3–7]. This offers the opportunity of increasing the storage density and making the memory nonvolatile. The inclusion of defect states into multiferroics can change their conductivity and the latter can be controlled by domain walls [8, 9] in an electric field. The conductivity of semiconductors can be governed by a magnetic field, even in the paramagnetic state [10–15]. In the magnetically ordered state, the conductivity over defect states depends on the type of a magnetic order [16–19].

According to the results of the LDA calculation [20], cobalt-substituted bismuth ferrite forms impurity states near the valence band. As the substitution concentration increases, the electron density in the t2g shell grows; this leads to the degeneracy, which can be eliminated by the strong electron correlations [21] or the orbital ordering of electrons. This leads to the conductivity change, which was observed in semiconductors with the orbital ordering [22–26]. The coexistence of the magnetoelectric effect and conductivity makes it possible to manipulate the current by magnetic and electric fields [27–28].

Substitution of holmium and strontium ions for bismuth in the BiFeO3 compound [29, 30] causes the metal-type conductivity of the material. The formation of defect states in the oxygen subsystem ensures the hopping-type conductivity. Substitution of 3d metals for iron ions leads to the partial filling of the impurity band and increases the conductivity. In [31], the IR radiation-induced diode effect in the BiMnxFe1–xO3 films was found, which is related to the asymmetry of quantum wells and band bending at the film surface.

In the BiMexFe1-xO3 (Me = Mn, Co, V) films, areas with the electric polarization can be retained and the conduction channels controlled by magnetic and electric fields can exist. An internal electric field creates the prerequisites for the photocurrent and photoinduced effects [32].

The aim of this study is to investigate the possibility of the illumination and magnetic-field control of the current in the BiCoFeO3 films with electron doping obtained by replacing trivalent iron by cobalt ions with an additional electron in the t2g shell.

2. Experimental part and discussion

The cobalt-doped bismuth ferrite films were synthesized from the $BiFe_{0.8}Co_{0.2}O_3$ solid solution precursors on object glasses by the ignition technique. Sputtering was performed in a vacuum system at a pressure of 10^{-3} Pa in the reaction chamber; the temperature of a tantalum evaporator was kept around 2000°C. The substrate temperature was varied within 250–300°C. The film was 13 × 17 mm in size and 980 nm thick. According to the X-ray diffraction data, the BiFe1-xCoxO3 (BFCO) compound has a rhombic structure (R3c sp. gr.) with crystal lattice parameters of a = b = 0.558 nm and c = 1.380 nm (figure 1). The X-ray diffraction patterns contain, along with the reflections of the main BFCO phase, the reflections of Bi_xFeO_{39} with a 5% sillenite structure, which is paramagnetic.

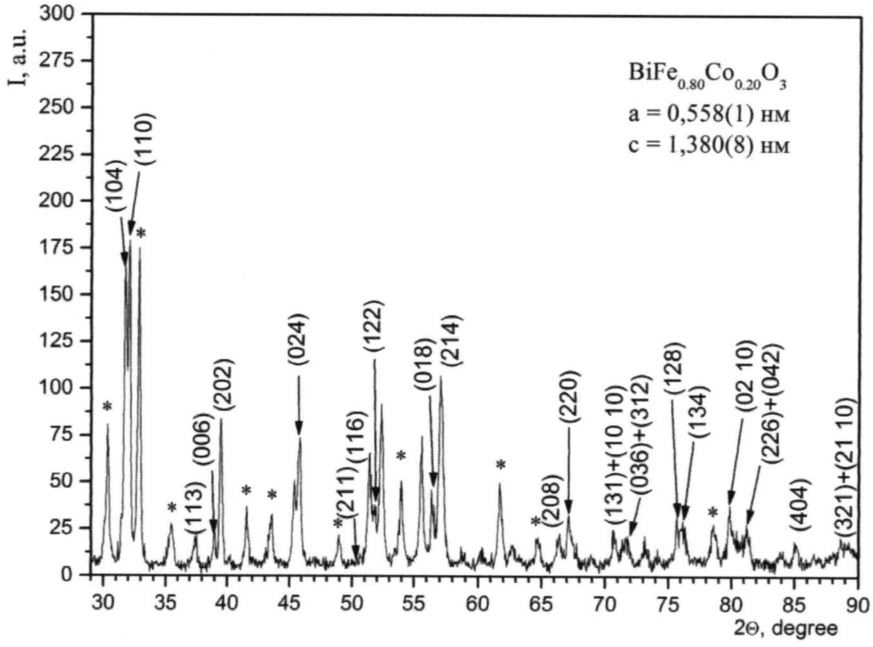

Figure 1. XRD pattern of the $BiCo_{0.2}Fe_{0.8}O_3$ at T=300K.

Figure 2 shows the temperature dependence of the magnetization measured in a magnetic field of 8.6 kOe by the Faraday method. Upon heating, the magnetization smoothly increases, passes through a maximum at 360 K, and tends to zero at 600 K, which was attributed by us to the occurrence of a magnetic phase transition. Above 300 K, the temperature behavior of the magnetization is satisfactorily described in the molecular field approximation by the power law $M = M_0(1 - T/Tc)1/2$ with Tc = 600 K. The deviation from the theoretical power dependence observed below 300 K is possibly due to the noncollinear arrangement of spins caused by the competition of exchange interactions between the spins of iron ions and the cobalt spins. The more than half-filling of the t2g shell facilitates the ferromagnetic exchange in cobalt clusters. The degenerate state occurring in the t2g shell can be eliminated by the orbital ordering, the formation of which leads to a change in the magnetic structure [33]. The degeneracy elimination by the Jahn–Teller channel causes the formation of a strong electron–phonon interaction, which leads to the four-spin interaction and a decrease in the magnetic moment on a site [34–40].

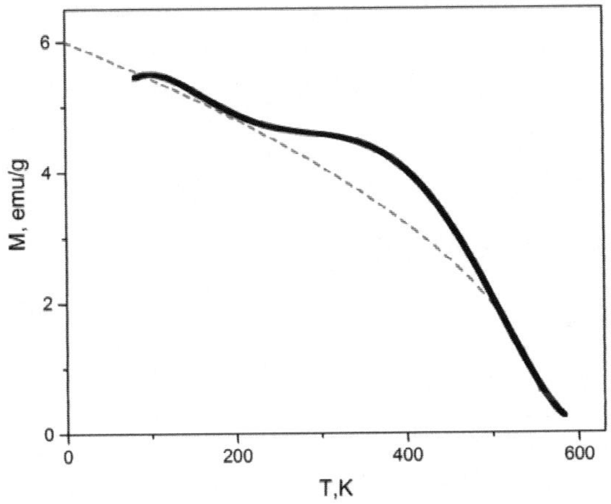

Figure 2. Temperature dependence of the magnetization for the BiCo0.2Fe0.8O3 film in a magnetic field of H = 8.6 kOe. Fitting function M=M0 (1-T/Tc)1/2 (dotted line).

We establish the effect of light on the resistive characteristics and current from the I–V characteristics measured without light and under laser illumination in the blue range at a wavelength of 405 nm. The I–V characteristics are linear up to 400 K; upon further heating, they become nonlinear (figure 3). This is related to the impurity states formed by cobalt ions and oxygen vacancies, which contribute to the conductivity. At low voltages, the carriers injected from the electrodes fill shallow traps. As the voltage increases ($U > U_c$), the shallow traps are filled and the transitions of electrons (holes) to the higher-lying vacant states start contributing to the resistance. Under illumination, the current slightly (by 2–5%) grows. The voltage dependence of the photocurrent is shown in figure 4.

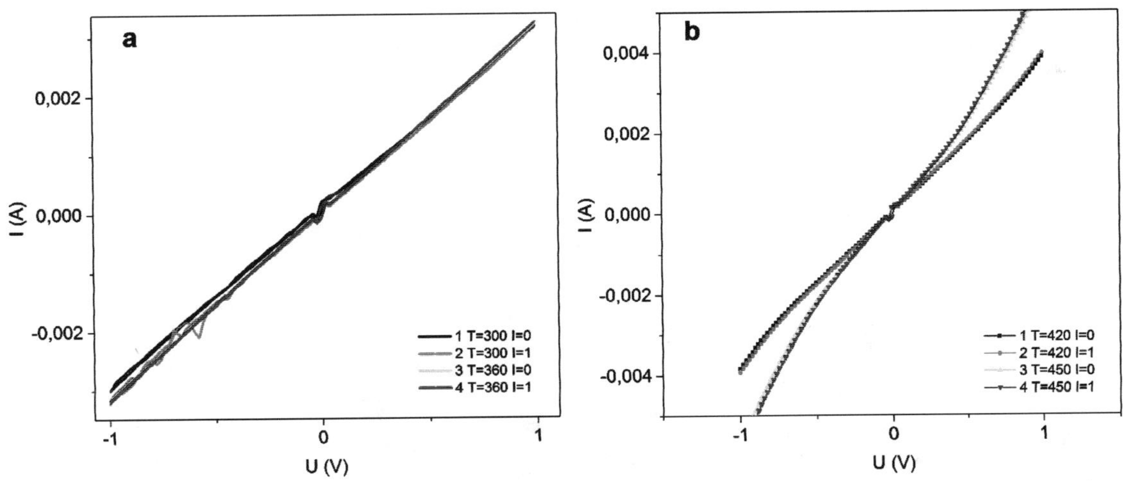

Figure 3. VAX of the BiCo0.2Fe0.8O3 film without illumination (1,3,5,7) and with λ=405 nm (2,4,6,8) at T=300 K (1,2), 360 K (3,4) (a), T=420 K (5,6), 450 K (7,8) (b).

At temperatures of up to 360 K, the relative current variation (I(IR) – I(0)/I(0)) increases with a decrease in the external electric field; above this temperature, it is almost voltage-independent (figure 4). In the film with the high (x = 0.2) cobalt concentration, an impurity subband is formed from manganese ions. The top of the valence band and the bottom of the conduction band change in space with the smooth band bending caused by structural defects and oxygen vacancies. When the film is illuminated by light quanta with an energy of hv = 3 eV, the majority carriers are holes in the valence band and electrons in the conduction band. The low photocurrent is explained by the disappearance of the electric polarization. The photocurrent is proportional to the density of carriers passed from the valence to conduction band upon absorption of a photon. At room temperature, the photocurrent asymmetry and the diode effect are 3%; above 360 K, they vanish.

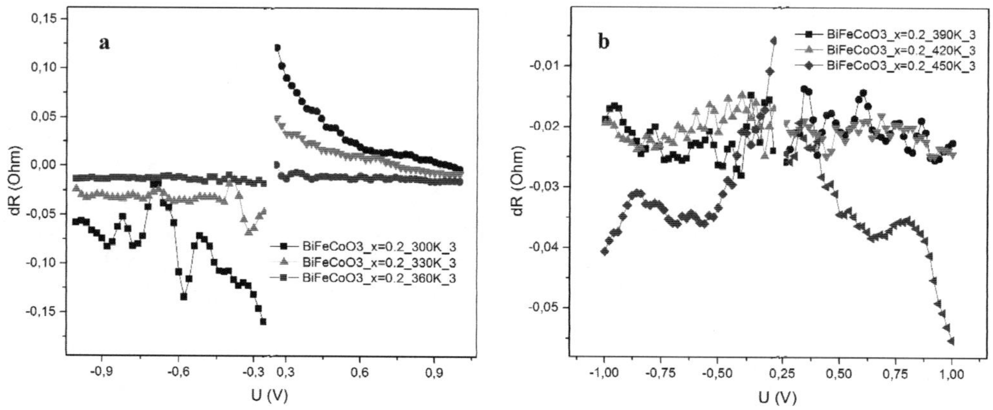

Figure 4. Relative photocurrent dIph= (I (0) −I(IR))/I(IR) from voltage for X=0.2 with illumination λ=405 nm at T=300 K (1), 330 K (2), 360 K (3) (a), T=390 K (4), 420 K (5), 450 K (6) (b).

We find the effect of the magnetic field on the conductivity from the field dependence of the conductivity at constant temperatures (figure 5). As the temperature of the magnetic phase transition is approached, the conductivity increases in a magnetic field. The increase in the conductivity is caused by electrons hoppings over iron ions. The hopping probability increases when the magnetic moments are parallel and decreases at other magnetization directions. The magnetic field dependence of the conductivity is described within the ferron model [41] as (1) where B is determined by the combination of the energies of the Coulomb interaction of electrons in the impurity region, H is the external magnetic field, and ξ is electron localization radius. Upon heating, the ferron radius increases and the potential barrier for electron hoppings between ferrons decreases.

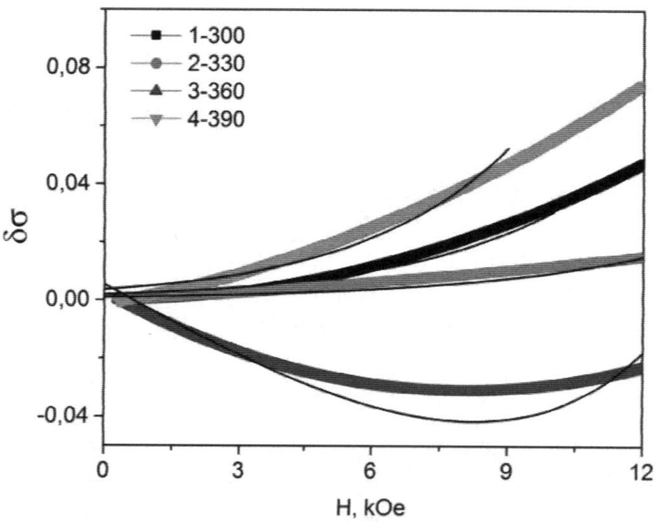

Figure 5. Field dependence of the magnetoconductivity $\delta 6 = \frac{6(H)-6(0)}{6(H)}$ for film BiCo0.2Fe0.95O3 at (1)300K, (2)330K, (3)360K and (4)390K. The solid line shows fitting by Eq. (1).

$$(\sigma(H) - \sigma(0))/\sigma(0) = 1 - \exp(-BH\xi/k_BT) \quad (1)$$

In the $BiCo_xFe_{1-x}O_3$ (x = 0.05) film, the voltage dependence of the current is linear (figure 6a). In a magnetic field of 12 kOe, the current and conductivity increase. The magnetic field dependence of the conductivity shown in figure 6b is satisfactorily described within the electron hopping model with regard to the spin polarization. Since the distance between cobalt ions at x = 0.05 is much shorter than for the composition with x = 0.2, an increase in the potential barrier leads to a decrease in the effect of the magnetic field on the conductivity. The two methods for measuring the conductivity lead to the conductivity growth in a magnetic field upon approaching the temperature of the magnetic phase transition.

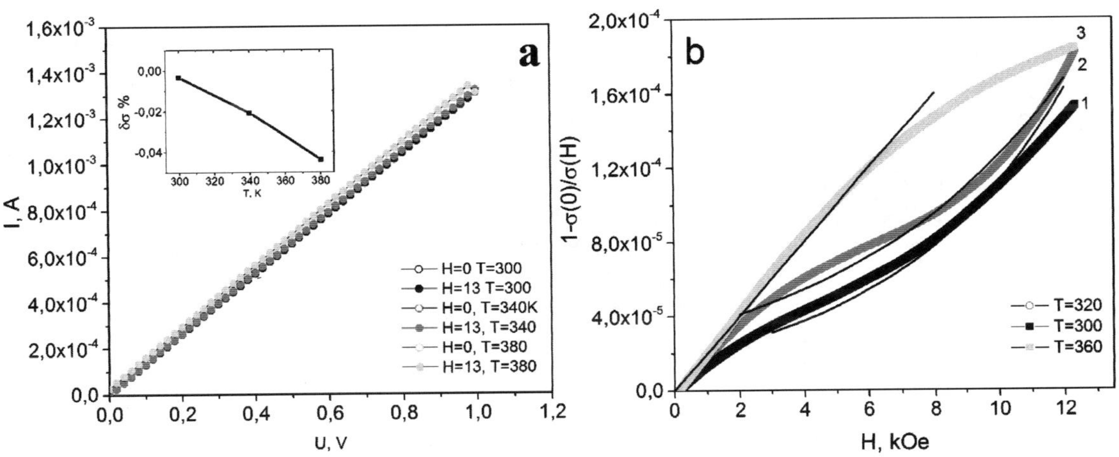

Figure 6. I-V characteristics measured in zero magnetic field and 12 kOe field for BiCo0.05Fe0.95O3 film at T = 300K (1), 340K (2), 380K (3). Insert: temperature dependence of magnetoconductivity $\delta_H = \frac{6(0)-6(H)}{6(0)}$. b) field dependence of normalized resistance at 300K (1), 320K (2) and 360K (3). The fitting functions by equation (1) (solid line).

3. Conclusions

In the $BiFe_{1-x}Co_xO_3$ (x = 0.2) films, the photo-effect and the weak diode effect at room temperatures were found. A linear voltage dependence of the current below 400 K was observed. The conductivity growth in a magnetic field at the two cobalt concentrations was established; it was found that the effect of the magnetic conductivity increases with the concentration. The conductivity growth was explained using the model of electron hoppings over ferrons with regard to the spin polarization of electrons.

Acknowledgments
This work was supported by the Grant of the President of the Russian Federation № MK-620.2021.1.2.

References

[1] Fiebig M 2005 *J. Phys. D. Appl. Phys.* B **8** R123
[2] Scott J F 2012 *J. Mater. Chem.* **11** 4567
[3] Aplesnin S S *et al.* 2020 *JETP Letters* **112** 636
[4] Udod L V *et al.* 2020 *Eur. Phys. J. Plus* **135** 776
[5] Aplesnin S S *et al.* 2021 *J. Mater. Sci.: Mater. Electron.* 10.1007/s10854-020-05121-9
[6] Aplesnin S S *et al.* 2019 *Mater. Res. Express* **6** 116125
[7] Aplesnin S S *et al.* 2019 *JETP Lett.* **110** 223
[8] Kulikova D P *et al.* 2018 *Phys. Status Solidi RRL* **12** 1800066
[9] Logginov A S *et al.* 2007 *JETP Lett.* **86** 115

[10] Aplesnin S S *et al.* 2015 *Phys. Status Solidi (B) Basic Research* **252** 1792
[11] Aplesnin S S and Sitnikov M N 2014 *JETP Lett.* **100** 95
[12] Aplesnin S S *et al.* 2013 *J. Condens. Matter Phys.* **25** 025802
[13] Aplesnin S S *et al.* 2004 *Phys. Solid State* **46** 2067
[14] Aplesnin S S *et al.* 2016 *Phys. Status Solidi (B) Basic Research* **253** 1771
[15] Aplesnin S S *et al.* 2017 *Journal of Applied Physics* **121** 075701
[16] Aplesnin S S *et al.* 2015 *Phys. Solid State* **57** 1627
[17] Aplesnin S S *et al.* 2012 *Phys. Solid State* **54** 2005
[18] Aplesnin S S *et al.* 2011 *IEEE Transactions on Magnetics* **47** 6027598
[19] Aplesnin S S *et al.* 2010 *Bull. Russ. Acad. Sci.: Phys.* **74** 708
[20] Deepak A Sharma and Basantakumar H Sharma 2019 *Integr. Ferroelectr.* **203** 81
[21] Aplesnin S S *et al.* 2005 *Phys. Rev. B Condens. Matter.* **71** 125204
[22] Aplesnin S S *et al.* 2007 *Phys. Solid State* **49** 2080
[23] Aplesnin S.S *et al.* 2004 *Solid State Commun.* **129** 195
[24] Aplesnin S S *et al.* 2009 *Phys. Solid State* **51** 698
[25] Aplesnin S S *et al.* 2008 *J. Exp. Theor. Phys.***106** 765
[26] Petrakovskiĭ G A *et al.* 1999 *Phys. Solid State* **41** 1520-1524.
[27] Aplesnin S S *et al.* 2019 *Phys. Status Solidi B* **256** 1900043
[28] Aplesnin S S *et al.* 2017 *Phys. Solid State* **59** 2268
[29] Ju Hong Miao *et al.* 2009 *Am. Ceram. Soc.* **92** 2762
[30] T. Durga Rao *et al.* 2014 *Phys. B Condensed Matter* **448** 267
[31] Aplesnin S S *et al.* 2019 *Semicond. Sci. Technol.* **34** 095007
[32] Romanova O B *et al.* 2019 *Int. J. Appl. Phys.* **125** 175706
[33] Aplesnin S S and Moskvin A I 2008 *J. Condens. Matter Phys.* **20** 325202
[34] Aplesnin S S and Piskunova N I 2005 *J. Condens. Matter Phys.* **17** 5881
[35] Aplesnin S S 2004 *Phys. Lett. A* **333** 446
[36] Aplesnin S S 1998 *J. Condens. Matter Phys.* **10** 10061
[37] Aplesnin S S 1996 *Phys. Solid State* **38** 1031
[38] Aplesnin S S 2000 *Phys. B Condensed Matter* **61** 6780
[39] Aplesnin S S 2003 *J. Exp. Theor. Phys.* **97** 969
[40] Aplesnin S S 2003 *Phys. Lett. A* **313** 122
[41] Kagan M Yu and Kugel K I 2001 *Phys. Usp* .**44** 553

Research of physical and mechanical properties of plywood based on thermally modified veneer

A I Krivorotova, A A Orlov, Yu A Korchuk[1] and A A Popov

Reshetnev Siberian State University of Science and Technology, prospect "Krasnoyarsky rabochy "31, Krasnoyarsk, Russia

[1]E-mail: korchuk75@mail.ru

Abstract. The article presents the results of a study of the physical and mechanical properties of plywood based on thermally modified birch veneer. The implementation of a full-factor experiment in the temperature range from 160 to 200 °C with a duration of 120 to 240 min made it possible to establish the nature of the dependence of the change in strength in static bending and strength in shearing along the adhesive layer on the veneer heat treatment mode. The parameters of veneer heat treatment are determined, which do not lead to a decrease in the strength of plywood, but allow to reduce the indicators of water absorption and swelling by an amount from 20.4 to 33.5%. Plywood based on thermally modified veneer acquires higher strength and water resistance, which significantly expands the scope of application of glued wood based on urea-formaldehyde resins.

1. Introduction

The scientific foundations of heat treatment of wood were laid in the 30s – 40s of the twentieth century in Germany and the USA. Further research carried out in Germany in the period from the 50s to the 70s became the basis for more modern research carried out in the 90s in Finland, France and the Netherlands [1]. At the same time, the historical roots of the phenomenon of thermal modification of wood can be traced back to hundreds of years. Our ancestors used fire to increase the durability of wood. For example, even in the days of the Vikings, elements of the enclosing structures were processed over an open fire.

Currently, in many countries, various methods of thermomodification of wood are patented and implemented. The first industrial technology Thermowood was developed in Finland in 1997. In France, the Retification technology is being implemented, developed in the 70s of the last century, and which was commercialized several decades later. The technology involves the use of nitrogen instead of superheated steam. The Westwood method (USA) is based on the "3D heat wave" technology, when the steam in the closed space of the equipment is circulated forcibly by means of fans.

A distinctive feature of the Plato method (Holland) is a two-stage heating and cooling of the stack. The first heating is carried out under high pressure in a steam-water medium to a temperature of 150 to 180 °C, followed by cooling and drying, then the second stage of heating to 190 °C is performed.

The analysis of the works of Russian scientists dealing with the thermodification of wood is considered in the works [2-4].

The generalizing element of most technologies for thermal modification of wood raw materials is its heating in various environments in the temperature range from 150 to 280 °C. The main positive changes acquired by wood are an increase in resistance to the biological effects of fungi and insect larvae, a decrease in water absorption and moisture absorption, the acquisition of stability of geometric dimensions during operation, and an increase in thermal insulation characteristics. Heat treatment saturates the color of the wood, makes its structure brighter and more effective. Due to this, the wood of low-value species acquires an external resemblance to the wood of valuable hardwood species. [5-6].

As negative aspects of the process of thermal modification, researchers note a decrease in a number of mechanical and operational properties of wood, which limits the scope of its application. In [7], a decrease in the ultimate strength at rupture is noted as a result of a change in the heat treatment temperature, which at the same time has practically no effect on the elastic modulus. In the studies of F. Kollmann, violations in the microstructure of heat-treated spruce wood are described due to a decrease in the volume of wood in the process of heat treatment [8]. Heating wood to 180–200 °C in the presence of moisture leads to a decrease in resistance to shock loads, elastic modulus, tensile strength and compressive strength [9].

In general, the analysis of works devoted to the thermal modification of wood shows a multidirectional change in the properties of the processed material: the strength characteristics of wood decrease, but the characteristics improve, allowing the material to be used in conditions of high humidity.

The positive results of changes in the physical and mechanical properties of wood in the process of its thermal modification have become the basis for further research to study the effect of the thermal modification process on the properties of glued wood materials of various types. Today, plywood based on rotary cut veneer is one of the most widespread and promising glued materials for internal and external use. Significant strength in the longitudinal and transverse directions, high wear resistance and elasticity are the distinctive advantages of this material, which ensure versatility of use.

A number of works of the Kazan National Research Technological University are devoted to the study of the properties of plywood products made on the basis of thermally modified veneer [10-14].

For the manufacture of plywood with increased water resistance and moisture resistance, the veneer sheets were placed in a heat treatment chamber on the press plates, after which the chamber was sealed with a lid, and the press plates were compressed. The high temperature of the press plates up to 280 ° C caused thermal modification of the veneer in 10-15 minutes [15]. In the process of processing, the odd rows of plates carried out reciprocating movements and there was a constant "smoothing" of the veneer sheets by the pressure rollers. It is assumed that at such a temperature regime of processing, molecular changes occur in the wood material, in which the wood turns into a completely new material in comparison with what it was before processing. It was found in [15] that high-temperature processing of veneer leads to a decrease in hygroscopicity, a change in color, density, weight and cellular structure of veneer samples. The rational processing temperature is determined in the range from 190 to 210 ° C.

Chimiris Yu. V. in his work [16] found that to refine (obtain a light-yellow color) wood veneer with a thickness of 1 mm, it is necessary to heat up to a temperature of 190 ° C for 10 minutes and hold at the same temperature for 90 minutes. The test results showed that the color of the workpiece depends on the thickness of its cross-section and the holding temperature. With an increase in the thickness of the workpiece, an increase in the heating duration to the required temperature and an increase in the holding time are required. In this case, the color of the workpiece is in direct proportion to the holding temperature: the higher the temperature, the darker the color.

In [17], the thermal modification of veneer was carried out by the method in a rarefied environment in the temperature range from 413 to 533 K. The change in the color gamut was most noted at temperatures above 473 K. Considering that in the process of thermal modification, many properties of the veneer change from color to cellular structure, an indicator of the degree of thermal modification was applied - the change in the density of veneer in the process of heat treatment at different temperatures and duration of exposure relative to its initial density. It has been established that the swelling pressure of modified veneer samples decreases according to an increase in the processing temperature of the wood material, with the main effect on the degree of processing of the wood material is exerted by temperature and duration.

2. Methods

At the Department of Woodworking Technology, Siberian State University M.F. Reshetnev, an experimental installation for thermal modification of wood was mounted and tested. In the course of preliminary experiments, a heat-shielding structure was developed and constructed, designed to ensure a uniform and stable arrangement of samples and their protection from thermal radiation from the surface of the cabinet walls. For the experiment, the temperature range was used, which is considered the most optimal for the thermal modification of wood from 150 to 230 °C [4]. Preliminary experiments have shown that for birch veneer with a thickness of 2.0 mm, the temperature range can be taken from 160 to 200 °C, since an increase in temperature above 200 °C leads to a significant change in the appearance and color of the veneer. Thermal modification of veneer consisted of the following stages: heating, heat treatment for a given duration, cooling-conditioning. After the end of the thermomodification mode, the veneer was removed, placed in bundles in batches, and kept for 24 h under a slight surcharge to equalize possible stresses. At the end of the aging process, the veneer was formed into plywood packages according to the standard general-purpose plywood assembly method and pressed according to the mode shown in table 1. The pressing value for all plywood samples corresponded to that adopted for birch plywood and ranged from 9.8 to 10.6%.

An experiment to study the influence of the parameters of thermal modification of veneer on the physical and mechanical properties of plywood products was carried out in accordance with the B3 plan. The required accuracy was ensured by 5-fold duplication of measurements. Factors and levels of their variation are given in table 2. The output parameters were static bending strength and shear strength along the adhesive layer. The results of the experiment are presented in table 3.

Table 1. Pressing mode of plywood based on thermally modified veneer.

Factor name	Designation	Parameter value
Pressing pressure, MPa	P	1.8
Specific pressing time, min / mm	τ_{ud}	0.52
Pressing temperature, ^{0}C	T	130

Table 2. Factors and levels of variation.

Factor name	Designation	Variation levels		
		upper	main	lower
		+1	0	-1
Duration of heat treatment, min	τ/Factor_A	240	180	120
Heat treatment temperature, ^{0}C	T/ Factor_B	160	180	200

Table 3. Experiment planning matrix.

Duration of heat treatment, min	Heat treatment temperature, ^{0}C	Static bending strength, MPa	Shearing strength along the adhesive layer, MPa
Factor_A	Factor_B	Var_1	Var_2
1	1	67	0.99
-1	1	110	2.05
1	-1	91	1.75
-1	-1	123	1.71
0	1	80.5	2.12
0	-1	115	1.93
1	0	86.5	1.97

The influence of the investigated factors (temperature and duration of heat treatment) on the strength of plywood under static bending (figure 1) and when chipping along the adhesive layer (figure 2) was assessed by graphic interpretation of the regression equation and graphs of the effects of factors and the effects of their interactions.

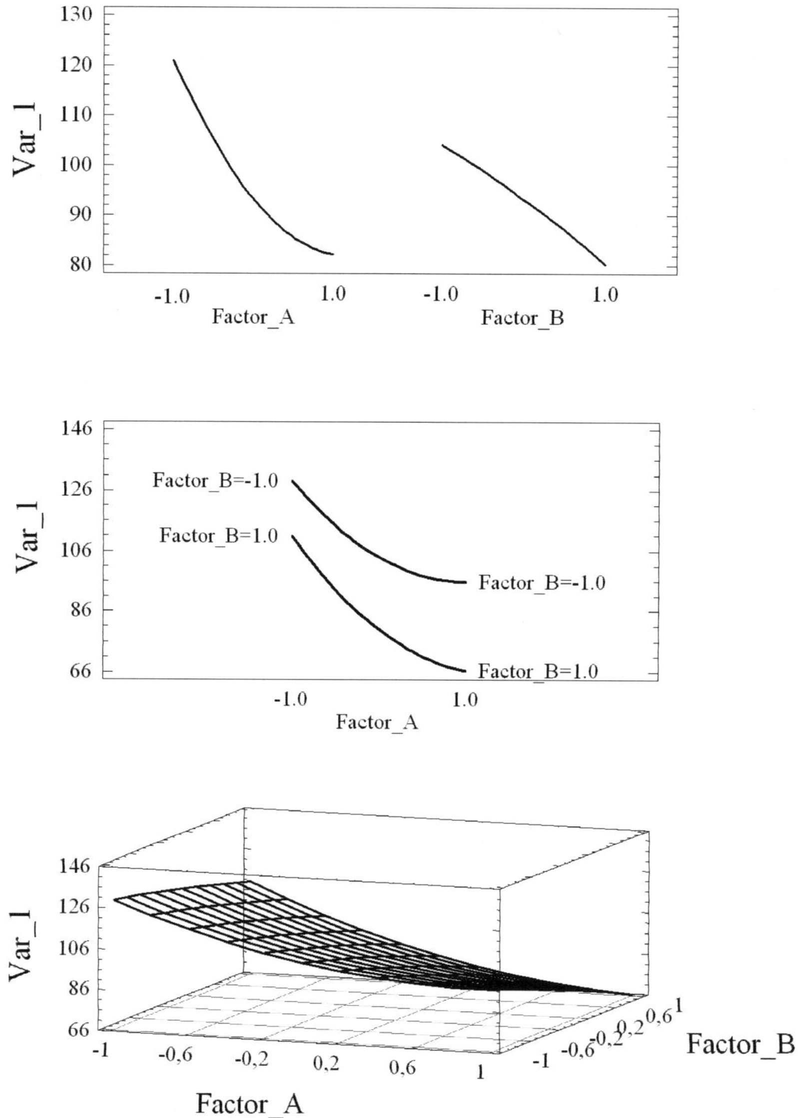

Figure 1. Results of the study of strength in static bending.

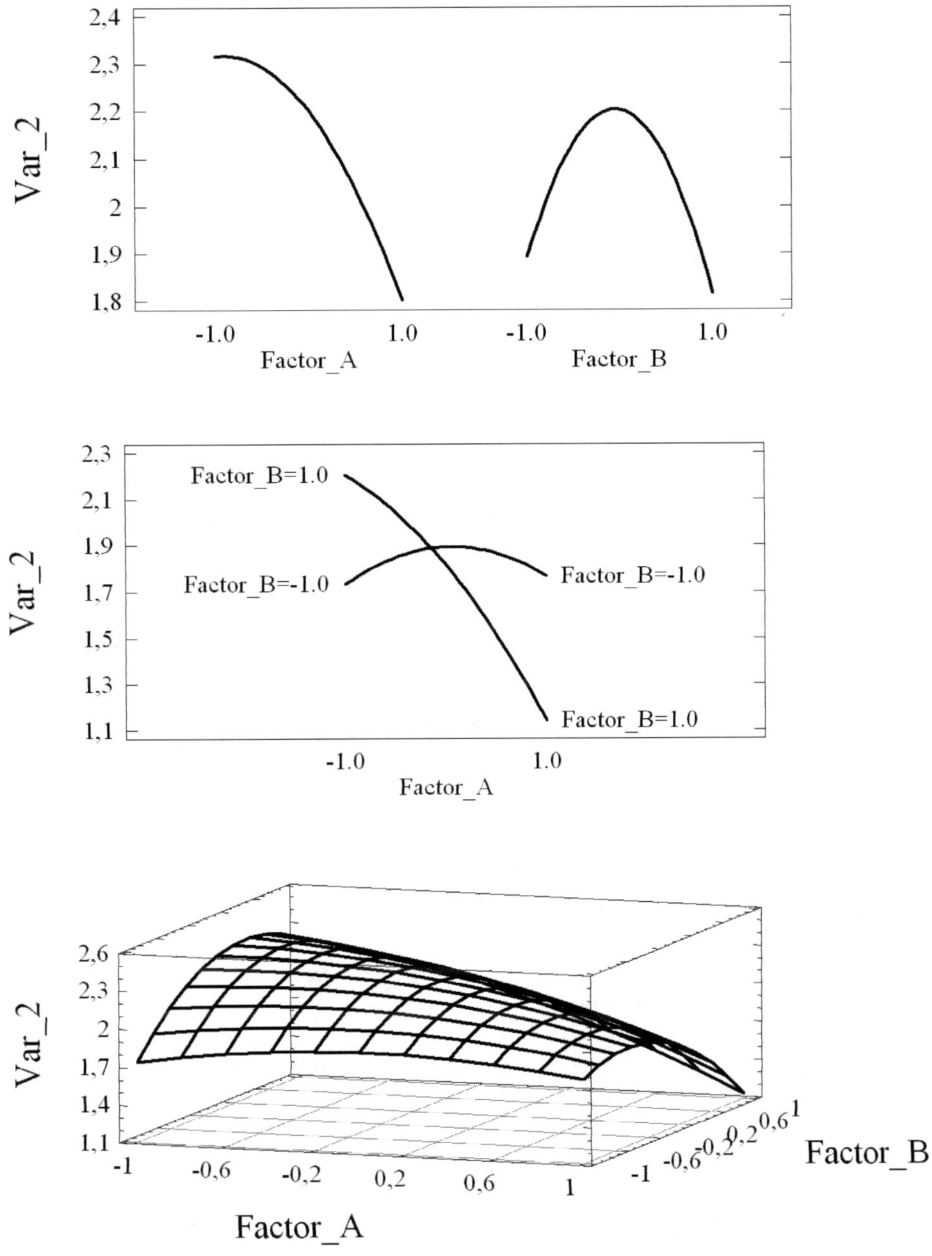

Figure 2. Results of a study of the shear strength along the adhesive layer.

The given graphical dependencies show the presence of directly proportional relationships between the duration and temperature of heat treatment and the strength of plywood under static bending. With an increase in the duration of heat treatment, the strength of plywood decreases more significantly compared with a decrease with an increase in the heat treatment temperature. At the same time, an increase in the duration at the maximum temperature of heat treatment (200 °C) leads to a more significant decrease in strength than when processing at the minimum temperature.

Analysis of the data when testing plywood for the index of shear strength along the adhesive layer shows a decrease in strength similar to that discussed above with an increase in the duration of heat treatment. In the case of a change in the heat treatment temperature, a different dependence is observed. An increase in the heat treatment temperature to 180 °C leads to an increase in the strength of plywood when chipping along the glue layer. This is probably due to a certain loss of veneer density during heat treatment, which contributes to a deeper penetration of the adhesive into the veneer and the formation of a single monolithic veneer and glue structure, which, as a consequence, leads to an increase in chipping strength. A further increase in the heat treatment temperature reduces the strength of the plywood when chipping due to a more significant loss of its weight by the wood veneer. In addition, a denser layer is formed on the veneer surface, which prevents the surface from wetting with glue, leading to a decrease in the adhesive contact between the glue and the wood.

These assumptions are confirmed by the appearance of the fracture surfaces of plywood samples during chipping. The destruction of samples from veneer, processed at temperatures up to 180 °C, is of a mixed nature, subsequent samples have a cohesive nature of destruction along the adhesive layer or along the wood. It should also be noted that an increase in the duration of heat treatment at the minimum temperature has practically no effect on the strength characteristics of plywood samples, and at the maximum temperature, an increase in the duration of heat treatment leads to a significant decrease in strength.

As a result of mathematical processing of experimental data, regression equations were obtained that adequately describe the dependence of the ultimate strength of plywood on the heat treatment mode:

- with static bending:

Var_1 = 93.222 - 19.416 x Factor_A - 11.918 x Factor_B + 8.417 x Factor_A ^ 2 - 1.083 x Factor_B ^ 2.- 2.75 x Factor_A x Factor_B;

- when chipping along the adhesive layer:

Var_2 = 2.202 - 0.257 × Factor_A - 0.038 × Factor_B - 0.143 × Factor_A ^ 2 - 0.348 × Factor_B ^ 2 - 0.275 × Factor_A × Factor_B

A certain number of works on thermal modification of wood is devoted to reducing the values of water absorption and swelling of wood [10, 12, 13]. Therefore, water absorption and swelling values were also determined for the manufactured plywood samples.

The water absorption of thermally modified samples in comparison with general-purpose plywood is reduced by 5.4 to 26.6%. The swelling is reduced by an amount from 8.8 to 43.4%. The maximum value of water absorption and swelling is observed in samples made of veneer, thermally modified for 120 minutes at a temperature of 160 °C (62.1 and 16.6%, respectively). The minimum value for samples made of modified veneer at a temperature of 240 °C for a duration of 120 minutes (40.2 and 9.9, respectively). It should be noted that water absorption and swelling are more pronouncedly reduced when using higher heat treatment temperatures.

Thus, plywood based on thermally modified veneer acquires higher water resistance, which significantly expands the scope of plywood based on urea-formaldehyde resins.

3. Conclusion

The use of thermally modified veneer for gluing plywood is one of the methods of directionally changing its physical and mechanical properties.

The implementation of the PFE in the work made it possible to establish that the strength characteristics of plywood are to a greater extent determined by the temperature of the heat treatment of the peeled veneer. The duration of heat treatment more significantly reduces the strength characteristics of plywood at maximum temperatures and practically does not affect the strength at minimum temperatures.

It has been established that heat treatment of veneer, which does not lead to a decrease in the strength of plywood, makes it possible to reduce water absorption and swelling by an amount from 20.4 to 33.5%.

Thus, plywood based on thermally modified veneer acquires higher strength and water resistance, which significantly expands the scope of application of glued materials based on urea-formaldehyde resins.

References

[1] Guide to heat treatment of wood Retrieved from: www.thermowood.fi
[2] Safin R G, Akhmetova D A and Stepanova T O 2015 Thermal modification of wood materials *Technological University Bulletin* **22** 112-8
[3] Safin R G, Akhmetova D A and Stepanova T O 2015 Review of Research on Thermal Modification of Wood *Woodworking industry* **4** 28-34
[4] Safin A V, Khasanshin R R and Safin R R 2015 Review of domestic and foreign research in the field of heat treatment of wood filler in the production of composite materials *Kazan Technological University Bulletin* **3** 194-8
[5] Calonego W F, Severo E T and Furtado E L 2010 Decay resistance of thermally-modified Eucalyptus grandis wood at 140 degrees C, 160 degrees C, 180 degrees C, 200 degrees C and 220 degrees C *Bioresour. Technol.* **101** 9391-4
[6] Boonstra M J, Tjeerdsma B F and Groeneveld H A 1996 Thermal modification of non-durable wood species technology *IRG* **40123** 13
[7] Giebeler E 1983 Dimensional stabilization of wood by moisture-heatpressur *Holz als Roh - und Werkstoff* **41** 87-94
[8] Kollmann F 1969 Gefuge – und Figenschaft – Sandesangen im Holz durch mechanische und thermische Beanspruchung *Holz als Roh – und Werkstoff* **1** 407-25
[9] Korkut S 2009 Effect of high-temperature treatment on the mechanical properties of rowan *Drying technology: an international journal* 1240-7
[10] Khasanshin R R 2012 Improving the performance of composite materials based on thermally modified wood *Kazan Technological University Bulletin* **7** 57-61
[11] Safin R R 2008 Investigation of thermal modification of pine wood in the conditions of vacuum-conductive devices *Furniture design and production* **2** 29-31
[12] Khasanshin R R 2013 Influence of the processing temperature on the water resistance of thermoplywood *Woodworking industry* **4** 28-30
[13] Khasanshin R R 2013 Improving the performance of glued materials based on heat-treated veneer *Technological University Bulletin* **13** 87-98
[14] Safin R R 2012 Development of technology for creating moisture-resistant plywood *Technological University Bulletin* **20** 64-5
[15] Safin R R, Galyavetdinov N R, Ibragimova A I and Talipova G A 2014 Distinctive properties of moisture-resistant plywood obtained by gluing thermally modified veneer *Actual problems of the forestry complex* **38** 237-9
[16] Chimiris Yu V 1995 Pat. of the Russian Federation No 2099180 C1 appl. 05.10.1995, publ. 20.12.1997
[17] Ziatdinov R R 2013 *Technology for the production of moisture-resistant plywood from thermally modified veneer* (Kazan)

Electrical resistance and impedance of manganese sulfides, substituted by lutetium

A M Kharkov, M N Sitnikov, O B Begisheva, A V Gelgorn, O S Nikitinskiy and N A Cheremnykh

Reshetnev Siberian State University of Science and Technology, 660037, Krasnoyarsk, Russia

E-mail: khark.anton@mail.ru

Abstract. The effect of electron doping in $Lu_xMn_{1-x}S$ and cationic vacancies in $Mn_{0.9}S$ manganese sulfide on the formation of phonon modes of vibration, the type of conductivity, and the impedance characteristics is studied. The IR spectra in the frequency range 450-7500 cm^{-1}, the I-V in the temperature range 300-500 K, and the impedance in the frequency range 10^2-10^6 Hz were measured. The qualitative difference in the temperature behavior of the impedance, the activation character of the relaxation time, and the change in the activation energy of the impurity conductivity in $Lu_xMn_{1-x}S$ at high temperatures are found.

1. Introduction

Semiconductors with electronic [1-3] and magnetic phase transitions [4-6], both in temperature and concentration, are promising materials for creating the element base of microelectronics. The transport of current carriers can be controlled both by action on the electron charge and the magnetic moment [7-12], in the presence of interaction between the electrons and the orbital magnetic moments [13-18]. In electrically inhomogeneous systems, the transport of current carriers in a magnetic field can be controlled on the basis of the Maxwell-Wagner effect [19-21] and the orbital ordering [22]. On the other hand, in systems with a random distribution of the electric potential and potential wells, it is possible to control the tunnel effects due to the external electric field [23-26] and the spin-phonon interaction [27-31].

To create electrically inhomogeneous states in manganese sulfide, the manganese ion is substituted by a trivalent lutetium ion. Nonstoichiometric substitution can lead to cationic defects [32]. To clarify the role of cationic defects, the $Mn_{0.9}S$ system is studied and its properties are compared with the $Lu_xMn_{1-x}S$ solid solution. It is possible that substitution will lead to areas with increased conductivity and to the formation of a metallic bond. The presence of defective states is established from the IR spectra and electrical resistivity on direct and alternating current.

2. Materials and discussion of results

2.1. IR spectra

The IR spectrum was measured by the VERTEX80v Fouriertrans device with a resolution of 0.5 cm^{-1} in the frequency range of 450-1200 cm^{-1} and temperatures of 80-460 K. Figure 1a shows the IR absorption spectrum for $Mn_{0.9}S$, and figure 1b shows the IR spectrum of $Lu_xMn_{1-x}S$ solid solutions for three concentrations. Below 700 cm^{-1}, we distinguish phonon bands at frequencies of 586-592 cm^{-1} [33-34] for all three concentrations, and the intensity of the absorption bands decreases with increasing concentration. Weak absorption bands in the range of 515-527 cm^{-1} and with increasing concentration, new absorption peaks appear at a frequency of 470 cm^{-1}. The phonon absorption frequencies are equal to the frequencies of 647 cm^{-1}, ω = 531 cm^{-1} in $LaMnO_3$ [35-37], which are associated with the oscillation modes of octahedra. It is possible that lutetium leads to deformation of octahedra containing manganese ions. At higher concentrations, the deformation of the octahedra decreases. In the $Mn_{0.9}S$ sample, phonon modes of oscillations are absent in this frequency range.

In the frequency range 1000-1200 cm^{-1}, there is an absorption band that varies slightly with concentration and in $Mn_{0.9}S$ it appears in the form of a kink. The existence of bands in the mid-IR range corresponds to localized electronic states and is observed in transition metal oxides and is considered as Jahn-Teller polarons, since vibronic coupling and the Jahn-Teller effect take place in the Mn-O octahedron. The polaron band was found in the $La_{0.9}Sr_{0.1}MnO_3$ manganite at an energy of ~ 0.15 eV [38, 39].

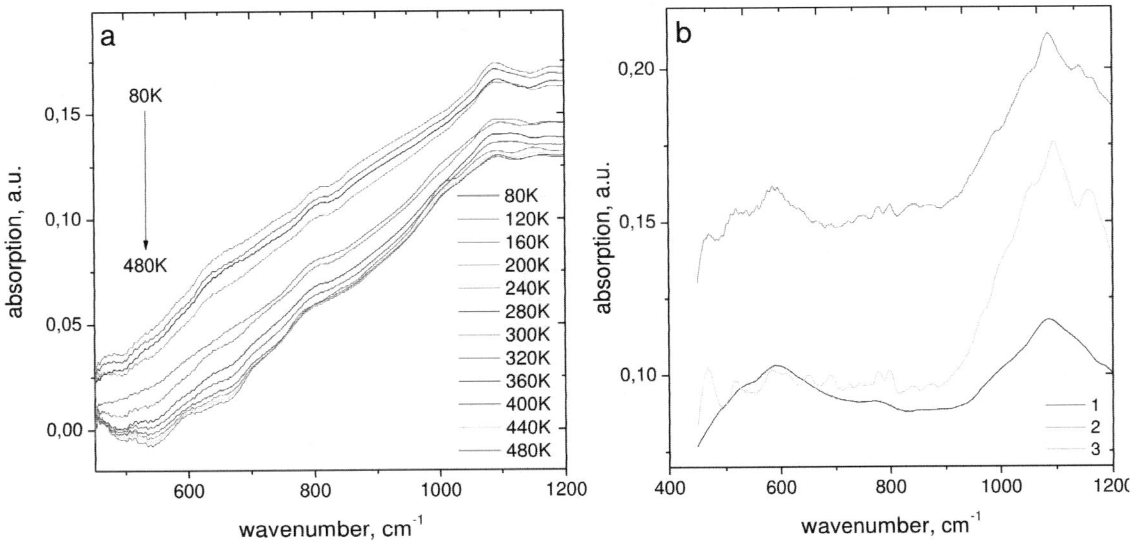

Figure 1. (a) IR absorption spectrum for $Mn_{0.9}S$ and (b) $Lu_xMn_{1-x}S$ c x = 0.05 (1), 0.1 (2), 0.2 (3).

2.2 I-V and electrical resistance

The presence of electrically inhomogeneous states is determined from the I-V. The dependence of current on voltage in semiconductors can be described by Ohm's law, a power function, and an exponential function. In a strong electric field, the height of potential barriers can change, and the I-V is described by the Poole-Frenkel law [40]. In charge-ordered semiconductors, the conductivity is described in the model of currents bounded by a spatial charge [41]:

$$j = \frac{9}{8}\tau_{\mu}\sigma_0\mu\frac{U^2}{L^3}S \quad (1)$$

where, j is the current density, τ_μ is the Maxwell relaxation time, σ_0 electrical conductivity in the volume of the material, μ is the carrier mobility, U is the applied voltage, L is the sample thickness, S is the sample area.

Figure 2 shows the I-V for $Lu_xMn_{1-x}S$. In a wide range of electric fields from 5 V to 200 V, the I-V in logarithmic coordinates lnI-lnU is described by a linear function lnI=A+nlnU with n = 1.16 at T = 300 K, n = 1.3 at T = 450 K for a concentration x = 0.1. With increasing concentration, the nonlinearity decreases and n = 1.08 at T = 300 K, n = 1.16 at T = 450 K. For concentrations is less than the percolation concentration $X_c = 0.17$, the conductivity is described by the hopping type and approaches to the Mott law. If lutetium ions percolate through the lattice, the conductivity becomes ohmic.

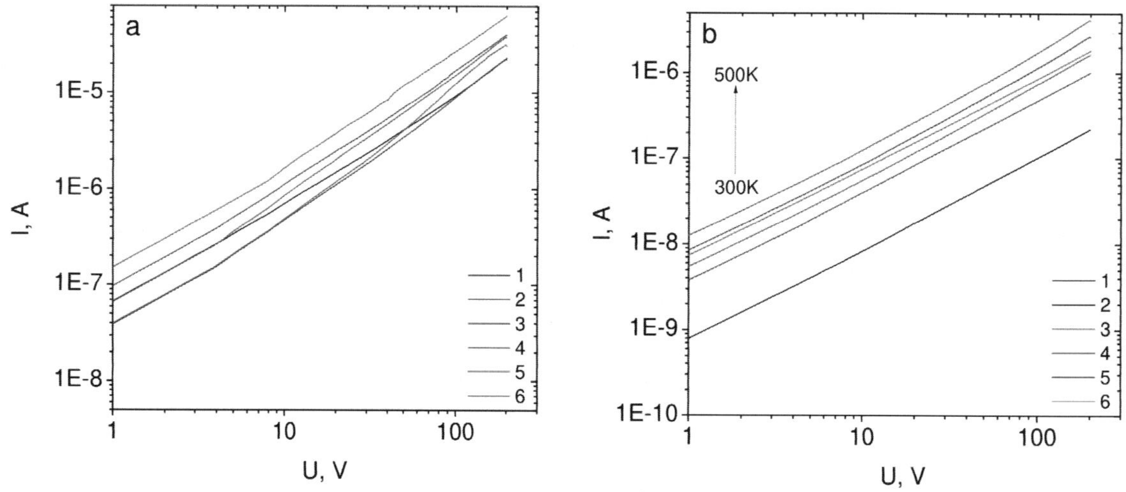

Figure 2. I-V of $Lu_xMn_{1-x}S$ solid solutions at temperatures T = 300K (1), 340K (2), 380K (3), 420K (4), 450K (5), 500K (6) for x = 0.1 (a), x = 0.2 (b).

In a cationic non-stoichiometric sample with cation defects in $Mn_{0.9}S$, the I-V is a linear function with n = 1.05, the index does not depend on temperature (figure 3). The conductivity in the $Mn_{0.9}S$ sample is ohmic. Manganese sulfides are Mott semiconductors with a charge gap formed by sulfur ions. Therefore, cationic vacancies do not significantly affect the conduction mechanism.

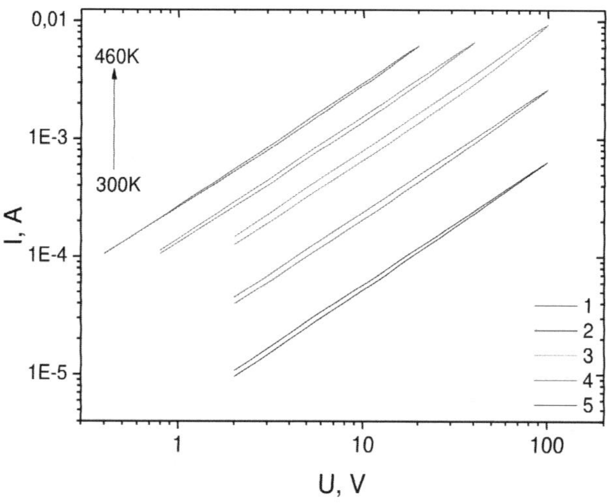

Figure 3. I-V of $Mn_{0.9}S$ solid solutions at temperatures T = 300K (1), 340K (2), 380K (3), 420K (4), 460K (5).

In manganese sulfide, the band gap is 3 eV [42, 43]. Electron doping of t_{2g} electronic states will lead to the formation of impurity states and, with increasing concentration, to the formation of an impurity subband [44, 46]. The energy interval between the valence band and the impurity state is found from the temperature dependence of the conductivity from the inverse temperature. For impurities semiconductors $R=R_0\exp(\Delta E/kT)$ and the activation energy is determined from $\ln R=\ln R_0+\Delta E/kT$. Figure 4 shows the dependence of the electrical resistance on the inverse temperature.

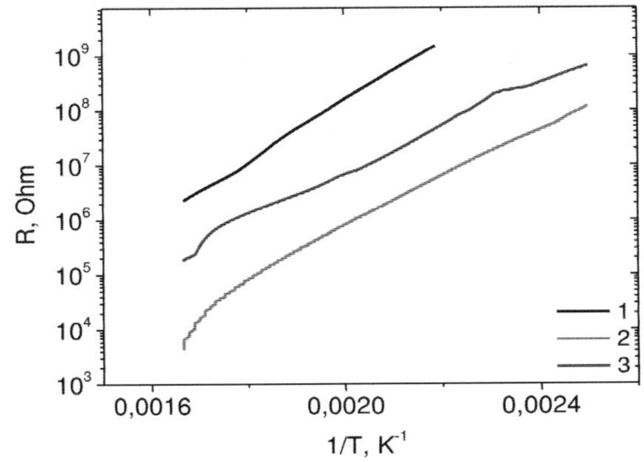

Figure 4. Electrical resistance $Lu_xMn_{1-x}S$ from the inverse temperature for x = 0.05 (1), x = 0.1 (2), x = 0.2 (3).

With increasing concentration, the activation energy decreases from ΔE = 0.93 eV for x = 0.05, ΔE = 0.86 eV for x = 0.1 to ΔE = 0.7 eV for x = 0.2. Perhaps this is due to the increasing of the impurity subband, and the increase in resistance above the percolation concentration is associated with a decrease in the mobility of current carriers. In the temperature range of 500-534 K, the activation energy changes. This can be caused by the pinning of lattice polarons or the formation of a charge density wave. Pinning of polarons is accompanied by lattice deformation. To verify this statement, it is necessary to carry out structural studies. Anomalies in the impedance spectrum can use as an indirect confirmation.

2.3 Impedance

Figure 5 shows the temperature dependences of the real part of the impedance and the normalized impedance Z(T)/Z(T=90 K) for $Lu_xMn_{1-x}S$ with x = 0.1. In the vicinity of the temperature T = 350 K, there is a maximum in the real part of the impedance, which disappears at high frequencies. This may be due to the formation of the charge ordering, which is accompanied by the deformation of the lattice. In the temperature range of 480-530 K, there is an electronic transition, which is accompanied by a maximum in the real part of the impedance and a maximum of the capacitance. The impedance abruptly decreases at the temperature of the magnetic phase transition as a result of the magnetoelastic interaction.

In a system with $Mn_{0.9}S$ cationic defects, the temperature behavior of the impedance is qualitatively different (figure 6). When heated, the impedance decreases by 4-5 orders of magnitude above 200 K. A small maximum is observed at T = 443 K, which does not depend on the temperature. The components of the impedance absolute magnitude have a maximum at this temperature. Apparently, this is due to local deformation near cationic vacancies [47, 48]. The change in the impedance from the temperature has an activation character $Z(T)=Z_0\exp(\Delta E/kT)$ with an activation energy ΔE = 0.11-0.13 eV in the range of 250-500 K.

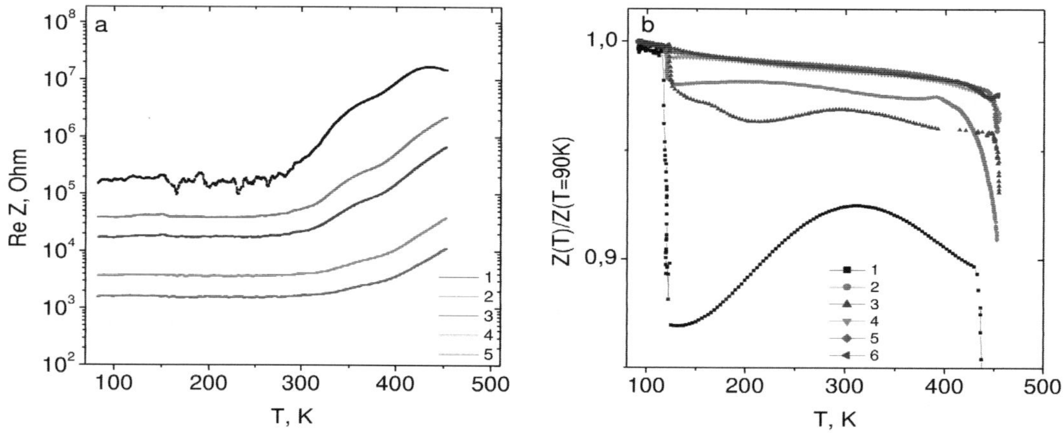

Figure 5. (a) – The real part of the Re(Z) impedance for $Lu_xMn_{1-x}S$ with x = 0.1 at the frequencies ω = 1 kHz (1), 5 kHz (2), 10 kHz (3), 50 kHz (4), 100 kHz (5) and (b) – the normalized impedance Z(T)/Z(T=90 K) at the frequencies ω = 1 kHz (1), 5 kHz (2), 10 kHz (3), 50 kHz (4), 100 kHz (5), 300 kHz (6) for $Lu_xMn_{1-x}S$ with x = 0.1.

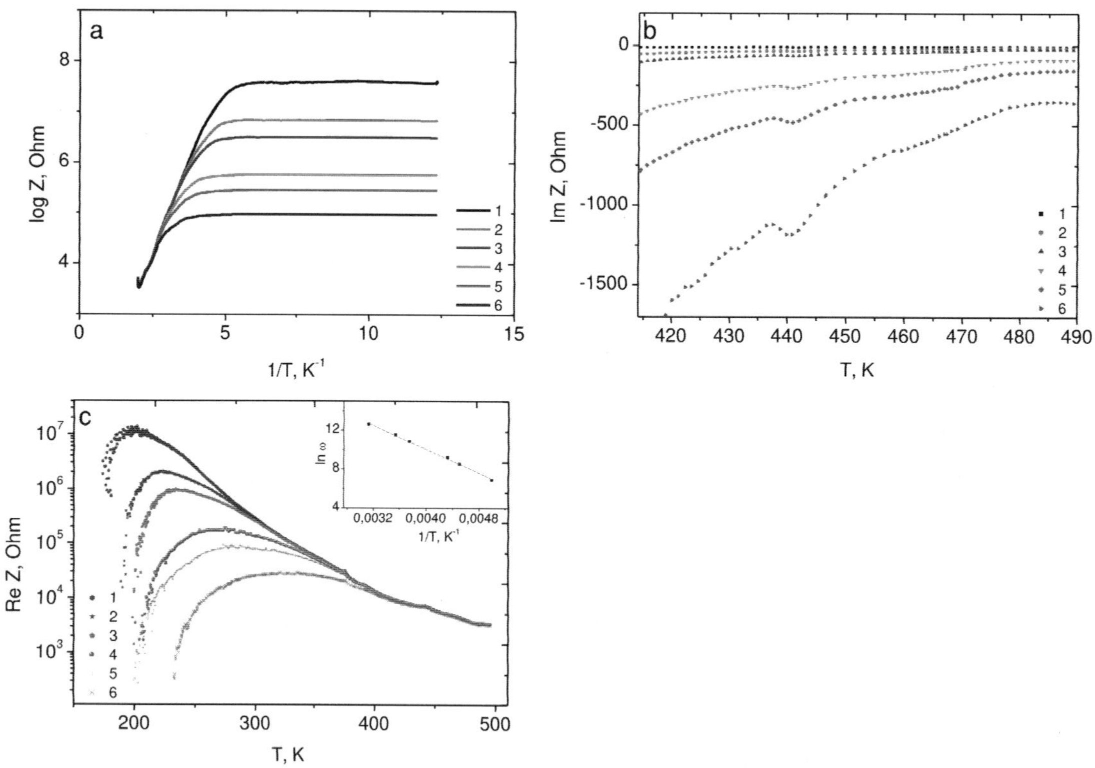

Figure 6. Logarithmic dependence of the impedance from the inverse temperature for $Mn_{0.9}S$. Temperature dependence of the imaginary (b) and real (c) parts of the impedance for $Mn_{0.9}S$ at frequencies ω = 1 kHz (1), 5 kHz (2), 10 kHz (3), 50 kHz (4), 100 kHz (5), 300 kHz (6).

The temperature behavior of the impedance components at room temperature and below is described in the Debye model:

$$\mathrm{Re}\, Z(\omega) = \frac{A}{1+(\omega\tau 1)^2} \qquad Im\, Z(\omega) = \frac{A\omega\tau 1}{1+(\omega\tau 1)^2} \tag{2}$$

where A is the constant parameters and τ_l is the relaxation time the temperature dependence of which is described by the exponent $\tau_l = \tau_{l0}\exp(\Delta E_l/kT)$ with activation energies of $\Delta E_l = 0.6$ eV.

3. Conclusion

In solid solutions of $Lu_xMn_{1-x}S$, cationic vacancies are absent at the non-stoichiometric substitution. Phonon modes of octahedron oscillations, which are absent in $Mn_{0.9}S$, were found in the $Lu_xMn_{1-x}S$ compound. Electronic doping leads to a hopping type of conductivity and to a deviation of the current-voltage dependence from the Ohm's law. In manganese sulfide with cationic defects, Ohm's law is fulfilled. The impedance characteristics are qualitatively different. The active part of the impedance in $Mn_{0.9}S$ increases at cooling below room temperature, passes through the maximum. The relaxation time of charged particles has an activation character. In $Lu_xMn_{1-x}S$, the maximum resistance at the alternating current in the region of the electronic transition is found.

Acknowledgments
The reported study was funded by Grant of the President of the Russian Federation № MK-620.2021.1.2.

References
[1] Aplesnin S S 1998 *J. Phys. Cond. Mat.* **10** 10061
[2] Aplesnin S S 2000 *Phys. Rev. B* **61** 6780
[3] Aplesnin S S *et al.* 2017 *J. Appl. Phys.* **121** 7
[4] Aplesnin S S *et al.* 2012 *Phys. Sol. St.* **54** 1374
[5] Decker D L and Wild R L 1971 *Phys. Rev. B* **4** 3425
[6] Szuszkiewicz B et al. 2005 *Phys. St. Sol. C* **2** 1141
[7] Aplesnin S S *et al.* 2004 *Sol. St. Com.* **129** 195
[8] Aplesnin S S *et al.* 2009 *Phys. Sol. St.* **51** 698
[9] Aplesnin S S *et al.* 2015 *Phys. St. Sol. B* **252** 1792
[10] Aplesnin S S and Sitnikov M N 2014 *JETP Lett.* **100** 95
[11] Volkov N V 2012 *Phys. Usp.* **55** 250
[12] Aplesnin S S *et al.* 2004 *Phys. Sol. St.* **46** 2067
[13] Peters R and Kawakami N 2011 *Phys. Rev. B.* **83** 125110
[14] Aplesnin S S *et al.* 2008 *J. Exp. Theor. Phys.* **106** 765
[15] Petrakovskii G A *et al.* 1999 *Phys. Sol. St.* **41** 1520
[16] Aplesnin S S *et al.* 2016 *Phys. Sol. St.* **58** 19
[17] Hotta T and Dagotto E 2002 *Phys. B: Cond. Mat.* **312** 700
[18] Aplesnin S S *et al.* 2007 *Phys. Sol. St.* **49** 2080
[19] Aplesnin S S *et al.* 2021 *Phys. Sol. St.* **63** 242
[20] Parish M M and Littlewood P B 2008 *Phys. Rev. Lett.* **101** 166602
[21] Aplesnin S S *et al.* 2010 *Bull. Rus. Acad. Sci.: Phys.* **74** 708
[22] Aplesnin S S and Moskvin A I 2008 *J. Phys. Cond. Mat.* **20** 325202
[23] Loseva G V *et al.* 1997 *Phys. Sol. St.* **39** 1267
[24] Aplesnin S S *et al.* 2011 *IEEE Tran. Mag.* **47** 4413
[25] Romanova O B *et al.* 2020 *Mater. Sci.: Mater. Electron.* **31** 7946
[26] Aplesnin S S *et al.* 2020 *JMMM* **513** 167104
[27] Aplesnin S S 2003 *J. Exp. Theor. Phys.* **97** 969
[28] Aplesnin S S 2003 *Phys. Lett. Section A* **313** 122
[29] Aplesnin S S 1996 *Phys. Sol. St.* **38** 1031

[30] Son J *et al.* 2019 *npj Quantum Mater.* **4** 17
[31] Kilian I and Jurgen S 2020 *Phys. Rev. B* **102** 054407
[32] Wang F *et al.* 2018 *npj Flex Electron* **2,** 22
[33] Udod L *et al.* 2020 *Eur. Phys. J. Plus* **135** 776
[34] Aplesnin S S *et al.* 2020 *JETP Lett.* **112** 636
[35] Fedorov I *et al.* 1999 *Phys. Rev. B* **60** 11875
[36] Aplesnin S S *et al.* 2019 *Semicond. Sci. Technol.* **34** 095007
[37] Aplesnin S S *et al.* 2019 *JETP Lett.* **110** 223
[38] Mertelj T *et al.* 2000 *Phys. Rev. B* **61** 15102
[39] Nekvasil V *et al.* 2009 *J. Appl. Phys.* **105** 113
[40] Connell G A *et al.* 1972 *Philos. Mag.* **26** 541
[41] Shklovsky B I and Efros A L 1979 *Electronic properties of doped semiconductors* (Moscow, Science)
[42] Aplesnin S S *et al.* 2012 *Phys. St. Sol. B* **249** 812
[43] Aplesnin S S *et al.* 2005 Phys. Rev. B **71** 125204
[44] Aplesnin S S *J. Exp. Theor. Phys.* 2007 **104** 766
[45] Aplesnin S S 2005 *JETP Lett.* **81** 66
[46] Kharkov A M *et al.* 2020 *J. Phys.: Conf. Ser.* **1614** 012103
[47] Adepalli K K 2013 *Adv. Funct. Mater.* **23** 1798
[48] Laredo E 1985 *Phys. Rev. B* **32** 8325

IOP Conf. Series: Materials Science and Engineering 1181 (2021) 012007 doi:10.1088/1757-899X/1181/1/012007

Carbon-containing materials for high-temperature friction units

M N Roshchin[1], A I Lukyanov[1] and A Yu Krivosheev[2]

[1] Mechanical Engineering Research Institute of the Russian Academy of Sciences, Moscow, Russia
[2] GAZ METALLOKOMPLEKT LLC, Taganrog, Russia

E-mail: roschin50@yandex.ru, gmk.taganrog@yandex.ru

Abstract. The paper considers frictional interaction of carbon-containing composites with steel at high temperatures. Material "Hardcarb-TPG" has better tribological characteristics than materials "Hardcarb" and "Hardcarb-TP". At a speed of 0.16 m/s and a load of 1.0 MPa, the friction coefficient of the Hardcarb-TPG material is 23% lower than that of the Hardcarb-TP material. Thus, at a temperature of 500 °C and a load of 1.0 MPa, the friction coefficient of the Hardcarb-TPG material at a speed of 0.16 and 0.25 m/s is 8 and 14% higher than at a speed of 0.05 m/s, respectively. The friction coefficient of the Hardcarb-TPG material at a speed of 0.16 m/s and a load of 1.0 MPa in the temperature range 300...600 °C varies from 0.13 to 0.30.

1. Introduction

The present period of development of metallurgy is characterized by a radical change in both the scale of production of high-quality and high-quality steel, and its share in total production and production methods.

High requirements for the quality of steel in modern metallurgy have led to the development of a large number of new technologies, which has significantly changed the state of affairs in the steel industry in recent years.

It is known that carbon-base materials are the materials for various purposes based on natural or artificial graphite. Carbon materials technology includes heat treatment processes. Structural graphites and electrodes, as a rule, are fired at 900-1100 ° C and graphitized at 2400-3000 ° C. To protect against oxidation during operation in oxidizing environments at t> 600 ° C, diffusion-reaction coatings (DRC) are applied to articles made of carbon materials: slip-firing carbide, boride and glass-silicide classes and are obtained by a chemical reaction from steam (for example, based on SiC). When operating carbon products with a coating in oxidizing environments at t> 1500 ° C, coatings of the carbide-boride class of the Hf (Zr) - B-Si-C system or the oxide-boride class of the Hf (Zr) - B-Si-O system are recommended. At an operating temperature of 1300 - 1500 ° C, glass-silicide class coatings based on borosilicate glass and Mo (W) Si2 are used.

Fields of application carbon materials with DRP: crucibles for melting metals and their alloys, heaters of electric furnaces, elements of metallurgical furnaces (roller tables, valves, etc.) operating at high temperatures in air, as well as heat-stressed parts of spacecraft (wing edges, nose cone), aviation gas turbine (elements of the combustion chamber, directional apparatus, turbine impellers) and ramjet engines.

Scope of carbon-base materials in steelmaking: carburizing steel (carbon-containing material with a fraction of 0-10 mm sits down during metal tapping); the use of a lump carbon-base materials as additional heat energy for the steel smelting process (supply of a lump carbon-base materials with a fraction of 25-50 mm to the convector allows to reduce the consumption of liquid iron for the steel smelting process and increase the proportion of filled up scrap, replacing part of the liquid iron with scrap provides a reduction in the cost of steel); use of carbon-base materials powder for slag foaming.

In the process of creating and operating new equipment capable of operating under extreme conditions, for example, at high temperatures, it is required to use materials in the friction units of the executive bodies that have a low coefficient of friction in friction pairs. Materials used in assemblies at high temperatures must ensure its performance and reliability. To rise the antifriction of the friction unit at high temperatures, high-temperature lubricants or solid lubricating coatings are used [1, 2]. When creating friction units for operation in aggressive atmospheric conditions and high temperatures, materials based on carbon-carbon compositions (CCCM) are used [3]. CCCM material has good characteristics in terms of specific gravity and strength. The production of CCCM material is constantly being improved [4]. Carbon nanotubes are introduced into the composition of the CCCM material to improve the mechanical properties [5]. At high temperatures CCCM is subject to oxidation [6]. Mechanical properties of CCCM at high temperatures are widely studied [7]. Tribological parameters of CCCM materials at high temperatures have not been considered to a full extent.

This work is being carried out jointly by the Institute of Mechanical Engineering named after A. A. Blagonravov of the Russian Academy of Sciences and Joint Stock Company "Scientific and Production Association named after S.A. Lavochkin". The research is being carried out in the expediency of using in the design of spacecraft new promising materials of friction units operating in the conditions of: open space, the atmosphere of Venus.

Structural carbon-containing composite materials are of interest in the space and aviation industries due to their low specific gravity, high strength characteristics and the ability to work at high temperatures. In friction units of various-purpose apparatuses, an additional requirement for the materials used is antifriction characteristics. High coefficients of friction lead to an increase in the power of the executive bodies, and this leads to an increase in the mass of spacecraft. The performance of CCCM friction units at high temperatures depends on the rate of graphite oxidation, which leads to a loss of mass and strength properties of parts. In this paper, the issues of tribological testing of promising materials are partially considered. Tribological research is associated with high costs, both material and financial. Carrying out large-scale research is not possible due to the impossibility of ensuring the operating conditions of the units. Basically, high-temperature tribological tests of carbon-containing materials are carried out to solve particular problems to increase the resource and reliability of friction units [8]. Carrying out high-temperature tribological tests of carbon-containing materials in the creation of spacecraft of the future is an urgent task. Among the problems of the future are the tasks of Venus exploration, which require the creation of devices for long-term operation on the planet's surface, both stationary and mobile platforms. The tribological characteristics of CCCM have been insufficiently studied. With a large amount of research on the effect of temperature on the strength properties of carbon composites, there are very few studies of the tribological properties of CCCM, especially in the field of high-temperature tests.

The purpose of this work is to test the tribological properties of carbon-containing composites at high temperatures during friction against 40X13 steel.

2. Materials and methods

The samples were made of volume-reinforced carbon-containing composite materials of the following brands: "Hardcarb", "Hardcarb-TP", "Hardcarb-TPG". Materials of the "Hardcarb" type have the following fiber composition:

- "Hardcarb" is made of carbon technical fabric of the T-0.5P-22N type (viscose).
- "Hardcarb-TP" is made of carbon fabric such as Serzha 2 / 2-1000-12K-400 (PAN fiber).

- "Hardcarb-TPG" is made of PAN fiber with graphitization.

Samples of heat-resistant corrosion-resistant steel 40X13 were used as a counterbody. Tribological studies were carried out on a modernized high-temperature test bench VTMT-1000 [9].

The test was carried out on specimens according to the "disc-finger" scheme in the temperature range of 300 - 800 °C at a load of 0.3…1.0 MPa and a sliding speed of 0.05; 0.16; 0.25 m/s. Which was installed discretely. The tests should simulate the operating conditions of a full-scale friction unit. The implemented test scheme "disk-finger" allows you to extend the results of bench tests to other interface schemes. In tribological tests of materials, on some test installations, the "disk-finger" test scheme is widely used. The test results of the samples during the experiment are easier to adapt to other mobile schemes of sliding pairs. The laboratory test unit can measure the moment of friction and temperature in the contact zone, the signals from the sensors are transmitted online to the control computer for processing.

3. Results and discussion

Tribological tests of the "Hardcarb-TPG" material at sliding speeds: 0.05; 0.16; 0.25 m / s, a load of 1.0 MPa paired with 40X13 steel at temperatures from 300 to 800 ° C, it was found that the coefficient of friction increases with increasing sliding speed (Figure 1, curves 1; 2; 3 at a speed, m/s-0.05; 0.16; 0.25- respectively). Thus, at a load of 1.0 MPa, a temperature of 500 °C and a speed of 0.16 and 0.25 m/s, the coefficient of friction is 8 and 14% higher than at a speed of 0.05 m / s, respectively.

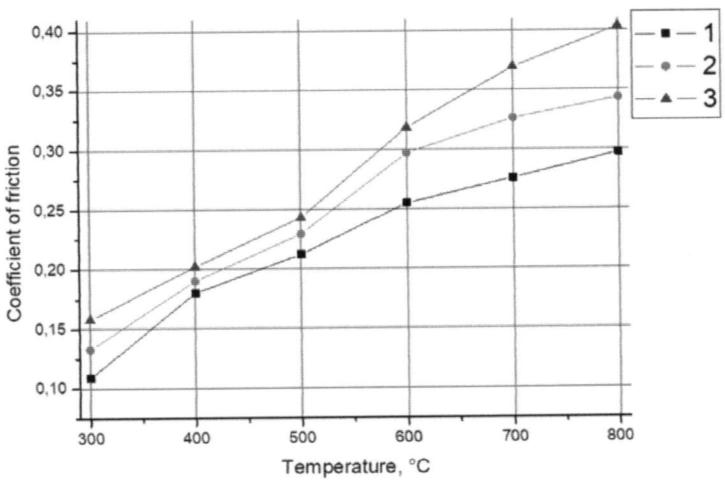

Figure 1. Temperature dependence of the coefficient of friction "Hardcarb-TPG" at a load of 1.0 MPa.

Figure 2 (at a load of 1.0 MPa and a speed of 0.16 m / s) shows the temperature dependence of the coefficient of friction of Hardcarb and Hardcarb-TPG materials at a load of 1.0 MPa and a speed of 0.16 m / s. In the temperature range of 300...600 °C the coefficient of friction of the Hardcarb-TPG material varies in the range of 0.13...0.30. With an increase in temperature above 300 °C, the coefficient of friction of the Hardcarb material increases. At a temperature of 400 °C, the coefficient of friction is 20% greater, and at a temperature of 500 °C, it is 38% greater than the coefficient of friction of the Hardcarb-TPG material. When testing Hardcarb and Hardcarb-TPG materials, with an increase in temperature above 300 °C, the mechanical characteristics, including hardness, decrease. The surface deformation component of the friction force increases with a decrease in hardness, therefore, an increase in the coefficient of friction is observed. The change in the friction coefficient for the Hardcarb-TPG material is shown in figure 3 (at a temperature of 500 °C and a speed of, m / s: 1-0,05; 2-0,16; 3-0,25). With an increase in the test load at a temperature of 500 °C, the friction coefficient for the Hardcarb-TPG material decreases. At a load of 1.0 MPa, the coefficient of friction at a sliding speed of 0.16 and 0.25 m/s is higher than the coefficient of friction at a speed of 0.05 m/s by 8 and 14%, respectively.

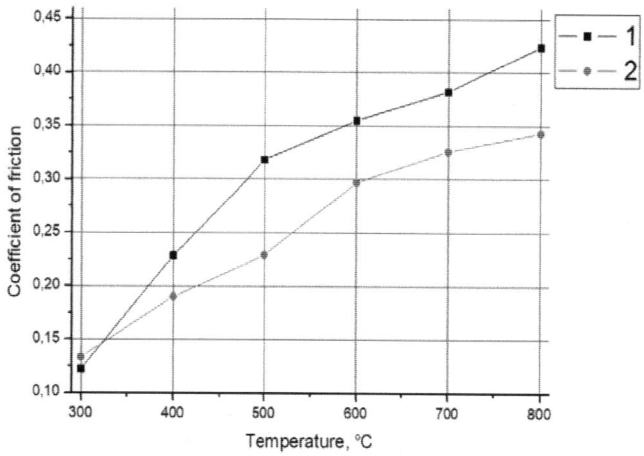

Figure 2. Temperature dependence of the coefficient of friction: 1- "Hardcarb"; 2- "Hardcarb-TPG".

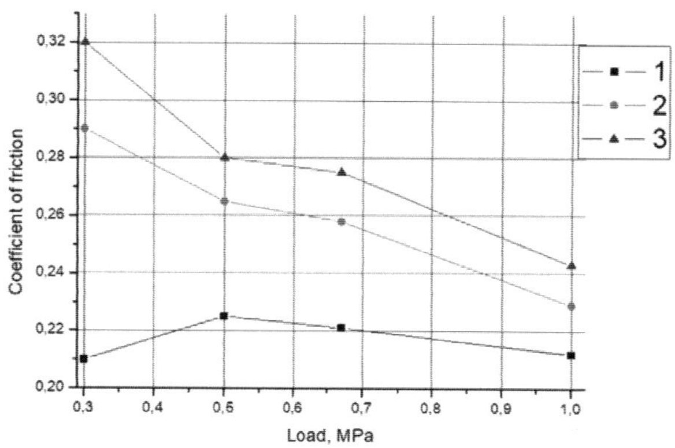

Figure 3. Influence of the load on the coefficient of friction "Hardcarb-TPG".

The change in the friction coefficient of the Hardcarb-TP and Hardcarb-TPG materials from the sliding speed is shown in figure 4 (load 1.0 MPa and temperature 500°C). The friction coefficient of the Hardcarb-TPG material in the sliding speed range of 0.05...0.25 m/s is lower than that of the material "Hardcarb-TP". At a speed of 0.16 m/s, the friction coefficient of the Hardcarb-TPG material is 23% lower than that of the Hardcarb-TP material. Technological innovations in the manufacture of the "Hardcarb-TPG" material with the use of graphitization lead to a reduced coefficient of friction than that of the "Hardcarb-TP" material.

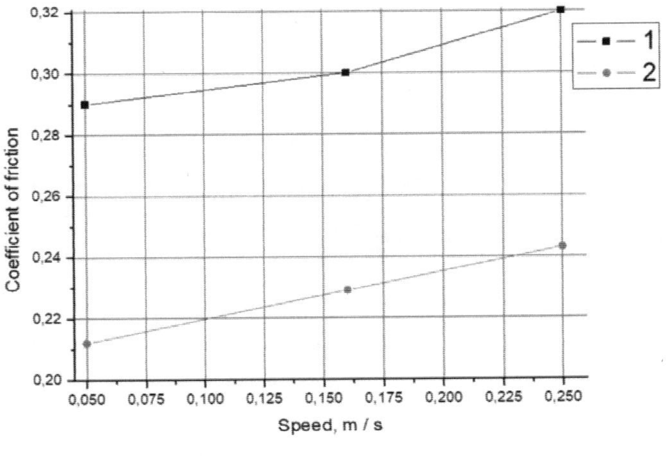

Figure 4. The effect of speed on the coefficient of friction: 1- "Hardcarb-TP", 2- "Hardcarb-TPG".

4. Conclusion

Material "Hardcarb-TPG" has better tribological characteristics than materials "Hardcarb" and "Hardcarb-TP". At a speed of 0.16 m/s and a load of 1.0 MPa, the friction coefficient of the Hardcarb-TPG material is 23% lower than that of the Hardcarb-TP material. Thus, at a temperature of 500 °C and a load of 1.0 MPa, the friction coefficient of the Hardcarb-TPG material at a speed of 0.16 and 0.25 m/s is 8 and 14% higher than at a speed of 0.05 m/s, respectively. At temperatures from 300 to 600 °C, a load of 1.0 MP and a speed of 0.16 m/s, the coefficient of friction of the Hardcarb-TPG material varies from 0.13 to 0.30. The experimental results obtained in the work can be used in the design of friction units operating at high temperatures.

References

[1] Buyanovskii I A, Tatur I R, Samusenko V D and Solenov V S 2020 *Journal of Friction and Wear* **41** 492-6
[2] Chen X H, Yuan X J, Xia J and Yu Z H 2015 *IOP Conf. Series: Materials Science and Engineering* **103** 012052
[3] Alam M, Parmigiani J P, and Kruzic J J 2017 *Eng. Fract. Mech.* **181** 116-29
[4] Zhiwei Sun, Xinhui Duan, Chandrasekar Srinivasakannan, Khalifa University 2018 *Science of Advanced Materials* **10(5)** 724-33
[5] Sharma S, Patel B and Patel R 2020 Friction **8(5)** 945–56
[6] Zhang S, Hao A, Nguyen N *et al.* 2018 *Carbon* **144** DOI:10.1016/j.carbon.2018.12.091
[7] Taylor R, Venkata Siva S B and Rama Sreekanth P S 2017 *Carbon Matrix Composites* In: Saleem Hashmi, Reference Module in Materials Science and Materials Engineering **13** (Oxford: Elsevier)
[8] Alisin V V, Roshchin M N, Lukyanov A I, Vorontsov V A, Bogachev V A and Markachev N A 2019 *Vestnik of NPO n. a. S A Lavochkin* **1** 61-5
[9] Roshchin M N 2020 *Journal of Physics: Conference Series* **1515** 042050

The type of current carriers and the influence of prehistory on the dynamic magnetic characteristics in $Ho_xMn_{1-x}S$

M N Sitnikov, S S Aplesnin, A M Kharkov and O N Bandurina

Reshetnev Siberian State University of Science and Technology, 660037, Krasnoyarsk, Russia

E-mail: kineru@mail.ru

Abstract. In a solid solution of $Ho_xMn_{1-x}S$, the influence of the prehistory of a sample cooled without a field and in a magnetic field on the magnetic characteristics is investigated. An increase in the inductance of a coil with a sample cooled in a magnetic field is found. The type of current carriers, the temperature of the change of the sign of the current carriers, the concentration and mobility from the Hall constant are determined. The results are explained in the model of spin and orbital-charge glass formation.

1. Introduction

The study of materials for microelectronics that can be controlled by an external electric field is an actual task [1-5]. This requires a coexictance of electric polarization and current carriers [6-11]. The formation of quasi-degenerate states, such as spin [12-19] or charge glass [20-27], allows the transport properties to be manipulated by an external electric field. The presence of glass states is determined from the magnetic permeability calculated from the inductance measured in the frequency range of 1-100 kHz when the sample is cooled in a zero and external magnetic field in the H=12 kOe [28-29]. The type of charge carriers is determined from Hall measurements, as a result of which we determine the position of the impurity subband relative to the valence band.

2. Materials and discussion of results

2.1 Influence of prehistory on dynamic magnetic properties $Ho_xMn_{1-x}S$

The magnetic permeability of $Ho_xMn_{1-x}S$ was determined from the inductance of a solenoid with an internal diameter of d = 2 mm and a length of 10 mm, inside which the sample was placed. The inductance of the coil with the sample (L_f) and without the sample (L_s) was measured, since the inductance of the solenoid is proportional to $L=n^2\mu\mu_0V$, then the magnetic permeability μr of the sample with the volume Vf is found as $\mu_r=L_f/L_s$. The normalized value of the permeability in $Ho_{0.05}Mn_{0.95}S$ for the five frequencies is shown in figure 1. In the region of the magnetic transition, the maximum permeability shifts towards high temperatures with increasing frequency and disappears at frequencies above 10 kHz (figure 1). Above 50 kHz, there are jumps at 160 K and 275 K, which may be associated with the formation of a charge-orbital glass [30-31].

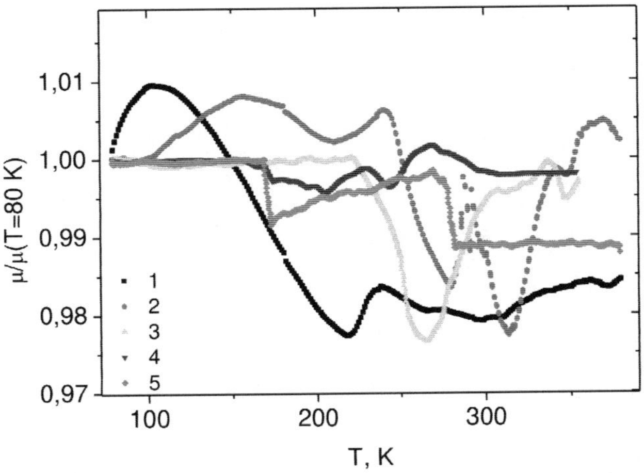

Figure 1. Normalized permeability value in $Ho_{0.05}Mn_{0.95}S$ for frequencies ω = 1 kHz (1), 5 kHz (2), 10 kHz (3), 25 kHz (4), 50 kHz (5) of temperature.

The magnetic susceptibility of $Ho_xMn_{1-x}S$ samples cooled in a magnetic field of 12 kOe below the Neel temperature at a frequency of ω = 1 kHz increases, since $\chi_H - \chi_0 \sim (\mu_H - \mu_0)$ increases figure 2a,b. This indicates the existence of spin-glass effects. For x = 0.05, the magnetic susceptibility in the high-frequency region decreases slightly, about one percent.

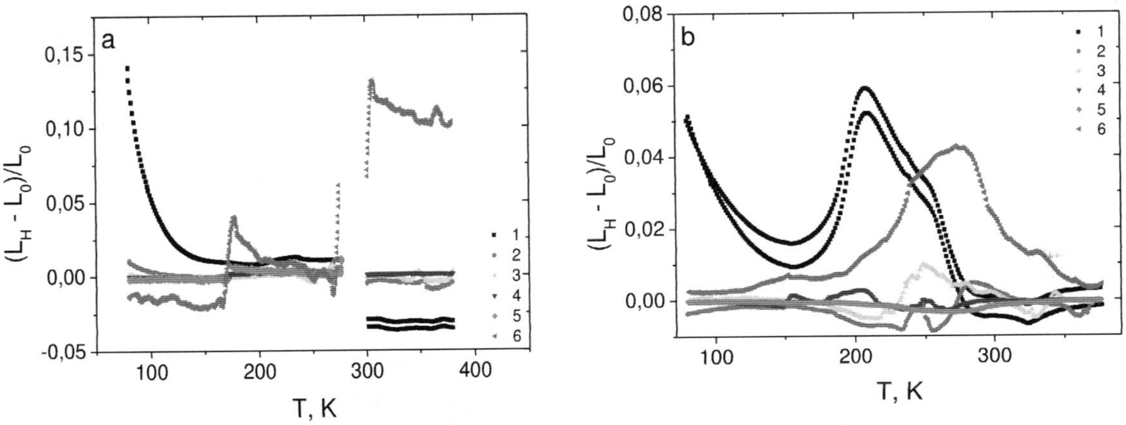

Figure 2. Relative change in the inductance of a coil with $Ho_xMn_{1-x}S$ cooled in a zero field (L_0) and in a magnetic field of 12 kOe (L_H), $(L_H - L_0)/L_0$ at frequencies ω = 1 kHz (1), 5 kHz (2), 10 kHz (3), 25 kHz (4), 50 kHz (5) and 100 kHz (6) for x = 0.05 (a), x = 0.1 (b) of temperature.

With increasing concentration, the influence of the sample prehistory on the inductance of the sample coil decreases several times. In the paramagnetic region, at T = 278 K, the sign of the susceptibility changes at the frequencies ω = 1 kHz and 100 kHz in $Ho_{0.05}Mn_{0.95}S$. For the concentration x = 0.1, the inversion of the magnetic susceptibility sign is absent during heating and a maximum $(L_H - L_0)/L_0$ is observed at T = 210 K, which shifts with increasing frequency to T = 275 K (figure 2b). This may be due to the formation of an orbital glass formed by the orbital angular moments of the electrons at the hopping between sites [32-35].

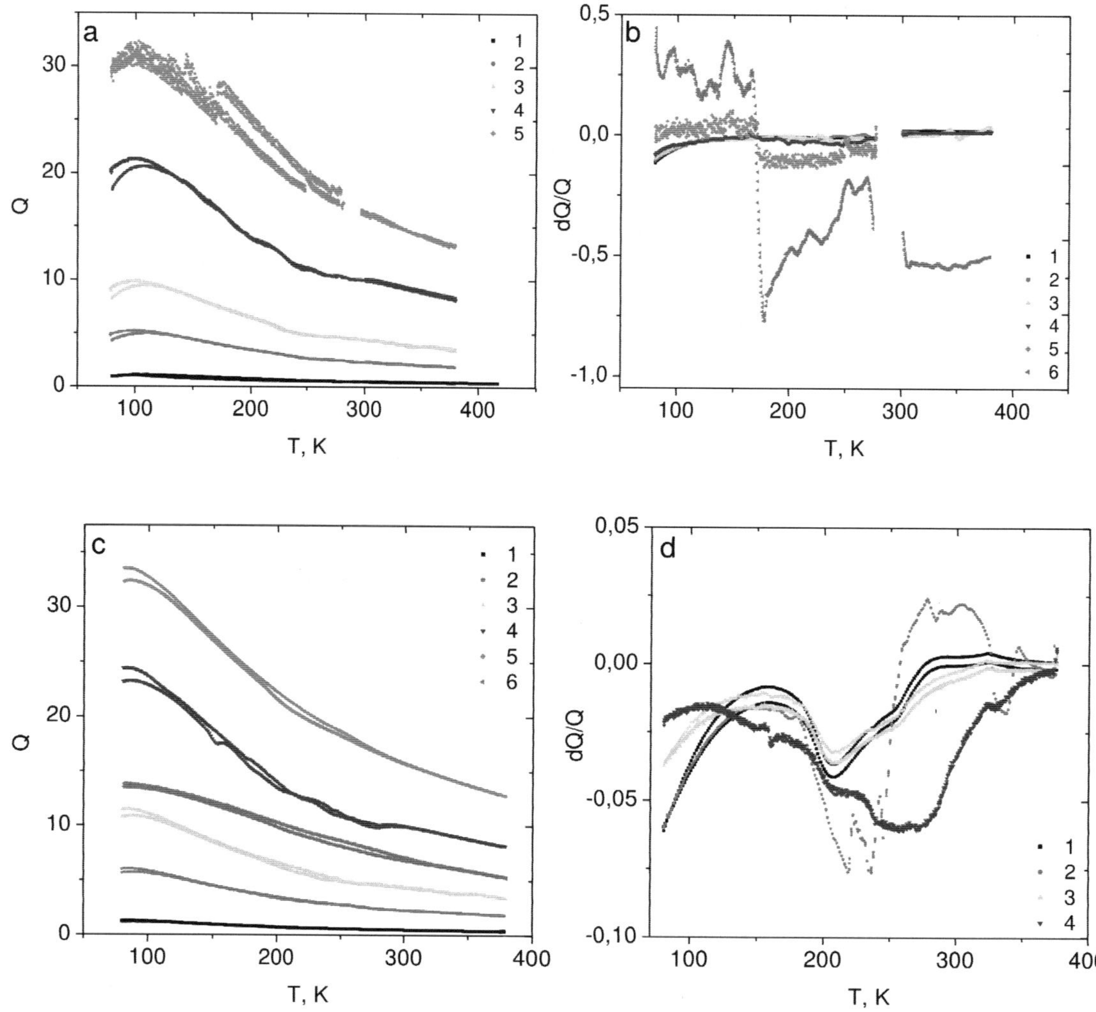

Figure 3. The Q-factor of the coil with $Ho_xMn_{1-x}S$ cooled in the zero field (Q_0) and in the magnetic field of 12 kOe (Q_H) (a, c) and $dQ/Q=(Q_H-Q_0)/Q_0$ (b, d) at frequencies ω = 1 kHz (1), 5 kHz (2), 10 kHz (3), 25 kHz (4), 50 kHz (5) and 100 kHz (6) for x = 0.05 (a, b), x = 0.1 (c, d) from the temperature.

The Q-factor of the electromagnetic oscillations of the coil with the sample decreases in the sample cooled in the magnetic field in the magnetically ordered state, with the exception of oscillations at the frequency ω = 100 kHz in $Ho_{0.05}Mn_{0.95}S$ (figure 3). The Q-factor also decreases in the region of the formation of the orbital glass. This effect is caused by an increase in magnetization when cooling in a magnetic field. So, the growth of the magnetization lead to increase of the hysteresis and to the magnetic loss.

2.2. Type of current carriers from Hall measurements

The Hall voltage was determined in a magnetic field of 12 kOe directed perpendicular to the current with a change in the polarity of the magnetic field. In figure 4, the resistance in $Ho_{0.1}Mn_{0.9}S$ drops exponentially when heated. The Hall voltage was measured at a fixed temperature at the averaging over 100 points (figure 5a).

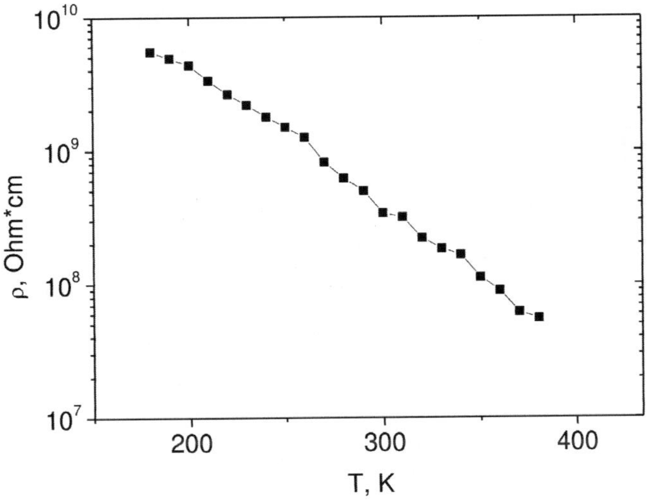

Figure 4. Electrical resistance of the solid solution Ho0.1Mn0.9S as a function of temperature.

The Hall voltage becomes positive below 230 K, with the exception of T=160 K and the temperature T=120 K in the vicinity of the magnetic phase transition. At T=230 K, the Hall constant changes its sign from positive to negative (figure 5b). The concentration of current carriers passes through the maximum at 250 K and the type of current carriers changes from holes to electrons at the heating (figure 6a). The mobility of the current carriers takes minimum in the range of 230-250 K (figure 6b).

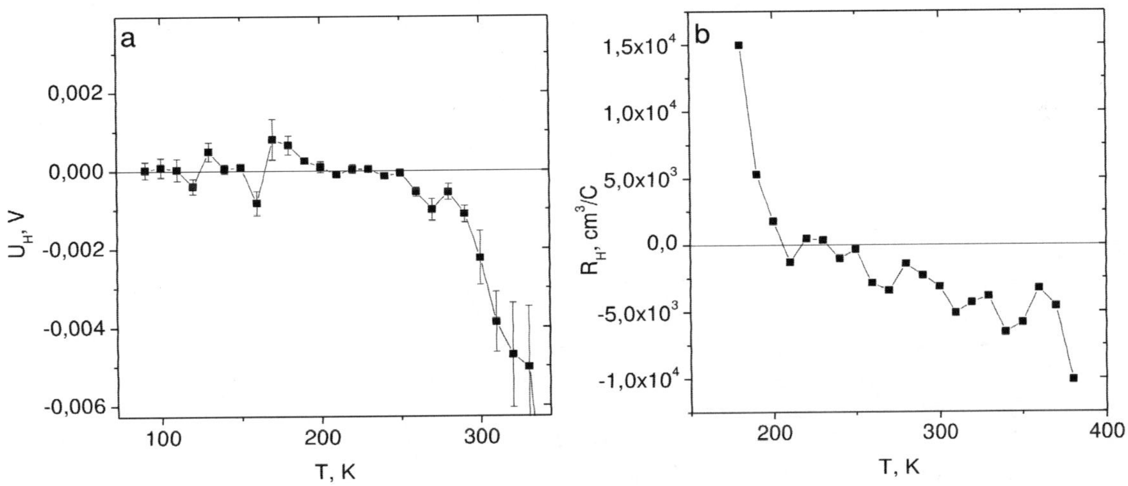

Figure 5. Hall voltage (a) and Hall constant (b) of the solid solution $Ho_{0.1}Mn_{0.9}S$ as a function of temperature.

The minimum of mobility is associated with the scattering of electrons at the orbital angular moments.

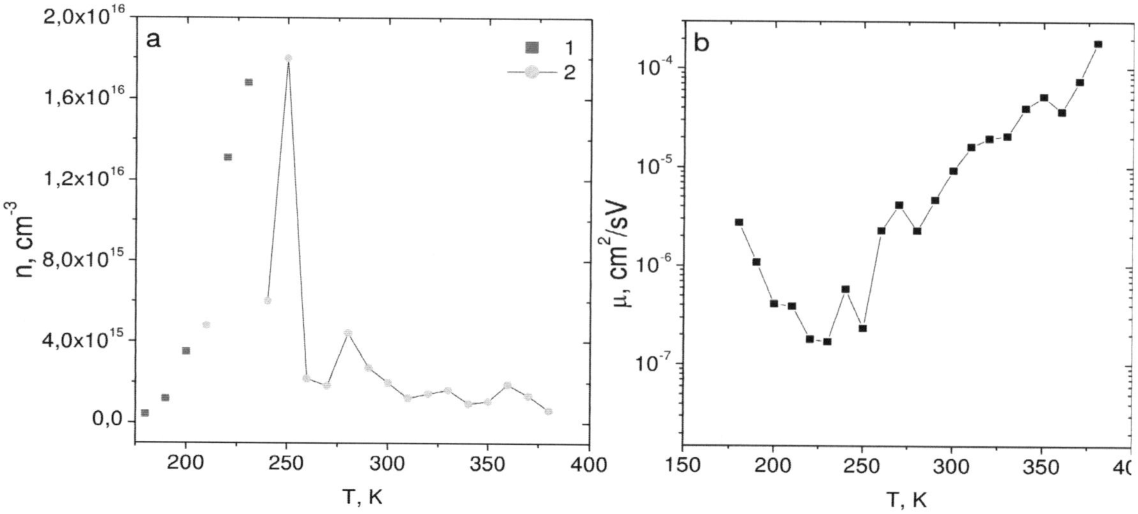

Figure 6. Concentration (a) and mobility (b) of charge carriers in a solid solution of $Ho_{0.1}Mn_{0.9}S$ as a function of temperature.

3. Conclusion

In $Ho_xMn_{1-x}S$, an increase in the inductance of a coil with a sample cooled in a magnetic field below the Neel temperature and a decrease in the Q-factor of the oscillations caused by an increase in the magnetic viscosity are found. This confirms the formation of a spin glass. With increasing concentration, the spin-glass effect decreases and the difference in inductance of coils cooled without a field and in a magnetic field shows a maximum in the temperature range of 200-250 K, which is explained by the formation of orbital glass. In this temperature range, the mobility of current carriers has a minimum and the type of current carriers changes from hole to electronic.

Acknowledgments

The research was funded by RFBR, Krasnoyarsk Territory and Krasnoyarsk Regional Fund of Science, project number 20-42-243002.

References

[1] Strukov D B *et al.* 2008 *Nature* **453** 80
[2] Liu J *et al.* 2017 *Neurocomputing* **266** 477
[3] Aplesnin S S *et al.* 2021 *J. Mat. Sci: Mat. Eletron* **32** 1946
[4] Romanova O B *et al.* 2020 *J. Mat. Sci.: Mat. Electron* **31** 7946
[5] Udod L V *et al.* 2020 *Eur. Phys. J. Plus* **135** 776
[6] Seidel J *et al.* 2010 *Phys. Rev. Lett.* **105** 197603
[7] Aplesnin S S 2008 *J. Exp. Theor. Phys.* **106** 765
[8] Aplesnin S S *et al.* 2018 *JMMM* **464** 44
[9] Khomskii D 2009 *Physics* **2** 20
[10] Aplesnin S S *et al.* 2019 *J. Exp. Theor. Phys. Lett.* **110** 223
[11] Farokhipoor S and Noheda B 2011 *Phys. Rev. Lett.* **107** 127601
[12] Petrakovskii G A *et al.* 2001 *Phys. Rev. B* **63** 18
[13] Aplesnin S S 1998 *Phys. Stat. Sol. B* **207** 491
[14] Aplesnin S S *et al.* 2009 *Phys. Sol. St.* **51** 698
[15] Petrakovskii G A *et al.* 1995 *JMMM* **140** 147
[16] Sherrington D 2013 *Phys. Rev. Lett.* **111** 227601
[17] Sherrington D 2014 *Phys. Stat. Sol. B* **1** 15

[18] Cannella V and Mydosh J A 1972 *Phys. Rev. B* **6** 4220
[19] Mydosh J A 2015 *Rep. Prog. Phys.* **78** 052501
[20] Aplesnin S S *et al.* 2005 *Phys. Rev. B* **71** 125204
[21] Udod L V *et al.* 2014 *Phys. Sol. St.* **56** 1315
[22] Aplesnin S S 1998 *J. Phys. Cond. Mat.* **10** 10061
[23] Aplesnin S S 2020 *JETP Lett.* **112** 636
[24] Nohara Y *et al.* 2006 *Phys. Rev. B* **74** 064417
[25] Solovyev I *et al.* 1996 *Phys. Rev. B* **53** 7158
[26] Biswas A *et al.* 2001 *Phys. Rev. B* **63** 184424
[27] Dho J *et al.* 1999 *Phys. Rev. B* **60** 14545
[28] Aplesnin S S *et al.* 2018 *Mat. Res. Exp.* **5** 115202
[29] Aplesnin S S *et al.* 2015 *J. Exp. Theor. Phys.* **121** 422
[30] Ritter C *et al.* 2000 *Phys. Rev. B* **61** 9229
[31] Aplesnin S S *et al.* 2021 *Ceramics Int.* **47** 1704
[32] Aplesnin S S *et al.* 2019 *Semicond. Sci. Technol.* **34** 095007
[33] Ramirez M O *et al.* 2009 *Phys. Rev. B* **79** 224106
[34] Mansuri I *et al.* 2011 *JMMM* **323** 316
[35] Kharkov A M *et al.* 2020 *J. Phys.: Conf. Ser.* **1614** 012103

To the design of the working body of the ground gun

I S Fedorchenko, M M Litvinova[1], V A Bakach and D A Belyaev

Reshetnev Siberian State University of Science and Technology, 31, Krasnoyarsky Rabochy Av., Krasnoyarsk, 660037, Russia

[1]E-mail: marg32883@gmail.com

Abstract. The article describes the invention for the creation of mineralized (reference) strips, as well as the justification of the parameters for the design of the grunt mill on the basis of experimental studies. The invention also solves the problem of improving operational performance when laying protective mineralized strips and extinguishing grass-roots fires with soil. This invention involves an increase in operational performance when laying protective mineralized strips and extinguishing grass-roots fires with soil.

1. Introduction

In solving the problem of prevention and control of forest fires, the main part of the time and money is occupied by grass-roots fires of various intensity. As practice shows, today various methods and equipment are used, among which, nevertheless, the use of manual labor and technical developments, which are obsolete and ineffective, prevails.

The analysis of the existing equipment and methods of extinguishing forest grass-roots fires [1] allows us to conclude that the most promising in this regard are prevention (the creation of mineralized, reference strips) and extinguishing with soil.

The invention [2] solves the problem of improving operational performance when laying protective mineralized strips and extinguishing grass-roots fires with soil.

The technical result is to ensure uniform spreading of the soil over the entire width of the mineralized strip.

The specified technical result is achieved by the fact that in a frontal forest fire grunt launcher containing a working body with throwers, a ground throwing mechanism, a control drive for the working body and a guide casing, the new one is that the working body is installed on the swing axis, carried out beyond the center of mass of the working body, closed by a guide casing connected through a lever with a loading hydraulic cylinder for lifting and lowering the working body, is made in the form of throwing knives, designed as a section of the spiral " rod», mounted on the milling disc at an angle α between the trailing edge of the knife and the plane of rotation of the milling disc, rotated relative to the trailing edge in the direction of movement at an angle β, and closed by a guide casing.

2. Materials and methods

The front-mounted forest fire grunt (figure 1) works as follows. The tractor dump removes the upper part of the soil layer, which may contain materials that support the gorenje process, as well as foreign bodies in the form of stones, etc. In this case, the front forest fire primer is buried in the treated soil to a certain depth h by its own weight, and if necessary, it is loaded with a hydraulic cylinder 7. The power plant 8, for example, a hydraulic motor operating from the hydraulic system of the base machine, transmits the torque to the shaft of the working body 3 of the frontal forest fire grunt. From the shaft of the working body, the torque is transmitted to the working body 3, which rotates and, when the frontal forest fire grunt moves forward, separates the soil chips from the array, and also throws and evenly distributes the cut volume of soil along the width of the mineralized strip. The uniform distribution of the soil over the width of the mineralized strip is carried out by throwing knives made in the form of a section of the spiral "rod", while the thickness of the soil chips is greater than the path traveled by the tractor during the cutting cycle. Throwing occurs as a result of giving the cut part of the ground different kinetic energy in different areas of the throwing knives. For the direction of the thrown soil, a guide casing 5 is used. When encountering an obstacle, the working body is raised above it, since it is mounted pivotally on the swing axis 4.

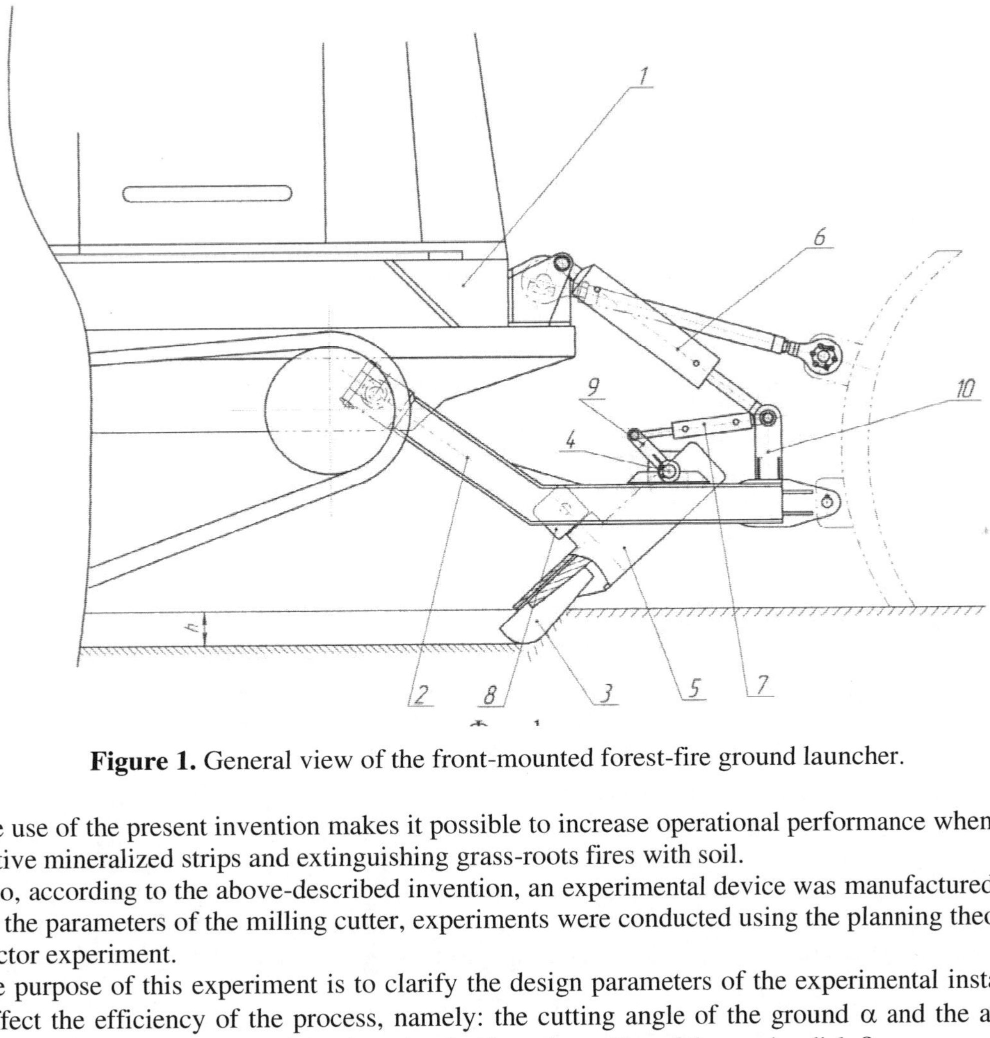

Figure 1. General view of the front-mounted forest-fire ground launcher.

The use of the present invention makes it possible to increase operational performance when laying protective mineralized strips and extinguishing grass-roots fires with soil.

Also, according to the above-described invention, an experimental device was manufactured [3], to justify the parameters of the milling cutter, experiments were conducted using the planning theory of a full-factor experiment.

The purpose of this experiment is to clarify the design parameters of the experimental installation that affect the efficiency of the process, namely: the cutting angle of the ground α and the angle of inclination of the cutting edge of the throwing knife to the radius of the carrier disk β.

Full-factor plan experiment 2^k, by k=2, that is, the regulation of two factors and at two levels is carried out. Y1 - the mass of soil per 1 m^2 of the bulk part at the near edge of the furrow being cut [4], Y2 - the mass of soil per 1 m2 of the bulk part in the middle part of the filling, Y3 – the mass of soil per 1 m^2 of the bulk part at the far edge of the furrow. The PFE matrix is shown in table 1.

Table 1. PFE 2^k, matrix, at k=2.

№	X_1	X_2	$X_1 * X_2$	Y_1	Y_2	Y_3
1	+	+	+	Y_{1cp1}	Y_{2cp1}	Y_{3cp1}
2	-	+	-	Y_{1cp2}	Y_{2cp2}	Y_{3cp2}
3	+	-	-	Y_{1cp3}	Y_{2cp3}	Y_{3cp3}
4	-	-	+	Y_{1cp4}	Y_{2cp4}	Y_{3cp4}

Determination of parameters of normalized variables by formulas (1-3):

$$z_{(i)n} = \frac{z_{(i)\max} + z_{(i)\min}}{2} \tag{1}$$

rationing step:

$$\Delta z_{(i)} = \frac{z_{(i)\max} - z_{(i)\min}}{2} \tag{2}$$

normalizing formulas:

$$x_{(1)} = \frac{z_{(i)} - z_{(i)н}}{\Delta z_i} \tag{3}$$

The levels and intervals of variation are shown in table 2.

Table 2. Levels and intervals of variation of factors.

Factor	Code	Factor level -1	Factor level +1	Variation step
Ground cutting angle α	x_1	45	55	5
Angle of inclination of the cutting edge of the throwing knife to the radius of the bearing disc β	x_2	25	55	15

The regression equation has the following form (4):

$$y = b_0 + b_1 \cdot x_1 + b_2 \cdot x_2 + b_{1,2} \cdot x_1 \cdot x_2 \tag{4}$$

All output data were previously checked for abnormality according to GOST 11.002-73 " Applied statistics. Rules for assessing the abnormality of observation results".

According to the method [5], the data obtained during the preliminary experiment were processed using the MS Office Excel application software package, and equations (5-7) were obtained showing the dependence of the soil mass on different sections of the bulk part of the mineralized (reference) strip (by its width) [6] from the cutting angle α and the angle of inclination of the cutting edge of the throwing knife to the radius of the bearing disc β:

$$y_1 = 6.025 + 0.055 \cdot x_1 + 1.195 \cdot x_2 + 1.125 \cdot x_1 \cdot x_2 \tag{5}$$

$$y_2 = 3.585 - 0.085 \cdot x_1 - 0.395 \cdot x_2 + 0.395 \cdot x_1 \cdot x_2 \tag{6}$$

$$y_3 = 2.645 - 0.405 \cdot x_1 - 0.305 \cdot x_2 - 0.295 \cdot x_1 \cdot x_2 \tag{7}$$

where y1 is the mass of soil per 1 m^2 at the near edge of the bulk part of the strip;
y2 – ground mass per 1 m^2 in the middle part of the bulk strip;

y3 – ground mass per 1 m^2 at the far edge of the embankment strip;

Based on the experimental data obtained, the graph shown in figure 2 was constructed.

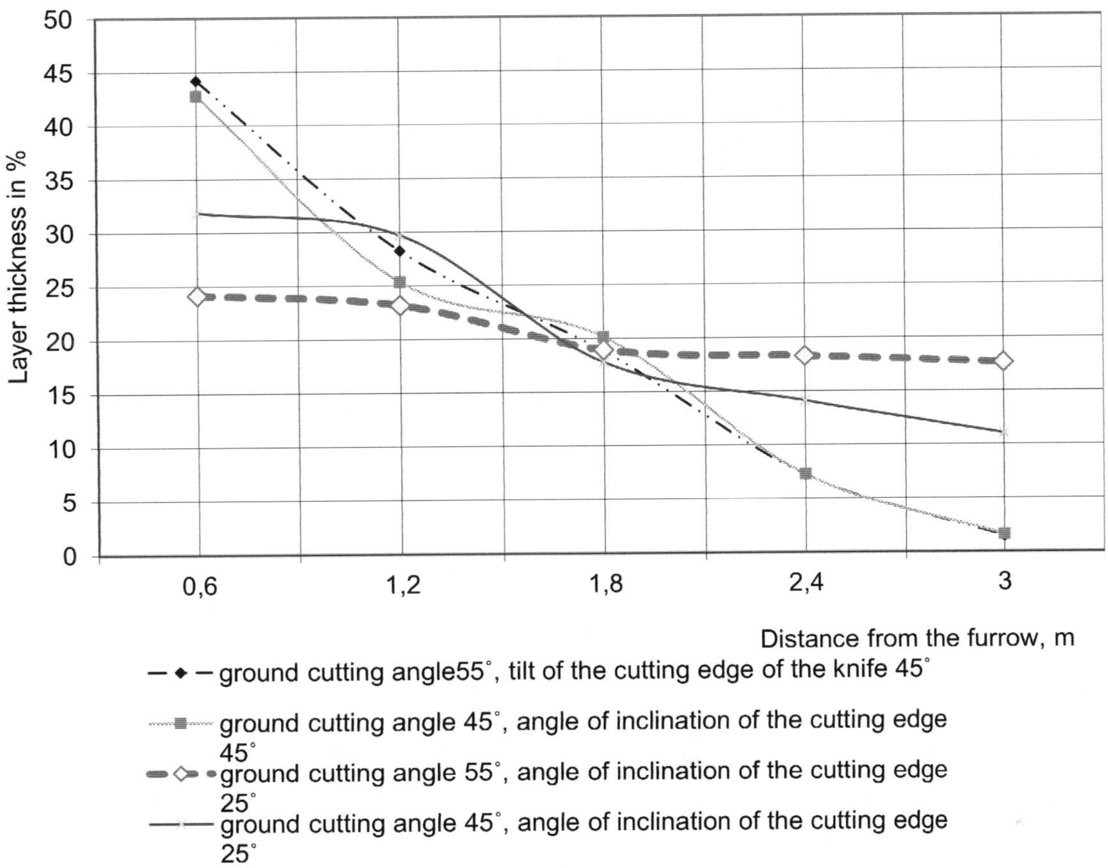

Figure 2. Distribution of soil over the width of the bulk part of the mineralized (reference) strip.

3. Conclusion

The graph shows that the best uniformity of the soil distribution is achieved at the cutting angle of the soil 55° and the angle of inclination of the cutting edge of the throwing knife to the radius of the bearing disc 25°, at which the minimum recommended soil mass of 4 kg/ m^2 is met [7].

As a result of the experiment, the design parameters of the experimental installation were clarified, namely: the angle of cutting the ground (55°) and the angle of inclination of the cutting edge of the throwing knife to the radius of the bearing disk (25°), which ensures the most uniform distribution of the soil over the bulk part of the mineralized (reference) strip [8]. The results of the study were used in the design of the working body of the plow of the forest combined PLC-5G for the modern forest fire unit LPA-521 of the Rubtsovsky Machine-building Plant.

References

[1] Fedorchenko I S and Maksimov E I 2009 The analysis of the existing equipment for suppression of forest fires by soil *The collection of works of the All-Russia conference "Young scientists in the solution of actual problems of a science"* **1** 192-4

[2] Maksimov E I and Fedorchenko I S Pat. of the Russian Federation RU 2009114066 C2 / No. 2009114066, 13.04.2009

[3] Fedorchenko I S and Maksimov E I 2009 *The experimental device for a soil throwing*

(Krasnoyarsk: Krasnoyarsk State Agrarian University) pp 234-9
[4] Orlovsky S N 2016 *Fighting forest, steppe and peat fires* (Krasnoyarsk: KrasGAU)
[5] Adler Yu P 1976 *Planning of experiment by search of optimum conditions* (M: Science)
[6] Orlovsky S N 2003 *Forest and peat fires. The practice of extinguishing them in Siberia* (Krasnoyarsk: KrasGAU)
[7] Ovchinnikov F M and Okhramets T I 2002 *Fire-prevention barriers in the woods of Siberia* 147
[8] Karpenchuk I V and Palubets S M 2010 Development of peat fire extinguishing tactics using specialized technical means *Bulletin of the Command and Engineering Institute of the Ministry of Emergencies Republic of Belarus* **2** 78-82

Filled polyaluminosilicate in LED emitting element design to increase its limit thermoelectric modes

A A Ivanov[1], V S Soldatkin[1] and A S Chermoshentseva[2]

[1] Tomsk State University of Control Systems and Radioelectronics, Tomsk, 634050, Russia
[2] Bauman Moscow State Technical University, Moscow, 105005, Russia

E-mail: alexchemtsu@rambler.ru

Abstract. The reliability of a LED lamp is conditioned by the need to increase its service life including the lamp in an inhabited spacecraft. The results of a filled polyaluminosilicate application as a heat-conducting intermediate layer, onto which LED chips are bonded, are considered in the article. The filled polyaluminosilicate application in the LED emitting element design allows increasing its operation temperature due to its specific properties.

1. Introduction

Reliability is the most important criterion for microelectronic and spacecraft devices. It is due to reliable materials application operating in a complex combination of force fields and being exposed to corrosive environment, extreme vacuum, and high pressure and temperature.

Highly efficient and reliable light sources are extremely important for interior space lighting in an inhabited spacecraft. Currently, LED light sources having such advantages as a long service life and high light intensity gradually displace the traditional light sources used in various industries. It is LED light sources that meet these requirements. Moreover, LED light sources have the longest service life (more than 80,000 hours) and the highest light output (more than 130 lm/W) compared with all other artificial light devices.

The main requirements for the modern LED engineering market are to reduce energy consumption and to increase luminous flux, service life, and heat removal. One can achieve the mentioned characteristics by using new materials and implementing modern technologies [1-2].

It is known [2] that an increase in a thermoelectric load on a LED intensifies the physicochemical degradation of a semiconductor structure and phosphor composition.

The following degradation mechanisms are known for LEDs and LED emitting elements: a change in the *p*-contact area resistance of a semiconductor structure due to interdiffusion processes and interfacial chemical reactions; mechanical stress relaxation in the semiconductor structure; hydrogen profile redistribution in the semiconductor structure; nitrogen vacancies formation due to Ga-N bonds breaking in the active chip area; indium and magnesium migration in the active semiconductor area; current contraction effect under the ohmic chip contacts; phosphor activator decay (Ce (cerium)) at temperatures above 120 °C.

The work aim is to use the filled polyaluminosilicate with a thermal conductivity of 142 W/m·K in the LED emitting element design to increase its limit thermoelectric operating modes.

The work relevance is due to the ever-increasing requirements for reliability and durability of the structural elements of microelectronics, lighting engineering, and spacecraft devices. In addition, the use of filled polyaluminosilicates in the LED emitting element design and the behavior study of the limit thermoelectric operating modes of LED light sources allow calculating, predicting, and increasing their service life.

A method to remove heat from the emitting element followed by dissipation in the environment with the use of the materials with high thermal conductivity, for example copper or aluminum, is used to manufacture high-power semiconductor emitters. When one develops a semiconductor emission source with the shape close to a filament lamp, it is practically impossible to use a massive radiator, since such a lamp has a closed structure, the so-called filament lamp. It makes it difficult to remove heat from the emitting semiconductor (chip) with subsequent dissipation in the environment. The fact that complicates the situation is that glass, which is a very poor heat conductor, is the main material by means of which the closed volume is created. Therefore, a problem to maximize the heat removal efficiency is very acute. It is known that the LED element operating temperature directly affects its operating mode and service life [3].

The standard filament lamp is a glass bulb, inside of which from four up to eight linears (LED filaments (figure 1)) are located. Twenty eight blue emitting semiconductor chips are located on the linears (in our case, in the LED emitting element design we have used a filled polyaluminosilicate applied on the aluminum base of the LED filament by means of 3D aerosol printing as an adhesive material). To obtain white emission the chips are coated with a layer of the optically transparent sealing polymer material with phosphor powder that converts the portion of blue emission into yellow one. A mixture of blue and yellow emissions produces white emission.

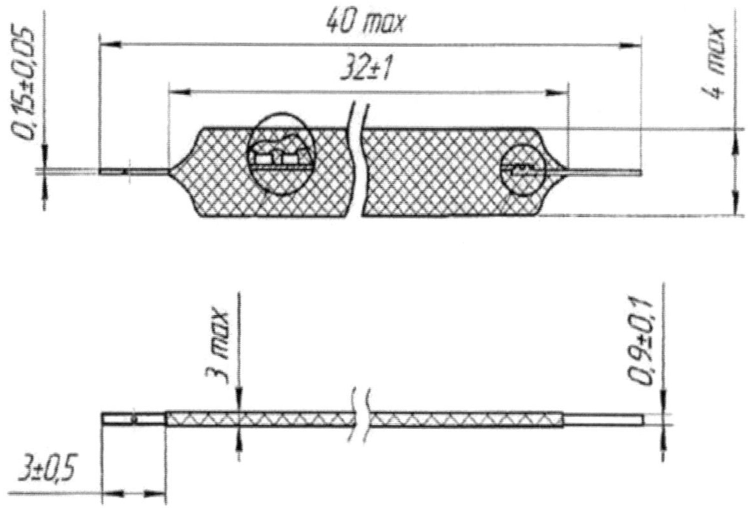

Figure 1. Dimension drawing of LED emitting element.

The space inside the lamp is filled with helium to increase the heat removal efficiency from chips and LED filaments [3].

The temperature effect on the performance and electrical characteristics of the LED emitting element has been studied to determine the possible conditions for the lamp operation.

2. Experiment

Experimental Obtainment of LED Filaments.

The AMG-6B aluminum has been chosen as a base material for the LED filament.

A filled polyaluminosilicate with a thermal conductivity of 142 W/m·K has been used as an intermediate dielectric layer to bond chips on the aluminum base. As the intermediate layer basis, the aluminosilicate meets the requirements for heat-resistant and wear-resistant coatings with high thermal conductivity provided that its macromolecules form such supramolecular formations, which in their turn form between them (interdomain) hollow areas capable of containing a large number of filler nano and microparticles. As the binder coating component, dendrimer aluminosilicate macromolecules with branched crown formation contain the maximum amount of fillers. The coating is capable of intense heat dissipation released in operating microelectronic and LED devices [1-7].

The developed composite material based on a highly filled dendrimer morphology polyaluminosilicate has been applied on LED linear bases by pneumatic 3D aerosol printing with a 3D printer. The highly filled dendrimer morphology polyaluminosilicate is pre-deagglomerated and placed in an aerosol generator in the Aerosol Jet 15EX printer of Neotech AMT company. The principle of pneumatic aerosol generation from a composite material used for the printer allows applying the dendrimer morphology polyaluminosilicates with various viscosities and the fillers with a wide range of particle size distribution. A working gas (nitrogen or air) is injected into the printer aerosol generator through a narrow hole under pressure to produce aerosol from the composite material. The pressure increase in the aerosol generator causes a rise of the material based on the highly filled dendrimer morphology polyaluminosilicate along the channel, and the aerosol is produced by the gas-material contact. The working gas pressure decreases at the exit from the narrow hole. As a result, a binder based on the dendrimer morphology polyaluminosilicate is sucked into the reduced pressure area through narrow channels from a chamber reservoir of the aerosol generator [8-9].

Primary aerosol formation. After the filled polyaluminosilicate meets a gas stream by the gas jet action, it breaks up into small particles ranging in size from 50 nanometers to 5 microns.

Secondary aerosol formation. The secondary aerosol forms after the primary aerosol particles collide with an obstacle and inhale to form even smaller particles than the primary aerosol ones.

A large proportion of the primary aerosol particles is deposited on the inner walls of the chamber and is drawn back in the aerosol formation. To focus the aerosol jet it is necessary to maintain a distance from the nozzle to the aluminum surface, on which the filled dendrimer morphology polyaluminosilicate material is applied. The distance is from 2 to 15 mm. Such aerosol jet focusing allows applying the composite material at 3D bases. This is carried out by moving the printing head along three axes and by inclinating the base along two axes.

28 Ga-N based blue emitting semiconductor chips are regularly located on the undried layer of the filled polyaluminosilicate applied on the aluminum base after the above-described operations. The chip clamping force to the aluminum base is visually controlled. There is no surplus of the material based on the filled polyaluminosilicate because the applied layer thickness is set in advance according to the previously obtained experimental results. After the emitting semiconductor chips have been glued, the obtained devices are exposed to heat treatment in an IR or thermal furnace at a temperature of 70 °C. Drying time is determined by the aluminum base thickness of the LED emitting elements and the number of the applied filled polyaluminosilicate layers. It varies from 15 to 25 minutes.

After the filled polyaluminosilicate drying we bond the chips by ultrasonic microbonding with the iBond5000 digital adjustable microbonding device.

The LED emitting element manufacturing is completed by filling them with a phosphor composition based on a two-component organosiloxane composition with the particles of an yttrium-aluminum garnet that converts the used chip emission into white colour.

3. Results and discussion

Operating (Nominal) Forward Current Determination of LED Emitting Element Samples.

Ten experimental samples of LED emitting elements with different in their chemical nature bases have been chosen for the research. LED chips have been bonded on the samples with various heat-conducting adhesives. Prior to the testing the current-voltage characteristics of the LED emitting elements (figure 2), the luminous flux dependence on the forward current of the LED emitting elements (figure 3), and the luminous efficiency dependence on the forward current (figure 4) have been measured with the Keithley 2410 source meter and FSH-1.0 photometric ball (the measuring range of the luminous flux is from 7 to 3000 lm, measurement inaccuracy is 10%).

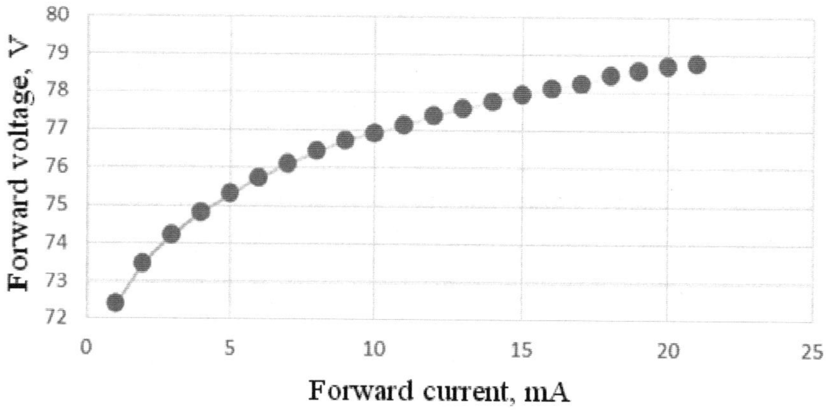

Figure 2. Current-voltage characteristic of LED emitting element.

Figure 2 shows that the current-voltage characteristic type of the LED emitting element with all the types of used bases and adhesives corresponds to the type of typical LED characteristic, and the slope is due to series-connected chips.

Figure 3 shows that the dependence type of the luminous flux on the LED emitting element forward current corresponds to the type of typical LED characteristic (for all ten studied samples). The luminous flux stops growing at a forward current of 15 mA.

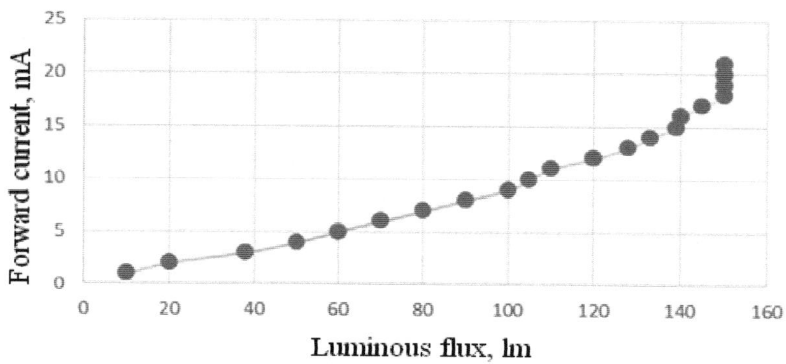

Figure 3. Luminous flux dependence on forward current of LED emitting element.

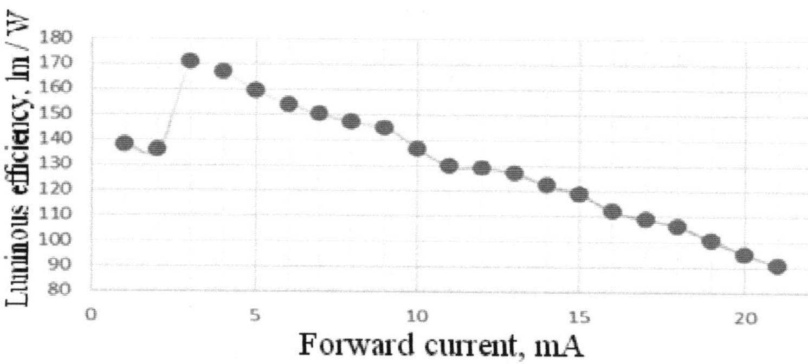

Figure 4. Luminous efficiency dependence on LED emitting element forward current.

IOP Conf. Series: Materials Science and Engineering 1181 (2021) 012010 doi:10.1088/1757-899X/1181/1/012010

Figure 4 shows a sharp increase in luminous efficiency at low forward current (2-3) mA. Then a decrease in luminous efficiency, which is linear, is observed. This is also typical for many types of LED emitting elements.

Basing on the dependency result analysis shown in Figures 2-4, we have decided to choose the working (nominal) forward current 10 mA for the experimental samples of LED emitting elements, which is 50% of the electric load margin for the limit forward current 21 mA.

Thermoelectric Load Test

A forward current of 10 mA has been found in the experimental samples of LED emitting elements by the Keithley 2410 source meter. A temperature of 30 °C has been set in the *SNOL* heat chamber (temperature maintenance range (30-350) °C ± 3 °C). The experimental samples of LED emitting elements have been installed in the heat chamber. The experimental samples of LED emitting elements have been kept in the heat chamber for five minutes for thermal stabilization. Then forward voltage has been measured by the Keithley 2410 source meter. Similarly, the measurements have been taken with increasing temperature in the heat chamber to the value, at which the experimental samples of LED emitting elements have failed. The failure has been fixed by a forward voltage of 0 V, which corresponds to an open electrical circuit. Ten experimental samples of LED emitting elements have been tested. The test results have shown that the limit thermoelectric loads for nine samples have not exceeded 194 °C. This is primarily due to the materials with insufficient thermal characteristics used in the LED emitting element construction to bond chips.

Figure 5 shows the test results of the limit thermoelectric loads for the LED emitting element sample, in the design of which a filled polyaluminosilicate has been used as the adhesive material for chips.

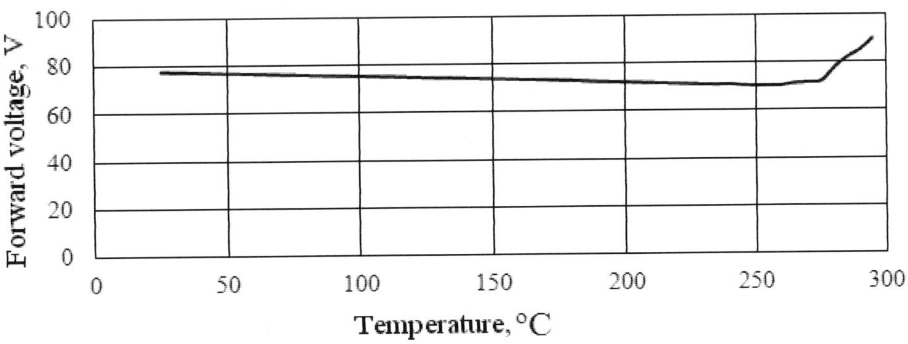

Figure 5. Forward voltage dependence on LED emitting element temperature.

Figure 5 shows that the forward voltage decreases linearly with increasing temperature, which is typical for LEDs. When the temperature in the heat chamber is 275 °C, there is a sharp increase in the forward voltage from 70 V to 90 V. This process is characterized by a thermal breakdown. When the temperature in the heat chamber is 299 °C, the experimental samples of LED emitting elements fail. Figure 6 shows a photograph of one of the experimental samples of LED emitting elements after testing.

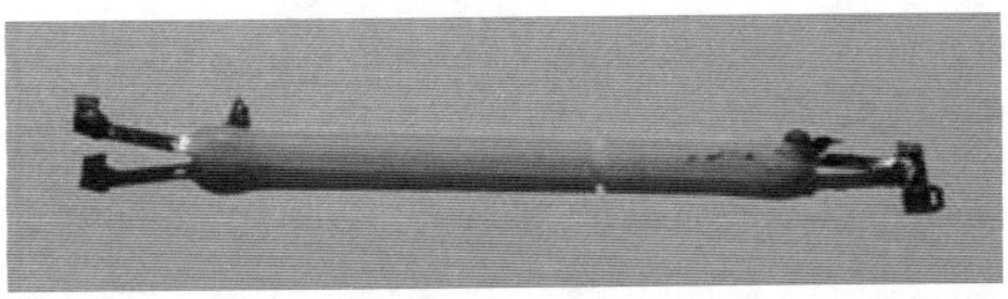

Figure 6. Tested LED emitting element sample containing filled polyaluminosilicate in its design after thermoelectric load test.

Photo 6 shows mechanical damage to the phosphor composition in the form of a crack. This is due to the loss of the highly elastic state of the organosiloxane binder used to prepare the phosphor composition, as well as the expansion of the LED emitting element supporting structure with increasing temperature. The crack formation of the phosphor led to a mechanical break in the circuit between chips, which are bonded with a thin gold wire by ultrasound.

4. Conclusion

Basing on the obtained results, one may conclude that the used combination of the base material and intermediate dielectric layer (a filled polyaluminosilicate has been used as the layer) has allowed increasing the temperature, at which the LED emitting element continues operating. This is due to the specific properties of the filled polyaluminosilicate, such as adhesive strength to the base material, thermal linear expansion factor, thermal conductivity, and elastic modulus.

Acknowledgement

The reported study was funded by RFBR according to the research project № 18-29-11018\20.

References

[1] Peter Marx 2009 LED-Anwendungstechnik *Licht* **3** 184-8
[2] Ivanov A A and Chermoshentseva A S 2019 *IOP Conf. Series: Materials Scienc and Engineering* **684**
[3] Ivanov A A, Afonin K N and Chermoshentseva A S 2020 *IOP Conf. Series: Journal of Phisics* **1488**
[4] Kang B D 2006 *KSAE* **28(2)** 82
[5] Takahashi K *et al.* 2006 *Int. J. of Adhesion and Adhesives* **26** 639-43
[6] Tomalia D A and Durst H D 1993 *Top. Curr. Chem.* **165** 193-313
[7] Tomalia D A, Naylor A M and Goddard W A 1990 *Angew Chem.* **102** 119-57
[8] Tomalia D A *et al.* 1986 *Macromolecules* **19** 2466-8
[9] Barton K, Mishra S, Shorter K A, Alleyne A, Ferreira P and Rogers J 2010 *A Desktop Electrohydrodynamic Jet Printing System Mechatronics* **20(5)** 611-6

Physicochemical regularities in highly filled polyaluminosilicate applied on aluminum base by 3D aerosol printing

A A Ivanov[1] and A S Chermoshentseva[2]

[1] Tomsk State University of Control Systems and Radioelectronics, Tomsk, 634050, Russia
[2] Bauman Moscow State Technical University, Moscow, 105005, Russia.

E-mail: alexchemtsu@rambler.ru

Abstract. A modified technique has been used to synthesize a highly filled polyaluminosilicate with a controlled supramolecular structure during its applying on an aluminium base by 3D aerosol printing. It has been found how the pH medium affects the polyaluminosilicate ability to form fractal supramolecular formations. Methods to control the highly filled polyaluminosilicate structure directly along the aluminium base surface have been found.

1. Introduction

In developed countries much attention is currently paid to the electronic industry development that leads to the need to create new materials and technologies to apply them on surfaces of different chemical nature. The materials should have not only the improved performance characteristics but also the controlled physicochemical properties during their application on various surfaces.

The current trend in technologies to apply the materials for microelectronic products and devices is reduced to the increasing use of 3D additive methods. 3D aerosol printing using a pneumatic aerosol generation principle is the most promising and developing method to apply materials on microelectronic products and devices. The pneumatic 3D aerosol printing method allows applying the materials with an accuracy of 50 nanometers that expands the scope of its application.

Therefore, the development of new approaches and scientific principles in the synthesis of new materials is basic to create fundamentally new microelectronic products. This is primarily due to the fact that when developing new materials one takes into account the technological features of their further use in production processes.

A sol-gel method is one of the methods to develop materials with specified properties that allows controlling their properties during the application by means of 3D aerosol printing. In our case, the aluminosilicates that form by hydrolytic polycondensation can produce macromolecules of different length, branching, and supramolecular formation. The gels with different supramolecular structures can form when the medium pH and number of unbound –OH groups in the aluminosilicates vary. By pH less than 3 the complete aluminosilicate hydrolysis leads to the formation of fractal supramolecular formations, the interdomain spaces of which are then closely packed with the particles of a dispersed filler. Consequently, a nonporous monolithic structure with new properties forms. At the system pH more than 7 the structure becomes more porous. The gel formation does not occur due to the incomplete precipitation of silicon ions and aluminum hydroxide dissolution [1].

The study aim is to research the structure formation effect of a sol-gel system *dendrimer morphology polyaluminosilicate - highly dispersed filler* on the properties of the dielectric layers on an aluminum base applied by 3D aerosol printing.

The following basic objectives have been set to achieve this aim:

- to study the possibility to control the supramolecular structure during the highly filled polyaluminosilicate sol application for its close packing;
- to determine a kinetic parameter effect of the highly filled polyaluminosilicate on the surface properties of the dielectric layers on the aluminum base obtained by 3D aerosol printing.

2. Experiment

The modern methods to synthesize materials, first of all, should be considered as a key problem to create the materials with specified and controlled properties. Of many liquid-phase methods to create ceramic polymer nanomaterials, such as chemical precipitation, ion layering, and hydrothermal synthesis, the sol-gel method allows regular controlling the physicochemical processes at some stages, which in its turn allows using the sol-gel method to create defect-free ceramic nanostructures [2-5].

It is necessary to take into account the stages of sol destabilization and aggregation in case of the sol-gel method use to synthesize highly filled polyaluminosilicates with their subsequent use for 3D aerosol printing (figure 1). It is also necessary to take into account the oligoaluminosilicate solubility in water, because the pneumatic 3D aerosol printing method involves two stages of aerosol formation. The lack of –OH groups in the supramolecular oligoaluminosilicate structures promotes the macromolecule enlargement, thereby complicating further fractal structure formation.

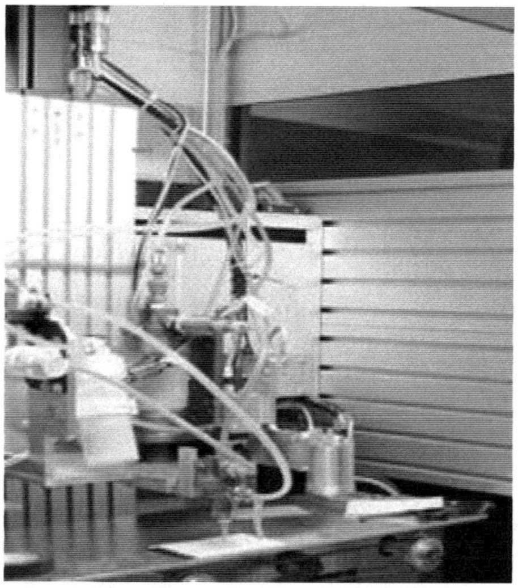

Figure 1. Pneumatic 3D aerosol printing unit in process of highly filled polyaluminosilicate topology applying on aluminum base.

When the aluminosilicate monomer concentration in a solution exceeds the equilibrium solubility, the solid phase, on which the dissolved aluminosilicate is precipitated, is absent. It complicates the formation of a closely packed structure after its precipitation on the aluminum surface by 3D aerosol printing.

When the oligomer particle sizes approach the lower limit of colloidal dispersion (in our case, the process begins after the material hits the aluminum base surface), homogeneous nucleation occurs (figure 2) at the aluminosilicate polycondensation stage. The dispersed polyaluminosilicate phase particles begin to grow on the nuclei according to the condensation mechanism. As the particles grow on the aluminum base surface, the aggregation of individual particles begins that determines the final

IOP Conf. Series: Materials Science and Engineering 1181 (2021) 012011 doi:10.1088/1757-899X/1181/1/012011

polyaluminosilicate structure (figure 3). Larger particles grow by absorbing smaller particles, their solubility is increased. When the process proceeds above 63°C and at pH more than 6, anhydrous particles form inside the supramolecular aluminosilicate formations that is confirmed by the thermogravimetric studies (figure 4).

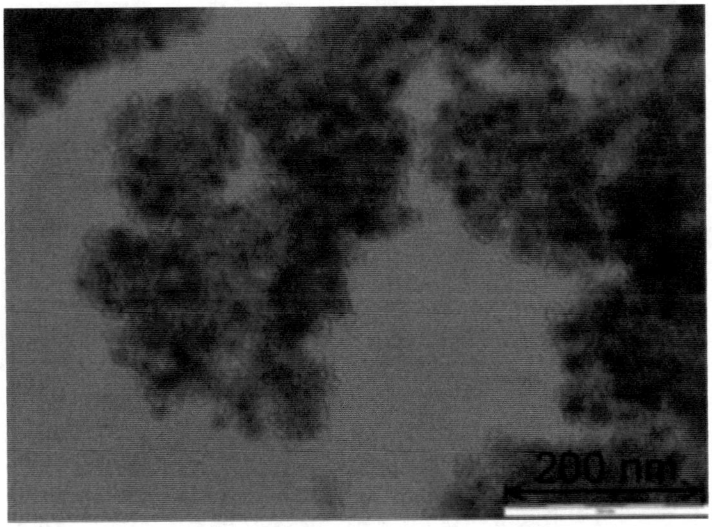

Figure 2. Microphotograph (transmission electron microscopy) of nuclei formation of highly filled polyaluminosilicate of fractal morphology.

The particles with chemically bound –OH groups, which form at pH more than 6, are negatively charged and repel one another. Their growth occurs not due to their aggregation but by polycondensation with particles from a true solution. Particles with a close supramolecular structure packing form by the appropriate control of the medium pH, which does not exceed 8, and interaction with the highly dispersed filler particles after their joint mechanochemical activation.

Obtaining the highly filled polyaluminosilicates with close packing requires a special approach to their drying, namely the environment temperature and humidity control should be provided. The lack of such control inevitably leads to a change in the polyaluminosilicate structure.

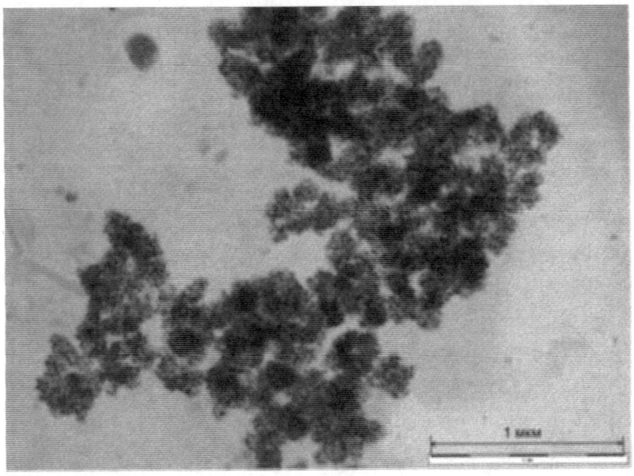

Figure 3. Microphotograph (transmission electron microscopy) of highly filled polyaluminosilicate of fractal morphology.

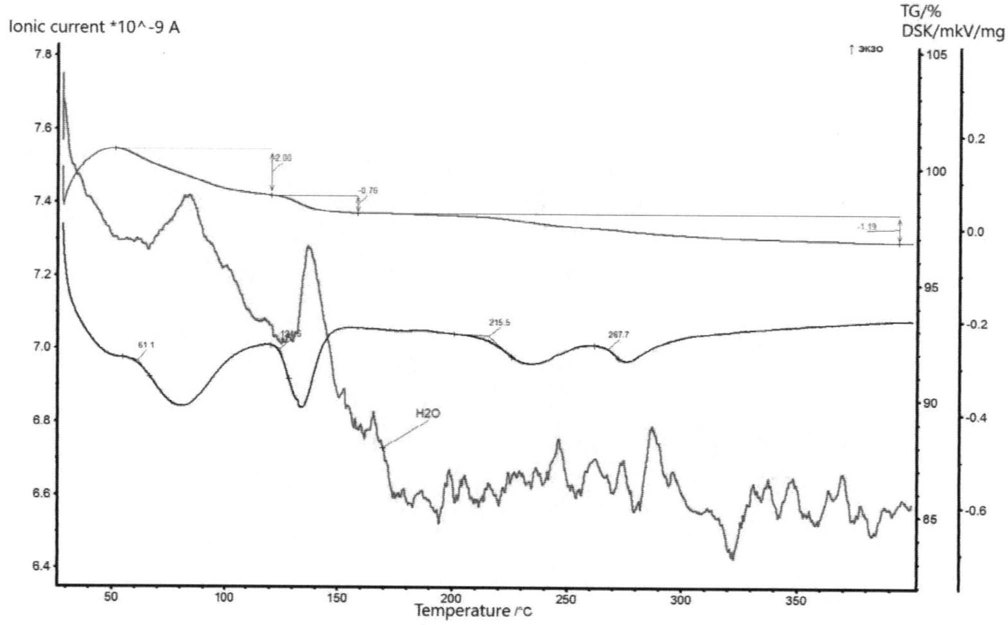

Figure 4. TG and DSC curves of dendrimer morphology polyaluminosilicate.

The thermogravimetric sample studies have been carried out by the NETZSCH STA 409 device and combined Q-600 TGA-DSC analyzer (*TA Instruments*, USA). The sample heating rate has been 10 gr./min in an argon atmosphere with a weighed analyzed sample from 8 to 10 mg. The gas medium composition has been controlled by mass spectrometry by means of the ProLab device (*Thermo Scientific*, USA).

The rate of chemically unbound water removal in interfractal and interdomain voids, which determines the final polyaluminosilicate macromolecular structure, is of great importance.

The first two peaks demonstrate water desorption in figure 4. The physically adsorbed water (moisture) desorption usually occurs at a temperature not lower than the adsorbate boiling point (in this case it is 100 °C). The fact that there is a desorption peak at 60 °C may indicate polycondensation processes occurring in the sample with water release. The chemically unbound water removal from the polyaluminosilicate matrix occurs due to capillary forces that transfer the liquid phase to the closely packed structure surface for evaporation. Therefore, it is important to select the optimal conditions to apply the highly filled polyaluminosilicate to the aluminum base surface by 3D aerosol printing.

The deep polycondensation of the highly filled polyaluminosilicate sol results in a target product formation with the closely packed supramolecular structure. It is carried out in an aluminosilicate sol phase directly on the aluminum base surface. Figures 4 and 5 illustrate the possibility to increase the polycondensation degree during heating. Water release occurs in two stages with maxima at 100 °C and 136 °C. At a temperature of 67-124 °C the sample loses six times more water than in the range of 129-147 °C, when all available OH-groups have already reacted (figure 4).

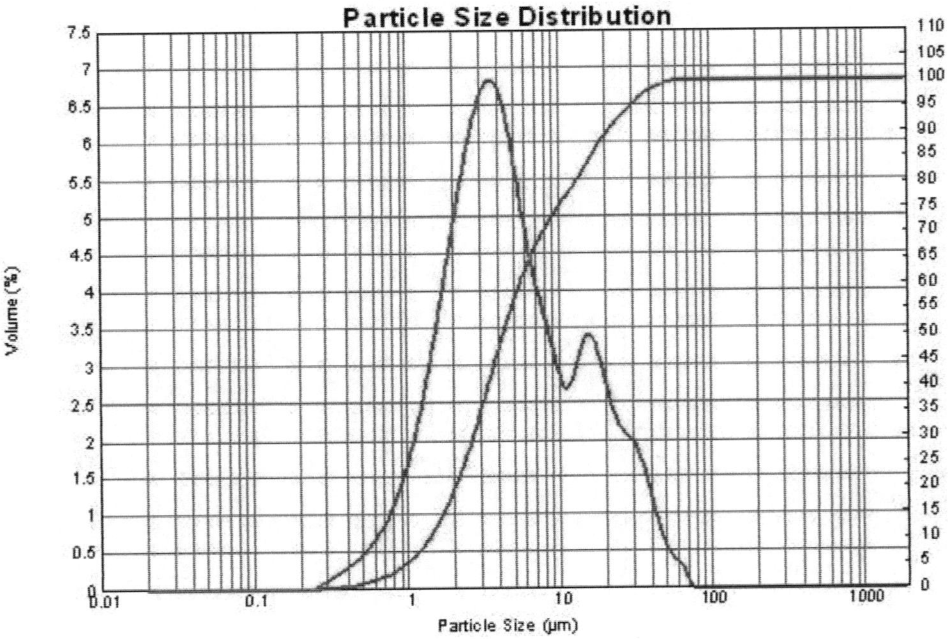

Figure 5. Distribution of highly filled polyaluminosilicate particle sizes (Mastersizer 2000 device).

3. Results and discussion

It is possible to control the supramolecular structure by the rate of its drying and polycondensation during the highly filled polyaluminosilicate sol applying. The chemically unbound water removal from the polyaluminosilicate matrix occurs due to capillary forces that transfer the liquid phase to the closely packed structure surface for evaporation.

The removal of chemically unbound –OH groups from the polymer polyaluminosilicate matrix occurs due to capillary forces that transfer the liquid phase to the supramolecular structure surface for evaporation.

Thus, one can control the failure stress, which directly affects the particle size and uniformity, by varying the drying temperature of the highly filled polyaluminosilicate. The high surface energy of the polyaluminosilicate is a driving force for this process.

The smaller is the size of the particles obtained by applying the highly filled polyaluminosilicate on the aluminum base by 3D aerosol printing, the larger is their specific surface area and the higher is their surface energy. Therefore, they should be dried at lower temperatures.

Thus, the study of a surface roughness degree of the highly filled polyaluminosilicate applied on the aluminum surface by 3D aerosol printing has shown that a sharp increase in temperature from room temperature to 70 °C at a heating rate of 5 degrees per second increases the surface roughness from 0.5 to 2.3 µm (figure 6). This is primarily due to excessively rapid hydrolysis deepening and polycondensation that affects the entire material structuring.

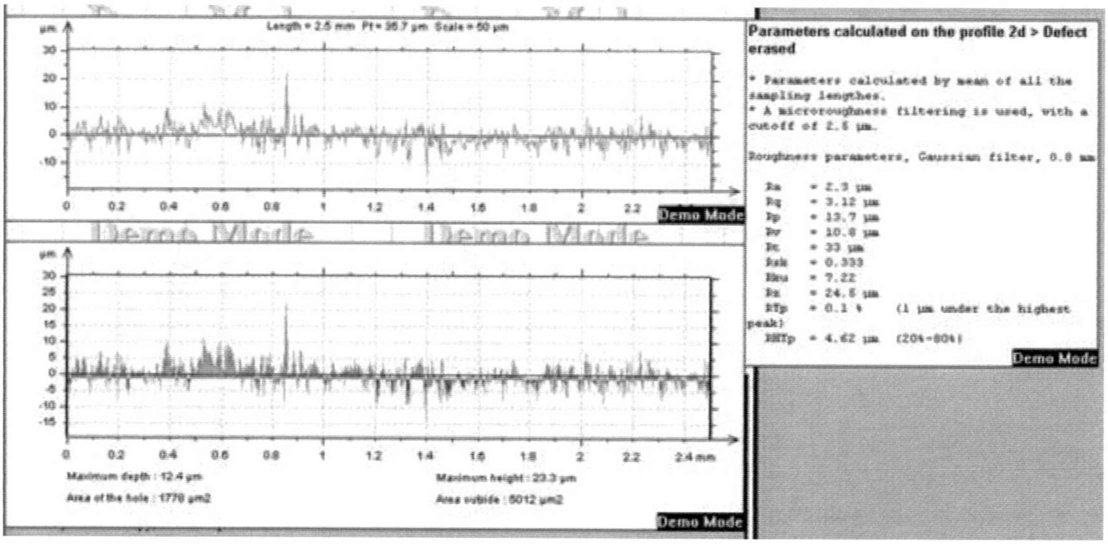

Figure 6. Surface profilogram of highly filled polyaluminosilicate applied on aluminum base by 3D aerosol printing.

Since the film forming properties of polyaluminosilicate sols directly depend on their drying temperature and affect their self-organizing supramolecular structure, a necessary condition to obtain a defect-free closely packed structure is to select adequate conditions to apply it on the aluminum surface and to control gelation kinetics, which in its turn directly depends on the starting sol components.

4. Conclusion

As a result, the modified methods to obtain the highly filled polyaluminosilicate and control its physicochemical parameters have allowed studying the effect of polyaluminosilicate gel structure formation on the ability to form defect-free dielectric layers with a given degree of surface roughness. This, in turn, has made it possible to control the cohesive interaction between the applied layers and its adhesion behavior by the contact with the substrate when applying the highly filled polyaluminosilicate on the aluminum base by 3D aerosol printing.

Acknowledgement

The reported study was funded by RFBR according to the research project № 18-29-11018\20.

References

[1] Ivanov A A 2020 *IOP Conf. Series: Materials Scienc and Engineering* **862**

[2] Walker Jr W J, Brown M C and Amarakoon V R W *J* 2001 *Europ. Ceram. Soc.* **21(1011)** 2031-6

[3] Selmi F A and Amarakoon V R W 1988 *J. Amer. Soc.* **71(11)** 934-7

[4] Hatano T, Yamaguchi T, Sakamoto W, Yogo T, Kikuta K, Yoshida H and Tanaka N 2004 *J. Europ.Ceram. Soc.* **24** 507-10

[5] Barton K, Mishra S, Shorter K A, Alleyne A, Ferreira P and Rogers J 2010 *A Desktop Electrohydrodynamic Jet Printing System Mechatronics* **20(5)** 611-6

Formation of a batch of parts in the conditions of one-off and small-scale production

B M Bazrov and N A Rodionova

Blagonravov Institute of Mechanical Engineering, Russian Academy of Sciences, Moscow, Russia

E-mail: modul_lab@mail.ru

Abstract. The article is devoted to increasing the efficiency of manufacturing a batch of parts by optimizing the sequence of their production. It is shown that the effect of the sequence of manufacturing parts in a batch on efficiency is manifested through the time spent associated with transitions to the manufacture of parts of different names. The criterion for the optimal variant of the sequence of manufacturing parts in a batch is the minimum time spent on changing technology during transitions to the manufacture of parts of different names.

The purpose of forming a batch of parts is to increase the efficiency of their manufacture by reducing the variety of technology and increasing its repeatability.

Under the formation of a batch of parts we mean the grouping of homogeneous parts and manufactured in a given sequence.

In practice, the formation of batches of parts is implemented using standard and group technologies [1-6].

The main the task of the typical technology is to reduce the labor intensity of the technological preparation of production.

In this regard, the grouping of parts is made in such a way that structurally similar parts with similar characteristics are included in one group, followed by the definition of a part - a representative [7], under the production of which a typical technological process is being developed, which is used in the development of technological processes for each part in the group.

The grouping of parts by design features is carried out by dividing them into a number of the following classes: shafts - V (smooth, stepped shafts, spindles, axles, pins, etc.); bushings - A (smooth bushings, stepped bushings, liners, sleeves, collets, etc.); discs - D (flat parts of rotation); plates - P (bodies, beds, frames, tables, strips, etc.); eccentric parts - E (eccentrics, crankshafts); levers - R, etc. Details of each class are subdivided into groups, subgroups and types. This classification is the basis for the typification of technological processes [1, 2].

Group technology pursues the goal of increasing the efficiency of operating technology by combining parts into groups containing the same surfaces, manufactured at one workplace.

Thus, the formation into groups is carried out according to the generality of processing workpieces in the corresponding operations. When forming a group, one should take into account the following [3, 4]:

- The generality of structural elements that form the configuration of the part (threads, chamfers, grooves, cylindrical outer and inner surfaces).
- Accuracy and roughness of surfaces.
- Homogeneity of the workpiece material, which allows the use of the same machining method and common tool for machining all parts of the group.
- The proximity of the dimensions of the workpieces, which allows processing on the same equipment using the same devices (group, readjustable and others).
- Serial production and labor intensity of parts processing.

Thus, typical technology includes structurally similar parts, and group technology includes structurally inhomogeneous parts, but containing groups of identical surfaces.

But in both cases, the parts are different both in design and in characteristics.

This leads to a difference in the schemes of basing their workpieces, fixtures, processing methods and processing tools.

In both cases, there remains a reserve for increasing the efficiency of their manufacture, which consists in establishing the optimal sequence for the manufacture of parts in a batch.

The influence of the sequence of parts in a batch on the efficiency of their manufacture affects the time spent on transitions to new technologies in the manufacture of other parts of different names.

For example, when switching to the manufacture of the second part in a batch, it will be necessary to change the machine tool to install the workpiece, and when switching to the manufacture of the third part, you will need the machine tool used in the manufacture of the first part [8].

Hence it follows that it is advisable to produce a third part after the first part, and not the second part.

In practice, the sequence of parts in a batch is usually determined by the type of the structural form of the parts, assuming that the change in technology during transitions to the manufacture of new parts will be minimal.

For example, body parts are made first, then parts such as caps, etc.

However, this sequence of parts to be produced may not be optimal.

It may be more efficient to manufacture a cap as a second part.

Therefore, depending on the sequence of production of parts of different names in the batch, the efficiency of the batch of parts produced can vary significantly.

And the wider the nomenclature of manufactured parts in a batch, the greater the difference in the production efficiency of a batch of parts, depending on the selected sequence of parts production.

This is mainly due to a change in the labor intensity of work associated with a change in the technology of transition from manufacturing parts of one name for the manufacture of parts of another name.

These works are associated with a change in certain elements of the manufacturing technology of parts.

So, the problem arises of lining up parts of different names in a row to minimize the labor intensity of work associated with the transition from the manufacture of a part of one name to the manufacture of a part of another name.

In connection with the foregoing, as a criterion for determining the sequence of manufacturing parts of different names in a batch, one should take not the constructive similarity of parts, but the minimum laboriousness of work associated with transitions to changes in the technology of manufacturing parts of another name.

When determining the sequence of manufacturing parts in a batch, the initial data should be drawings of parts in the batch, the machine tool park of the enterprise, technological equipment and technological processes for manufacturing parts of each name, and the laboriousness of changing technology elements associated with transitions to the production of parts of other names.

In table 1, as an example, an operational map of grinding the surface of a shaft is given, according to which the elements of the manufacturing technology of parts that can change during the transition from the manufacture of a part of one name to the manufacture of parts of another name include:

- change of devices for workpiece installation;
- changing the processing tool;
- changing the control and measuring device;
- changing the processing mode;
- setting the machine for the accuracy of the part.

Table 1. Operational chart of shaft surface grinding.

№	Transition	Cutting tool	Measuring tool	Estimated size, mm		Processing mode			
				Diameter	Length	t, min	S, mm / rev	n, rev/ min	V, m/ min
1	Install the shaft in the center								
2	Sand surface 1	Grinding wheel PL500×50×305 24A10PS2 35m/s 1kl A GOST2424-75	Indicator bracket D=50mm	35	10	0.2	0.2	380	35
3	Sand surface 2			30	6	0.25	0.2	380	35
4	Remove part								

Each of the listed technology elements is characterized by the corresponding time consumption (see table 2).

Table 2. Elements of technology that change during the transition to the manufacture of a new part.

№	Technology element	Time (t)
1	Machine change	T_s
2	Change of machine tool	T_p
3	Changing the processing tool	T_i
4	Change of control and measuring device	T_u
5	Changing the processing mode	T_r
6	Adjustment for machining accuracy	T_n

When determining the best sequence for manufacturing parts in a batch, there will be a sequence in which the total laboriousness of changing the above elements of technology in the manufacture of the entire batch will be minimal.

During transitions to the manufacture of parts of different names, various combinations of changing technology elements are possible. For example, in one case, you will have to change the machine tool, assign other processing modes, and in the other case, change the machine tool and part of the processing tools [9].

The task of determining the optimal sequence for the manufacture of parts of different names in a batch includes the following steps:

- determination of all possible options for the sequence of manufacturing parts of different names in the batch;

- calculation of the amount of time spent on changing the manufacturing technology of parts during transitions to the manufacture of parts of different names;
- determination of the sequence for manufacturing parts of different names, which has a minimum amount of time spent on changes in technology during transitions to the manufacture of parts of different names.

The problem of determining the optimal variant of the sequence of manufacturing parts of different names of a batch of parts is solved by enumerating options.

- the reserve for increasing the efficiency of manufacturing a batch of parts is the optimal sequence of manufacturing parts of different names;
- the sequence of manufacturing parts of different names in a batch determines the time spent associated with changes in technology during transitions to the production of parts of different names;
- as a criterion for the optimal sequence of manufacturing parts in a batch, it is necessary to take the minimum time spent on changing the technology during transitions to the manufacture of parts of different names;
- the optimal variant of the sequence of manufacturing parts in a batch is determined by enumerating all possible variants of the sequence of parts.

References

[1] Mitrofanov S P 1983 *Group technology of engineering production Organization of group production* **1** (Leningrad Branch: Mashinostroenie)
[2] Bukhalkov M I, Kuzmin M A and Pavlov V V 2010 Features of the design and organization of group production in mechanical engineering *Organizer of production* **4(47)** 27-32
[3] Mitrofanov S P, Kulikov D D, Milyaev O N and Padun B S 1987 *Technological preparation of flexible production systems* (Leningrad Branch: Mashinostroenie)
[4] Bazrov B M, Dilanyan R Z and Melnikov G N 2010 *Group and modular technologies* (Moscow: Publishing house of MSTU named after N E Bauman)
[5] Matalin A A 1985 *Technology of Mechanical Engineering* (Leningrad Branch: Mashinostroenie)
[6] Chistoserdov P S 1964 *Typification of technological processes and group methods of processing parts* (Minsk: Belarus)
[7] Bazrov B M and Rodionova N A 2021 The influence of the choice of a representative part on the efficiency of production of parts *STIN* **1** 22-6
[8] Kovalenko AV and Podshivalov R N 1986 *Machine tools* (Moscow: Mashinostroenie)
[9] Kozhevnikov D V, Grechishnikov V A, Kirsanov S V, Kokarev V I and Skhirtladze A G 2005 *Cutting tool* ed S V Kirsanova (Moscow: Mashinostroenie)

Research on the influence of deformable sheet aluminum alloy rheological behavior on stretch forming process limits

V A Mikheev, E A Nosova, Roberto de Alvarenga and M M Demidova

Samara National Research University, 34, Moskovskoye shosse, Samara, 443086, Russia

E-mail: vamicheev@rambler.ru

Abstract. Our research was aimed at modeling a symmetrical wrapping and stable control of the processes of shaping by wrapping a sheet blank along a wrapping punch having a double curvature surface and placed on the table of a virtual stretch-tightening press. The main task of modeling was to determine the deformed state with a uniform change in thickness in different regions of the sheet blank without folds and breaks, but without the influence of the rheological behavior of the deformable aluminum alloy. Here, the LS-DYNA software package is used with the use of an explicit or implicit solver for nonlinear problems of solid mechanics, characteristic of nonequilibrium transitions of the initial dissipative phase due to viscosity and heat transfer. Now it is possible to predict defect formation from a practical point of view, using the forming limit diagram (FLD), which makes it possible with sufficient accuracy to determine the degree of deformation of a given sheet material, at which any defects do not occur. The presented results of the study revealed the features of the influence of the rheological behavior of a wrought sheet aluminum alloy on the limits of the shaping process by a wrap.

1. Introduction

The development of new processes of shaping by a double-curvature shell is to a certain extent restrained by the poor knowledge of the rheology of the behavior of sheet aluminum alloy in the annealed, especially in the freshly hardened state. In addition, there is no information about the influence of a particular rheological state of the material on the stress-strain state in these processes, which determine the nature of the flow of a deformable sheet blank and the formation of the quality of the finished product. The rheological behavior of deformable materials is quite fully characterized by flow curves: σ_s - $\acute{\varepsilon}$, which reflect the interaction of strain hardening processes and dynamic softening processes [1]. Any deformable material can be represented as a set of elementary models: elasticity, viscosity and plasticity [2]. In our opinion, the behavior of a deformable body in all its diversity is most accurately reflected by a medium where a «spring» describing strain hardening is included in series with a viscous element that simulates diffusion-relaxation processes.

It is known that heat treatment of deformable aluminum semi-finished products, in particular sheet, has a significant effect on their structure and properties. A wide variety of structures that can be obtained in aluminum alloys D16 and 1163 after different heat treatment is the result of different degrees of deviation of alloys from the thermodynamic equilibrium state at room temperature [3]. The equilibrium (annealed) state of thermally hardened aluminum alloys is a solid solution with a low content of alloying elements (tenths of a percent). With this phase composition, the alloy has low strength and very high ductility.

The most unstable structure at room temperature in thermally hardened aluminum alloys is a supersaturated solid solution of alloying elements in aluminum, the concentration of which can exceed the equilibrium state by tens of times. With this structure, the aluminum alloy is also very ductile, but stronger than in the equilibrium state. The instability of the freshly quenched state of the alloy at room temperature and the tendency at the initial stage to premature decomposition of the supersaturated solid solution technologically limits the possibilities of shaping the sheet material by the skin. However, this intermediate structure, which corresponds to the initial stages of the decomposition of a supersaturated solid solution, attracts attention. There are results of stabilization of the freshly hardened state of an aluminum alloy under regulated heating for hardening and holding the freshly hardened sheet blank in the cold before the forming operation [4].

If the deformation of the sheet blank during shaping occurs under such conditions when the diffusion-relaxation processes are inhibited, then the work of the viscous element can be neglected and a model of an elastoplastic body with strain hardening can be considered. Strain hardening is accompanied by an increase in the nonequilibrium of the system, i.e. a forced process that occurs due to external forces. In turn, the spontaneous diffusion-relaxation process is inverse, removing the strain hardening.

The main goal of modeling the process of shaping by wrapping shells of double curvature is to select the technological mode of sheet shaping with minimization of rejection and maximization of the degree of shaping at an acceptable level of thinning of the sheet blank. Simulation of the process by the finite element method is carried out on a covering punch, with a surface oriented relative to the curvature lines and placed on the table of a virtual stretching press, which leads to symmetry during shaping by a covering [5]. In LS-DYNA, the modeling of sheet material in the process of shaping with a wrap has become reliable after the introduction of a number of algorithms. These include:

- adaptive rebuilding of the mesh of the workpiece finite elements; generation of a specialized mesh for contact sheet forming;
- models of material behavior for sheet forming, taking into account the anisotropy of the sheet, such as Barlat (MAT_036) and Hill (MAT_037);
- advanced post-processing for sheet forming processes, including building FLD.

Now it is possible to perform the necessary adjustment of the parameters of the control program based on the results of modeling a particular wrapping method using the ANSYS/LS-DYNA software package, using a number of pre-calculated finite element models for controlling the shaping of the wrapping on a stretching press. The development of an automated system for the production of bypass-forming elements of aircraft has become possible with the use of modern simulation tools. The operator of the software complex selects a model of material behavior, enters into the computational model the geometry of the tightening punch and the dimensions of the sheet blank, constants from the material database, selects the characteristics of friction. Next, the solver calculates stresses, strains and displacements based on the input data on materials, geometry, dimensions, boundary conditions. At the same time, the system predicts the moment of destruction or formation of folds of sheet material. Therefore, when designing an optimal shaping scheme with a wrap, it is necessary to have specific information on the current state of the sheet material, using the FLD diagrams of the limiting shaping of the sheet material during modeling.

At this stage of modeling the shaping process by the wrapping, we used the toolkit of the graphical user interface (Create and control RO-630) of the computer application of the kinematics control system of the virtual stretch-tightening RO-630-11. In practice, we have just started machine learning of the control system. In the course of the preparation, more than three dozen numerical calculations were carried out, as a result of which a stable kinematic model of the shaping process by a wrap was developed, which significantly reduces the degree of uneven deformation of the sheet blank.

A decrease in the thickness difference of the double curvature shell formed by the covering was carried out in the modeling process by manually adjusting the values of the controlled coordinates on the graphical control interface of the virtual model of the RO-630-11 stretch-tightening press based on the results of the analysis of the thickness values at the points indicated by us on the surface of the sheet blank. Two innovative wrap shaping methods have been proposed, each with a control parameter matrix. The preparation of geometric models of elements for calculation in the LS-DYNA processor was carried out in the LS-PREPOST program.

It was necessary to go through a cycle of machine learning of the control system using the graphical user interface toolkit (Create and control RO-630) to control the kinematic movements performed by the working bodies of the press. Using the example of a simple user, the control system allows you to accurately simulate the real kinematics of the RO-630-11 stretch-tightening press, using grounded kinematic methods of forming the cover. Matrices of press control parameters were generated in the form of animation for both methods. At the stage of calculations, the calculation results were sequentially checked and possible adjustments that had to be made to the animation in question were evaluated in order to send a new animation into the calculation, comparing the current results.

This took a lot of time from us and so far we decided to carry out mathematical modeling of one of the method options taking into account the influence of the rheological behavior of the wrought aluminum alloy under various annealing modes, hardening conditions and friction conditions of the sheet on the surface of the tightening punch. The main object for the analysis and optimization of the method is the graphs of deformation and thickness changes at characteristic or selected points on the surface of the sheet blank. To visualize the size and shape of the deformation zone in a particular method of wrapping, we used the data of the simulation results, drawn up in the form of graphs, histograms of thickness changes. Analysis of the histograms allows you to make a conclusion about the state of the process. By the simplicity of its construction, the histogram gives useful information about the spread of indicators (shell thickness), average values, accuracy and stability of technological processes, and the accuracy of the kinematics of technological equipment.

2. Application of FLD curves for sheet forming simulation

An important step in the production of aircraft skin sheet components is the development of forming limit diagrams (FLD), which are composed of forming limit curves (FLC). First introduced by Keeler [6] and Goodwin [7], FLC is an empirical relationship between critical (or limiting) strains in a plane that describe the deformability limits of a sheet material. FLCs distinguish combinations of major and minor deformations that are «safe», that is, for which a given combination of deformations in the FLD is not expected to fail. Major deformation occurs vertically (y-axis) and minor deformation occurs horizontally (x-axis) FLD [8]. Localized necking is usually of interest to the fracture mechanism itself, since the necking regions are potential sites for fracture (note that the key differences between «diffuse» and «localized» necking are detailed in Hosford and Cadell [9].

Efforts are currently under way to develop stress-based FLD. The stress-based approach has some advantages over the widespread strain-based approach, which may well lead to its becoming the preferred method in the future. Because experiments can be costly and time consuming, some effort has been made to predict strain-based FLD using various theoretical methodologies. The theoretical prediction of FLD for sheet materials relevant to the transport industry was based on one of three existing theories.

The first one belongs to Hill, and it is assumed that a band or neck of localized deformation is formed along the direction of zero tension, and there is an angle between the normal direction of the neck in the plane of the sheet and the direction of the main deformation: this is the so-called «hypothesis of zero elongation». However, Hill's zero elongation hypothesis is only applicable to the left side (LHS) FLD, since there is no direction of zero elongation in a positive slight strain stretched sheet, and of course, predicting localized necking is not possible.

The second theory is based on the hypothesis of the groove (or inhomogeneity of the initial thickness) by Marcinyak and Kuchinsky [10], also known as the «MK theory». It postulates that a localized neck arises from a geometric discontinuity or groove in the initial thickness of the sheet from which the localized neck develops. The obvious physical significance of the MK theory and its simple mathematical form have contributed to its widespread application for more than four decades. Unfortunately, MK models are overly sensitive to the assumed initial thickness inhomogeneity. In addition, several studies have been published that question the extent to which the underlying geometric discontinuity is a direct cause of local necking. For example, Zhang and Wang [11] showed that localized geometric softening at a certain strain state is in fact the cause of localized necking in 2036-T4 aluminum, rather than the initial geometric inhomogeneity.

The third theory is Storen and Rice's vertex theory [12], also known as the «STO theory». Here, it is assumed that the localized necking corresponds to the angle or vertex that develops on the yield surface at the loading point.

Theoretical prediction of the FLD of sheet alloys requires a yield criterion and a hardening model. Various non-quadratic tests developed recently are good candidates. F. Barlat [13] proposed high-order yield criteria known as Yld89, Yld91, Yld94, Yld97, Yld2000 and Yld2004, respectively. Hill's fluidity criterion, known as Hill's criterion since the 1950s, is still used today because it has a relatively simple mathematical form. Most theoretical methods based on the high-order yield anisotropy criterion do not account for in-plane anisotropy due to mathematical complexity. Predicting defect formation from a practical point of view makes it possible with sufficient accuracy to determine the degree of deformation of a given sheet material, at which no defects arise. In production, this possibility is extremely important, especially when shaping the skin parts of the aircraft. One way to determine the allowable deformation for sheet materials is to use a forming limit diagram (FLD).

Forming limit diagrams are constructed experimentally for each grade of sheet material based on test results under various stress states. The plane strain limit corresponds to the point where the FLC intersects the major strain axis. The data required to create the FLD is produced in a series of additional experiments on special equipment. The field on the surface of deformed samples is measured either with etched grids consisting of an array of circular elements or with digital image correlation, using an optical method of displaying the whole field.

The ordinate of the diagram shows the largest principal deformations, and the abscissa shows the smallest principal deformations. The resulting curve is called the ultimate shape change curve. The region of positive values of deformations δ2 corresponds to biaxial tension, at δ2 = 0, a plane deformed state is observed, in the region of negative values of δ2, uniaxial tension. The shape and location of the ultimate strain curve is unique for each material (figure 1).

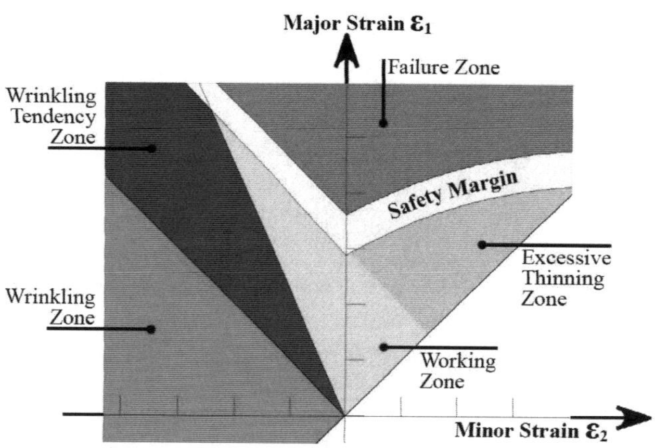

Figure 1. Diagram of ultimate deformations.

The forming limit diagram is valid in cases where the ratio of the main plastic strains is constant throughout the entire deformation process, that is, the deformation is monotonic. The diagram is divided into zones, the combination of the main deformations in which predicts a particular defect.

3. Different rheology behavior of aluminum sheet material

To compare the various rheologies of the behavior of aluminum sheet material, we use the results of modeling in the ANSYS/LS-DYNA software package of the process of forming a wrapped anisotropic sheet blank, primarily in the state of an elastoplastic body under strain hardening. As a rheological model of the material for the billet, the 3-Parameter Barlat Model was used - a model used to simulate aluminum sheet material under conditions of a plane stress state. For this material rheology, it was necessary to first create a local coordinate system taking into account the rolling direction of the sheet blank. The properties of the material of the workpiece according to the 3-parameter Barlat model are summarized in table 1.

Table 1. Material properties of the workpiece according to the 3-Parameter Barlat Model.

Parameter	Designation	Dimension	Parameter value
Density	DENS	kg/m^3	2700
Elastic modulus	EX	Pa	7.2E+010
Poisson's ratio	NUXY	–	0.35
Hardening criterion type	HardeningRule	–	1
Tangential module	Tang Mod	Pa	7E+008
Yield point	Yld Strs	Pa	1.12E+008
Barlat exponent value	Barlat Expon	–	6
(R00)	Width ratio-orig	–	0.613
(R45)	Width ratio-45 deg	–	0.818
(R90)	Width ratio-90 deg	–	0.639
Material local coordinate system number	CSID	–	2

However, when analyzing the simulation results, some inconsistencies with the cases observed in practice appeared, the breaks of the sheet blank in the process of its shaping. In the modeling, the results of which are given in [14-15], attention was paid, first of all, to the symmetric covering due to the external symmetry of the shaping punch relative to the main planes and directions in the «pole» of the double curvature shell. At the same time, the alignment of the directions of the lines of the main curvatures in the «pole» of the shell surface with the directions of anisotropy of the orthotropic sheet blank (directions along the RD (rolling direction) and across the TD (transverse direction) of the rolling direction of the sheet) was also ensured.

An EVP-model (Elastic Viscoplastic Thermal Model) is used as a rheological model of the material for the workpiece, which makes it possible to take into account internal friction with an irreversible transition of elastic energy into thermal energy. This is a complex rheological model of an elastic-viscoplastic medium. Low stresses during pressure treatment, high plasticity and formability easily explain the ability of the alloy under the considered rheological state to viscous flow. The EVP model of the medium can describe the rheology of the sheet material during shaping over a wide range of strain rates.

The irreversible transition of elastic energy to heat is a relaxation process that restores the disturbed equilibrium during deformation of the workpiece by activating diffusion. In annealed alloys in an equilibrium stable state, the mobility of atoms is uniquely determined by the energy of the interatomic bond, i.e. the energy of the ground state, where the diffusion activation energy is equal to a constant fraction of the binding energy. In a nonequilibrium unstable state, as a result of distortion of the crystal lattice, temporary energy barriers are created for atoms of different heights. The mobility of atoms and the activation energy in such a state depends on the values of these barriers, which are determined by structural features, i.e. depends on the energy of the transition state and is not directly related to the energy of the ground state.

As a result, the external force field from the working bodies of the press, spreading over the sheet material, can be scattered. When deformed, the material begins to absorb mechanical energy to one degree or another. The accepted model can describe the rheology of the elastic-viscoplastic flow of the material, which exhibits the effects of quasi-plastic deformation. They can be associated with the activation of diffusion along the boundaries or near the grain boundaries. From the rheological point of view, the elastic-viscoplastic flow of the material can be attributed to the quasi-liquid phase of the deformable material. The properties of the material of the elastic-viscoplastic behavior of the workpiece material are summarized in tables 2, 3, 4, 5 and 6.

Table 2. Properties of the workpiece material according to the EVP model of the elastic-viscoplastic behavior of the workpiece material.

Parameter	Designation	Dimension	Parameter value
Density	DENS	kg/m^3	2700
Coefficient of thermalexpansion	ALPX	m/m-C	2.1E-005
Elastic modulus	EX	Pa	7.2E+010
Iinitial yield stress	In Yld Strs	Pa	1.12E+008
Load curve ID	LCID1	–	1
	LCID2	–	2
	LCID3	–	3
	LCID7	–	7

The loading curve is a functional dependence of the parameters and is set by the ratio of tables with data on the values of these parameters. For a given material, four load curves are set:

- LCID1 - function of effective stress on effective plastic deformation.
- LCID2 - a function of the dependence of the elastic modulus on temperature.
- LCID3 - Poisson's ratio function on temperature.
- LCID7 - function of the dependence of the coefficient of thermal expansion on temperature.

Table 3. Load curve parameter values 1 (LCID1).

Parameter	Designation	Dimension	Value number					
			1	2	3	4	5	6
Effective stress	STRESS	MPa	0	340	375	405	423	438
Effective plastic strain	STRAIN	–	0	0.05	0.1	0.15	0.2	0.3

Table 4. Load curve parameter values 2 (LCID2).

Parameter	Designation	Dimension	Value number			
			1	2	3	4
Elastic modulus	EX	MPa	67000	66000	65000	52500
Temperature	TEMP_EX	°C	100	125	150	175

Table 5. Load curve parameter values 3 (LCID3)

Parameter	Designation	Dimension	Value number			
			1	2	3	4
Poisson's ratio	NUXY	–	0.3137	0.3195	0.3252	0.3478
Temperature	TEMP_NUXY	°C	100	170	240	310

Table 6. Load curve parameter values 4 (LCID7)

Parameter	Designation	Dimension	Value number			
			1	2	3	4
Coefficient of thermal expansion	ALPHA	m/m-C	2.1E-005	2.94E-005	4.85E-005	11.7E-005
Temperature	TEMP_ALPHA	°C	100	200	300	390

Aluminum alloys that undergo phase transformations during deformation can show, under certain conditions, elastic-viscoplastic behavior, independent of the deformation rate. However, it is impossible to establish how the rheological state of the material changes in the process of phase transformation, accompanied by a strong change in the structure upon deformation. In [16], it was concluded that islands of the «quasi-liquid» phase can appear for a short time in the process of phase transformation at the grain boundaries of the heterogeneous system. The existence of islands of the «quasi-liquid» phase at grain boundaries is substantiated from a thermodynamic point of view. The material may not deform as a viscoplastic medium, and then, for example, in the middle of the phase transformation, it may show a linear-viscous state and then will not return to its original state. This behavior of materials deformed during phase transformation was recorded during polymorphic transformation of pure iron [17].

4. Methodology, hardware and software

When investigating the influence of the rheological behavior of a deformable sheet aluminum alloy on the limits of the shaping process of a double curvature shell, it is necessary to have specific information on the deformed state of the sheet material using FLD diagrams of the limiting shaping of the sheet material during simulation. The construction of the ultimate strain curve is based on the ISO 12004-2 standard. FLD construction is carried out by two main methods: the Nakazima method and the Marchignac method. To accurately determine the FLD points, friction between the punch and the workpiece must be almost completely eliminated. The Nakazima method is based on the use of a hemispherical punch to deform the workpiece and on the use of a multi-layer lubricant between the punch and the workpiece. One FLD uses one specific multi-layer grease.

One of the following blanks is used: a round blank with a parallel fillet in the center; round blank with round cutouts along the edges (applied to biaxial tension). Examples of workpieces are shown in figure 2. By changing the fillet width, cut radius or workpiece width, various deformed states are obtained. Different strain states give different points on the ultimate strain curve. The samples were tested on a Zwick Roell BUP200 universal testing machine. Forming limit diagrams were constructed experimentally for a certain brand of sheet material based on the results of tests under conditions of various stress states (variability is achieved through the manufacture of blanks with different sizes of the bridge).

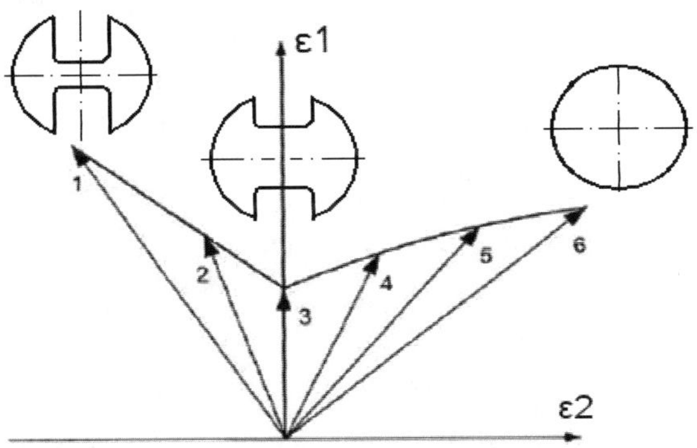

Figure 2. Deformed states depending on the width of the fillet of the workpiece.

Varying the width of the fillets and the radii of the cutouts in the samples is required to construct the FLD by points, which allows you to obtain various stress-strain state in the workpiece during deformation. The sample without side cutouts produces a point on the FLD curve that corresponds to bilateral tension $\delta 1 = \delta 2$; a sample with maximum notches allows obtaining a point that approximately corresponds to uniaxial tension $\delta 1 = -2\delta 2$. When testing blanks by the Nakazima method, the results were obtained for plotting the curve of the limiting shape change in the uniaxial tension zone, the value of $\delta 1$, at plane deformation and in the biaxial tension zone for the original sheet material of aluminum alloys 1163RDMV, 1.5 mm thickness. Four samples were taken for one point. Figure 3 shows the curves of the limiting deformation of each specimen and shows the average values in the zones of uniaxial and biaxial t The original sheet billet made of aluminum alloy 1163RDMV, 1.5 mm thickness and dimensions in terms of 1880x10500 mm x mm. Sheathing (B) sheet of grade 1163 alloy with regulated hard plating (HP), in annealed condition (M).

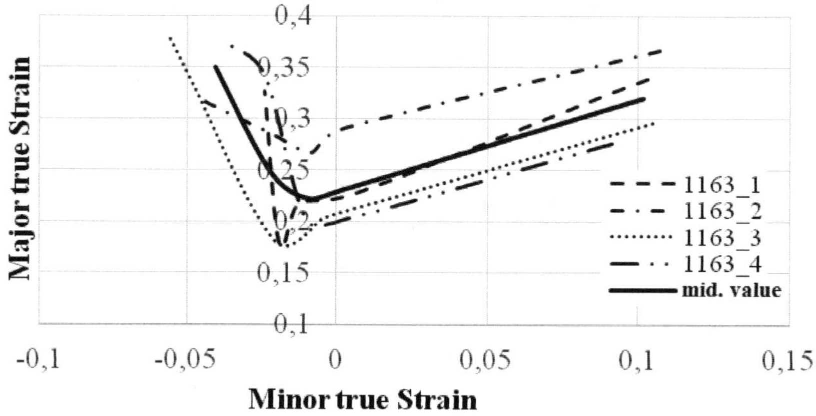

Figure 3. Experimentally obtained FLD diagram for sheet material 1163RDMV, 1.5 mm thickness.

For the main shaping operation, we give preference to the weakened state. According to the VIAM instruction No.904-67, in addition to the initial state, we will subject this alloy to five different anneals and one quenching:

- Full annealing: 380 ... 420°C., exposure for 30 minutes, cooling with an oven, then in air. Options: 1a - temp. 360°C., 1b - 400°C, 1c - 420°C.
- Reduced annealing: 360°C, holding for 30 minutes and cooling: 2a - air, 2b - water.
- Hardening: 500 ... 505°C., Holding for 30 minutes and cooling in water at room temperature.

The range of values of the coefficient of friction is quite wide, which is associated with the study of various rubbing pairs. The following values of the coefficient of friction were determined:

- 0.08 - fluoroplastic and lubrication with machine oil; rubber grade 7338 with grease lubrication; film B-118 and grease lubrication. In addition, later studies of rubbing pairs, which include a cast metal punching punch made of secondary aluminum, showed that an oxide film with fluorinated graphite and a plasma layer of aluminum bronze BrA10, even with dry friction, provides a friction coefficient of 0.04.
- 0.1 - this value is found in several rubbing pairs that are typical for production.
- 0.2 - this value most often refers to the friction of an aluminum sheet and a covering punch from a sand-glue mass.

The mechanical properties of sheets of alloy 1163RDMV after each annealing were obtained in the course of testing four samples and statistical processing of the results. The results of mechanical properties after testing and processing are summarized in table 7.

The preparation of geometric models of elements for calculation in the LS-DYNA processor was carried out in the LS-PREPOST program. All geometry is exported for further processing, which is presented in the form of creating a mesh, specifying the main parameters, generating an output calculation file, and so on. The necessary formatting for the installation and preparation of the initial data for the calculation is carried out in the graphical user interface (Create and control RO-630) of the computer application of the developed control system, which is used to perform actions.

Table 7. Mechanical characteristics of samples in the initial, freshly hardened state and various annealing.

Mode	sigma_в [MPa]	sigma_0,2 [MPa]	delta_n	mi_21 (0)	mi_12 (90)	mi_1 (45)
420	173	56	0.164	0.404	0.426	0.468
400	157	41	0.142	0.308	0.311	0.342
380	154	44	0.1846	0.325	0.280	0.308
360 2a	159	47	0.149	0.366	0.383	0.421
360 2b	155	40	0.0669	0.293	0.328	0.360
ishod	158	69	0.113	0.4	0.428	0.4708
sveshezak	357	189	0.199	0.258	0.364	0.383

The computer application of the control system consists of a simplified three-dimensional virtual model of the RO-630-11 stretch-tightening press and a graphical user interface designed to control the kinematic movements performed by the working bodies of the press, which makes it possible to accurately simulate its real kinematics using grounded kinematic shaping schemes. In addition to this, the interface includes control and positioning of each working body of the virtual press. As a result, it will be possible to customize the animation of the movements of the working bodies of the press for a specific shaping process. In figure 4, the virtual model of the RO-630-11 stretch-tightening press contains 10 controlled coordinates: LB1, LB2, LB3, LB4, STO1, STO4, PB1, PB2, PB3, PB4 (main and auxiliary hydraulic cylinders of the press).

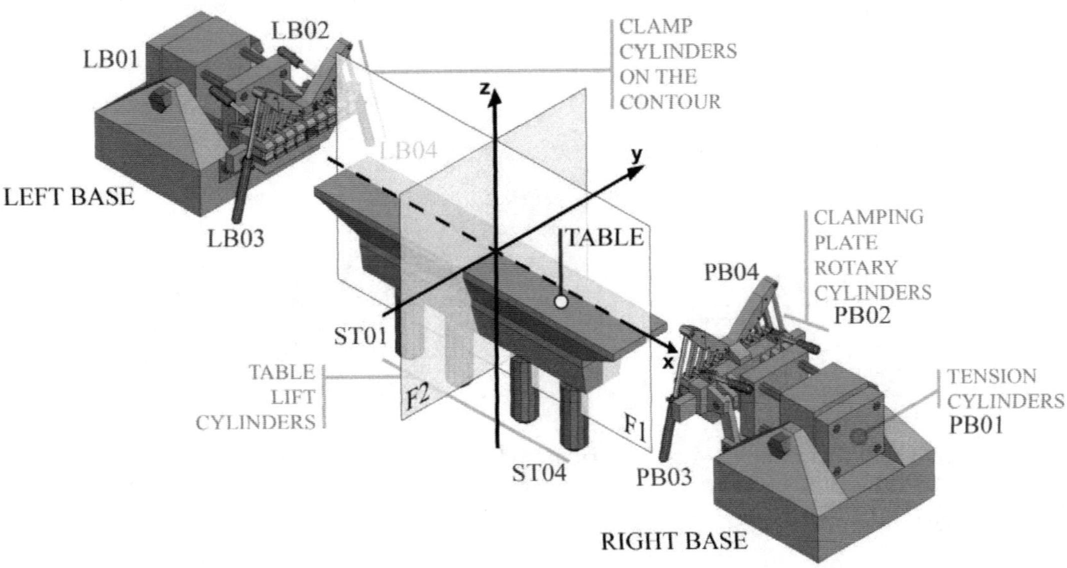

Figure 4. Three-dimensional virtual model of the stretching and drawing press RO-630-11.

To simulate the process using the computer application of the kinematics control system of the RO-630-11 virtual stretch-tightening press, a diagram of the step movements of the main and auxiliary working cylinders of the press was previously obtained using kinematic techniques in the presence of two effects. This made it possible to minimize the thickness difference of the simulated shell to the recommended limits. To create the conditions for symmetric wrapping, it is necessary to import the skinning punch surface (the model of the skinning punch in IGES format) into a curved coordinate system, to a double curvature surface attached to the grid of the main curvature lines with a pole point as the origin. As a result, we have left one version of the kinematic methods of shaping by the tight-fitting. So far, we decided to simulate one option, taking into account the influence of the rheological behavior of the wrought aluminum alloy. The features of the properties of sheet material under various modes of annealing, quenching and conditions of friction of the sheet on the surface of the tightening punch are given above.

The considered variant of the shaping method by the wrap provides a uniform change in thickness in different areas of the sheet blank without folds and breaks. It includes an innovative element at the outset where the slab was forced flat with rectilinear discrete clamping devices. This provides a plastic configuration of the middle part of the sheet blank without localizing the thinning deformation and sheet breakage in one of its free sections between the edge of the punch and the press clamps. The second innovative element is associated with the formation of the second curvature of the shell surface due to the position of discrete clamping devices along the contour after the plastic configuration of its middle part and unloading, providing, upon subsequent stretching of the shell, deformation of only the corner parts of the sheet blank.

5. Results of the influence of the rheological behavior of a wrought aluminum alloy on the limits of the shaping process

In figure 5, the results of calculations in the LS-DYNA processor of the values of the main deformations δ1 and δ2 are collected in separate files for their placement on FLD forming limit diagram of the sheet material of one of the annealing options. The minimum difference in thickness refers to the case when a sample of sheet material made of alloy 1163 passed the reduced annealing 360°C at a soak time of 30 minutes and was cooled in water. All samples were cut in the direction of sheet rolling. The mechanical characteristics of the samples for this case were: ultimate strength - 155 MPa; yield point - 40 MPa; uniform elongation - 6.69%; hardening index n = 0.386, tangential modulus - 440.2 MPa and anisotropy coefficients $\mu_{21} = 0.293$, $\mu_{12} = 0.328$ and $\mu_1 = 0.361$.

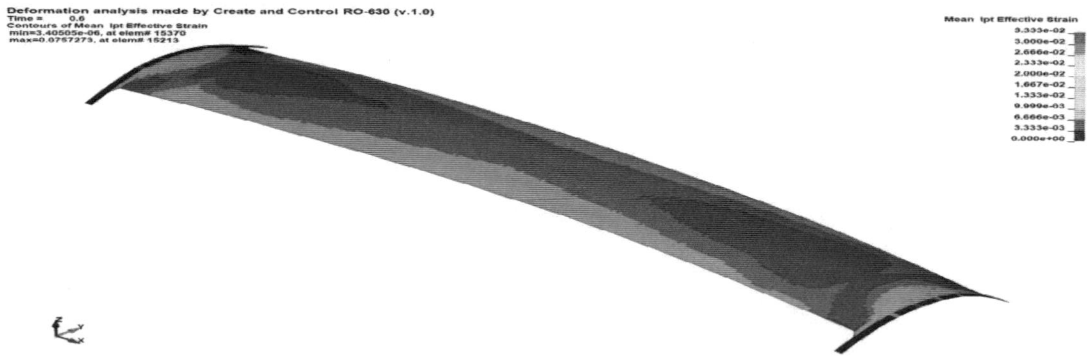

Figure 5. Deformed state (equivalent deformations) of the shell under study (calculation option - annealing at 360°C, holding for 30 minutes, air cooling and friction coefficient of 0.20). Calculations are performed in the LS-DYNA processor (finite element mesh is off).

The limiting deformation of the shaping process by the wrap was evaluated in the LS-DYNA postprocessor. The dialog for working with the diagram of ultimate deformations is called after clicking on the characteristic icon. Figure 6 shows the FLD post-processing for annealing at 360°C, holding for 30 minutes, cooling in air and with a friction coefficient of 0.20.

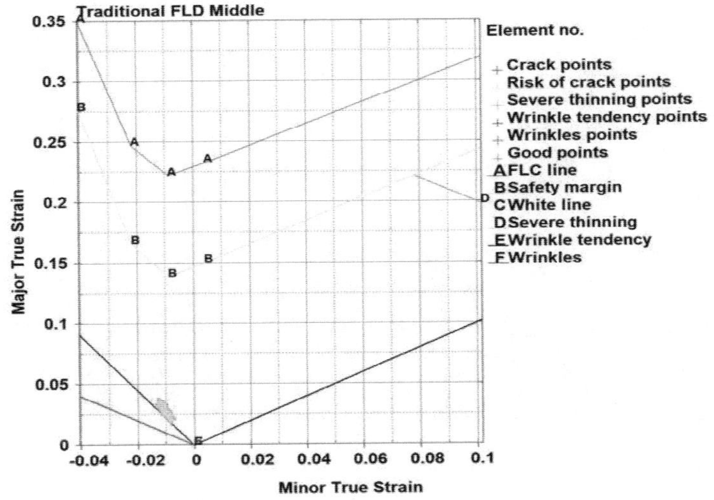

Figure 6. Post-processing FLD. The data source is taken from a separate file associated with the deformed state of the shell under study in figure 6.

The string fields «t =» and «n =» on the FLD diagram are responsible for the appearance of theoretical curves, one of which is displayed in the diagram by default in red and is responsible for fracture deformations, and the other is yellow, and shows the boundary of the theoretical safe deformation zone. The radio buttons «t_n» and «File» select the theoretical curve or the real one taken from the file. Figure 7 shows the result of processing the deformed state of the shell under study in the LS-DYNA postprocessor.

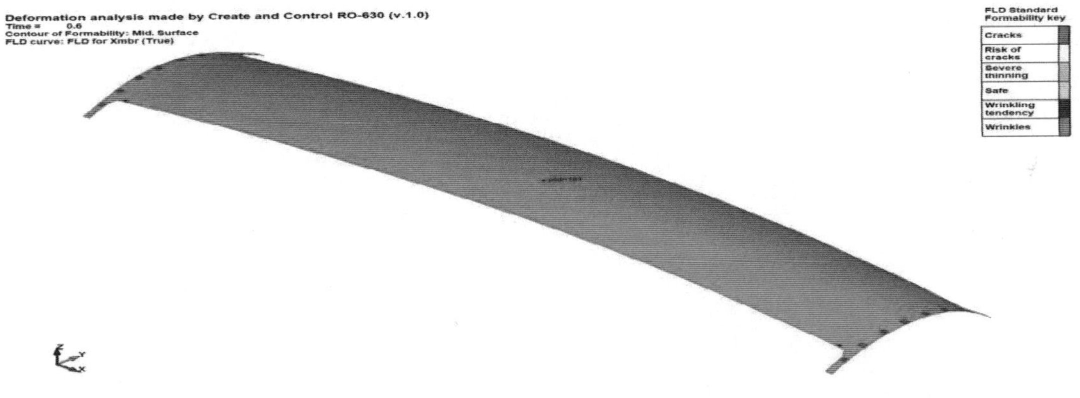

Figure 7. Result of processing the deformed state of the shell in the LS-DYNA postprocessor using FLD data (calculation option - annealing at 360°C, holding for 30 minutes, air cooling and a friction coefficient of 0.20). Compare with the colored area in the ultimate strain diagram (figure 1).

In figure 8, the results of calculations in the LS-DYNA processor of the values of the main deformations δ1 and δ2 are collected in separate files for their placement on forming limit diagram FLD of a material in a freshly hardened state.

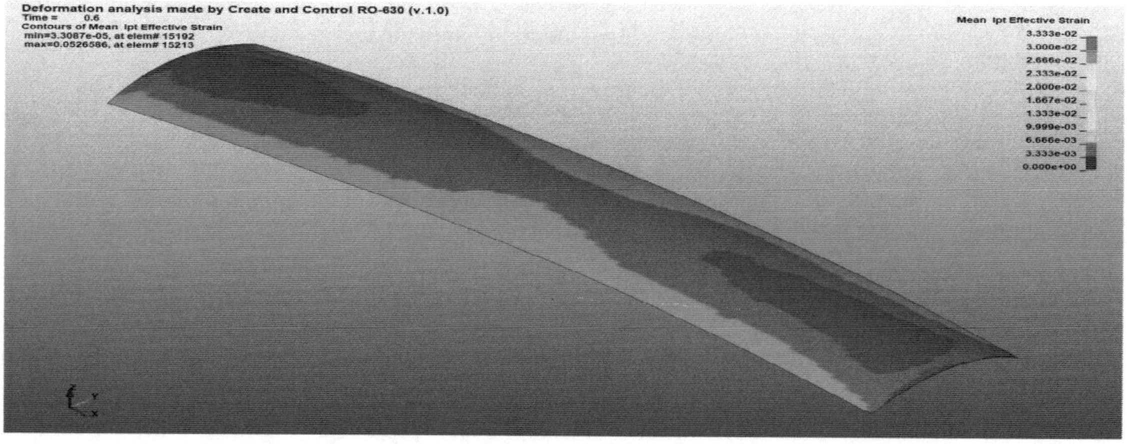

Figure 8. Deformed state (equivalent deformations) of the investigated shell MAT_sveshezak_VISCOPLASTIC_ftr0.20 (calculation option - hardening: 500 ... 505°C, holding for 30 minutes, cooling in water at room temperature and friction coefficient 0.20). Calculations are performed in the LS-DYNA processor (finite element mesh is off).

The limiting deformation of the shaping process by the wrap was evaluated in the LS-DYNA postprocessor. The dialog for working with the diagram of ultimate deformations is called after clicking on the characteristic icon. Figure 8 shows post-processing FLD for hardening mode 500...505°C, holding for 30 minutes and cooling in water at room temperature and with a friction coefficient of 0.20.

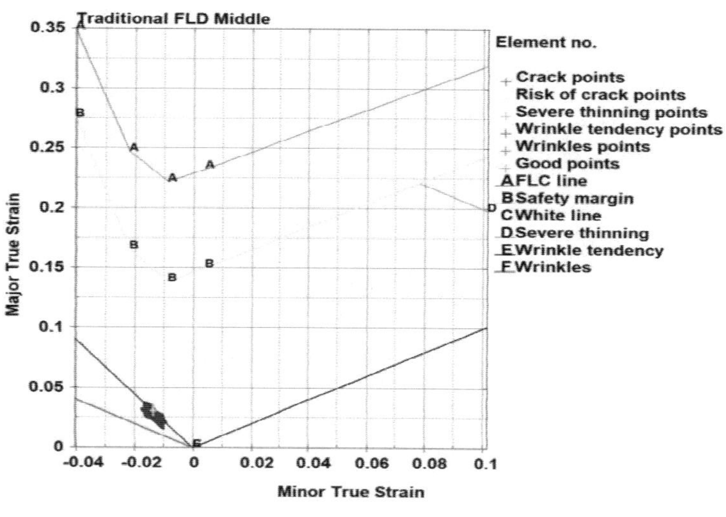

Figure 9. Post-processing FLD. The data source is taken from a separate file associated with the deformed state of the shell under study in figure 8.

Figure 9 shows the result of processing the deformed state of the shell under study in the LS-DYNA postprocessor.

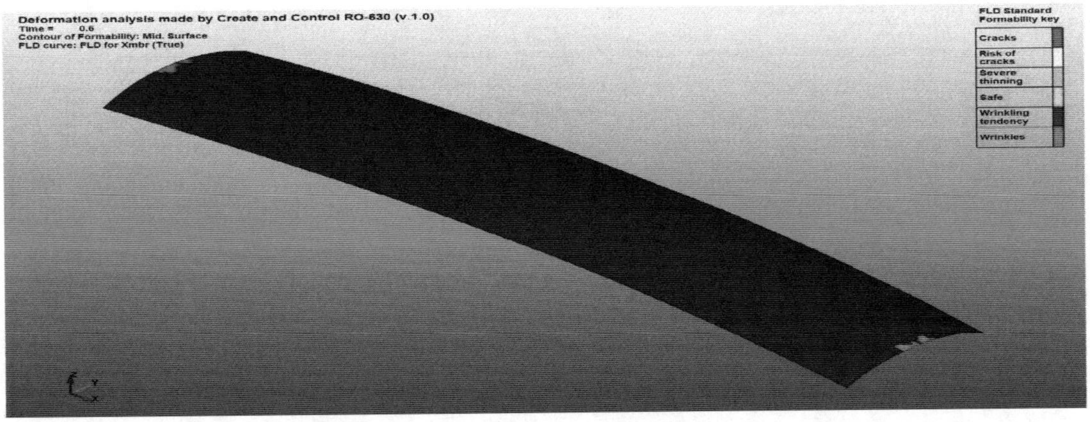

Figure 10. The result of processing the deformed state of the shell in the LS-DYNA postprocessor (calculation option - quenching: 500...505°C, holding for 30 minutes and cooling in water at room temperature, quenching at 505°C and friction coefficient 0.20). Compare with the colored area in the forming limit diagram (figure 1).

6. Summary

The main goal of modeling the wrapping process is to apply the graphical user interface toolkit (Create and control RO-630) of the computer application of the kinematics control system of the RO-630-11 virtual stretch-tightening press. A decrease in the thickness difference of the part formed by the stretch during the modeling process was carried out by manually adjusting the values of the controlled coordinates on the graphical control interface of the virtual model of the RO-630-11 stretch-tightening press based on the results of the analysis of the thickness values at the points designated by us on the surface of the sheet blank. The preparation of geometric models of elements for calculation in the LS-DYNA processor was carried out in the LS-PREPOST program.

At the computation stage, the calculation results were carefully checked and possible adjustments that had to be made to the animation in question were evaluated in order to send a new animation into the calculation, comparing the current results. Our research was aimed at creating conditions for symmetric wrapping and the implementation of stable control of the shaping processes of wrapping shells, close in geometric shape to the skin, with a uniform change in thickness in different areas of the sheet blank without folds and ruptures. Verification of material behavior models can be carried out on the basis of LS-DYNA. The only thing that remains to be introduced into the model is the criterion of destruction.

In production, the FLD chart can be used as this criterion, where the limit levels of deformation are indicated. In modeling, the main deformations in the model of the molded part are compared with the FLD: if the found deformations fall into the safe zone, then in production, the wrapping of the real metal will be successful. The result of processing the deformed state of the shell in annealed form in the LS-DYNA postprocessor using the FLD postprocessing data in our case indicates the absence of defects on the shell (the FLD zone is green). But in the case of processing the deformed state of the shell in a freshly hardened form, it indicates the danger of folds (FLD zone in blue).

We believe that the prediction of defect formation from a practical point of view makes it possible to determine with sufficient accuracy the degree of deformation of a given sheet material, at which no defects arise. In production, this possibility is extremely important, especially when shaping the skin parts of the aircraft from new sheet materials. The most unstable structure at room temperature in thermally hardened aluminum alloys is a supersaturated solid solution of alloying elements in aluminum, the concentration of which can exceed the equilibrium state by tens of times. In this state, the aluminum sheet is also very ductile, more precisely, very viscous, therefore, not resistant to shaping than in the equilibrium state.

References

[1] Dzugutov M Ya 1984 *Plasticity, its prediction and use in MF* (Moscow: Metallurgy)
[2] Grigoriev A K, Kolbasnikov N G and Fomin S G *1992 Structure formation during plastic deformation of metals* (St. Petersburg: SPGTU)
[3] Kishkina S I and Fridlander I N 1982 *Aviation Materials* (Moscow: ONTI)
[4] Mikveev V A 2004 *Development of processes of shaping by covering bypass-forming shells of aircraft with a minimum thickness variation*: dissertation of a Doctor of Technical Sciences: 05.03.2005 338
[5] Mikheev V A, Grechnikov F V, Roberto de Alvarenga and Bodrov V V 2021 Metric transformation of the surface of a tight punch for forming sheet shells of double curvature *IOP Conference Series: Materials Science and Engineering* **905** 12010
[6] Keeler S P 1965 Determination of Forming Limits in Automotive Stampings *Sheet Met. Ind.* 683-91
[7] Goodwin G M 1968 Application of strain analysis to sheet metal forming problems in press shop *Metall* (Ital) 767-74
[8] Wagoner R H, Chan K S and Keeler S P 1989 *On the Origins of the FLD, Forming Limit Diagrams: Concepts, Methods and Applications* 3-8
[9] Hosford W F and Caddell R M 2011 *Metal Forming: Mechanics and Metallurgy* (New York: Cambridge University Press)
[10] Marciniak Z and Kuczynski K 1967 Limit Strains in the Processes of Stretch Forming Sheet Metal *Int. J. Mech. Sci.* **9** 609-20
[11] Zhang L and Wang J 2012 Modeling the Localized Necking in Anisotropic Sheet Metals *International Journal of Plasticity* **39** 103-18
[12] Storen S and Rice J R 1975 *Localized Necking in Thin Sheets* 421-41
[13] Barlat F, Aretz H, Yoon J W *et al.* 2005 *Linear Transformation Based Anisotropic Yield Function* 1009-39
[14] Mikheev V A, Dementyev S G, Samokhvalov V P, Savin D V and Surudin S V 2013 Isometric conditions for the formation of a tight shell of double curvature with minimal thickness difference *Izvestia of Samara Scientific Center of the Russian Academy of Sciences* **15(6-1)** 161-6
[15] Mikheev V A and Surudin S V 2017 Fundamentals of the calculation of the process of forming tight thin shells of double curvature *Izvestia of Samara Scientific Center of the Russian Academy of Sciences* **19(1-3)** 555-62
[16] Chumachenko E N 2005 *Superplasticity: materials, theory, technologies* (Moscow: KomKniga)
[17] Smirnov O M 1979 *Processing of metals by pressure in the state of superplasticity* (Moscow: Mashinostroenie)

Crystal chemistry of anisotropy magnetic properties gas atomization powders of an alloy of the Fe-Nd-B system

V V Savin[1], A O Marukhin[1], A V Osadchy[2], L A Savina[1] and V A Chaika[1]

[1] Immanuel Kant Baltic Federal University, Nevskogo, 14A, 236016, Kaliningrad, Russia
[2] Prokhorov General Physics Institute of the Russian Academy of Sciences, Vavilova 38, 119991, Moscow, Russia

E-mail: VVSavin@kantiana.ru

Abstract. The magnetic material of Fe-Nd-B alloys is widely used in science and technology due to the unique physical and chemical properties of borides based on it. The functional properties of the material significantly depend on the structural-phase state in the alloy. The transition on the Melt Quenching technology (MQ) for its production required additional theoretical and experimental studies of the regularities of the formation and stability of the structural-phase state in multicomponent REM-containing alloys under conditions of nonequilibrium crystallization and solidification of a highly supercooled melt. The implementation of the industrial MQ technology by the method of Gas Atomization (GA) of medium pressure imposes additional significant requirements for such studies. The paper comprehensively studies and summarizes the available results on the relationship between the conditions for obtaining powder materials of Fe-Nd-B alloys by the MQ GA method and the functional properties of materials based on them. It is proved that the anisotropy features of their magnetic properties are related to the existing inhomogeneities in the distribution and morphology of $Fe_{14}Nd_2B$ phase grains in MQ GA powder materials. A crystal-chemical explanation of the occurrence of the identified features is proposed and the possibility of their practical application is shown.

1. Introduction

A number of mandatory requirements are imposed on the materials used for the production of high-energy permanent magnets: high residual induction; high coercive force; required temperature stability and corrosion resistance; sufficient mechanical properties; manufacturability of production and technical feasibility of subsequent processing; availability of raw materials; production cost and the cost of the finished product. Alloys of the Fe - Nd - B system and sintered powder magnets based on them have been invariably the most promising materials that meet the requirement for the price-property ratio over the past few decades [1-2]. During this long period, numerous targeted studies are being systematically conducted to develop new alloy compositions (alloying) and improve the technology for their production in order to increase the magnetic characteristics of the material and optimize the functional properties of the finished product (magnets and magnetic systems) [3-40]. It is generally accepted that the maximum magnetic properties of the material and the finished product are determined by the optimal structural-phase state of the alloy [1, 3, 6, 17, 19, 21, 24, 26]. The introduction of the liquid quenching (Melt Quenching; MQ) technology into the production of hard magnetic alloys of the

Fe - Nd - B system has significantly expanded the limits of the attainable level of magnetic characteristics of the material and the functionality of products [19, 20, 25, 32, 33, 35]. Especially with a combination of MQ technologies and hot plastic deformation [17, 18, 32, 40]. Of course, the design of magnetic systems, especially for special purposes, should be guided by the maximum magnetic characteristics of the material used. But this does not contradict another statement: for a given magnetic system (finished product), the magnetic properties of the material should not be maximum, but optimal. Moreover, taking into account the cost of the material and products from it. Such a method of MQ is undoubtedly the spraying of a melt jet by an inert gas stream (Gas Atomization; GA) or any other spraying medium that ensures the disintegration of the melt jet into droplets and guarantees the protection of the alloy from oxidation [18, 28].

The usage of MQ technology by the GA method allows us obtaining a powder material - a gas atomized powder (GAP) of a given chemical and fractional composition, which has increased corrosion resistance during long-term storage (up to 10 years) under natural conditions and at the same time retains its technological properties [28]. The use of the GAP for the production of magnets makes it possible to exclude the melting, crushing and grinding of the ingot into powder from the technological process, increases its manufacturability (reproducibility) and reduces the cost of producing a unit of production.

Academic discussion about the formation of high properties in magnets from GAP-alloys of the Fe - Nd - B system, arose from the moment of their preparation [4, 8, 17, 18, 19, 25, 28] and approbation for the production of magnets using standard powder technology [19]. Today, the main working theory is the model of recrystallization of non-equilibrium grains GAP-alloys during sintering by powder technology [42].

The grinding of a powder material of an average weighted fractional composition (+60, -640 microns) to particles (crystallites) with an optimal size (~6-12 microns) remains a mandatory technological operation in the production of magnets from GAP. But this only intensifies the process of recrystallization of grains in the alloy during sintering of powders due to an increase in the defect in the resulting crystallites. At the same time, the initial (weighted average) powder array always contains a certain number of $Fe_{14}Nd_2B$ phase grains close to the optimal one. As a result, under the optimal grinding mode, this leads to the formation of the necessary and sufficient number of crystallites with the morphology and size necessary for optimal magnetic properties. It is these crystallites that become the basis (centers) of recrystallization of the sintered powder and ensure the formation of the necessary grain structure in the magnetic material.

This effect is significantly enhanced when the technological processes of mechanical grinding (in certain types of ball mills or disintegrators) are combined with the process of hydrogen treatment (sorption - desorption of hydrogen) [7, 29, 31], including the imposition of thermal cycling in the temperature range of phase magnetic (310 °C) or structural (510 °C) transitions [27]. It should be borne in mind that in the case of the MQ quenching technology by the GA method for alloys of the Fe-Nd-B system, depending on the dispersion of the resulting powder and the mechanisms of interaction of the drops with each other in the sputtering flow, cooling rates are realized and supercooling of the melt sufficient for its amorphous solidification is achieved [19, 25, 35, 37, 41]. Moreover, the alloy contains an amorphous state (AS) of two compositions: AS_1 - the composition of the alloy; AS_2 is a composition with a high content of neodymium [41, 42]. Crystallization of these components during the thermal treatment of GAP or the production of magnets can significantly affect the recrystallization process and the formation of the magnetic properties of the finished product [17, 19, 20, 25, 35, 41]. However, the presence of an amorphous component in MQ alloys of the Fe-Nd-B system opens up fundamentally new opportunities for improving known technologies [20, 31, 32, 40, 42] and allows considering the use of fundamentally new technologies for the production of sintered magnets, including selective fusion [34] or additive growing [43].

2. Research objective

Based on the statement that the GSP of alloys of the Fe - Nd - B system, obtained by MQ using GA technology, are promising materials for the manufacture of permanent magnets for general and special purposes both according to traditional powder technology and innovative technologies of selective sintering, in work the research goal was formulated: to test the hypothesis that in each GAP powder of ternary and alloyed alloys, regardless of the fractional composition, there is always a dominant direction of grain orientation of the main magnetic phase - $Fe_{14}Nd_2B$. The indicated dominant direction is formed as a result of crystallization of a supercooled melt under conditions of rapid cooling and spherically symmetric heat removal of the drop during MQ by the GA method.

3. Materials and research methods

To achieve this goal, the following research methods were used: X-ray phase analysis (modernized DRON-3 device with Cu- and Co-irradiation); metallographic studies (scanning station based on the analyzer of fragments of microstructure of solids SIAMS-800 for panoramic applications, including metallographic microscope OLYMPUS BX-51 and SIAMS Drive System); scanning electron microscopy (JSM-6390LV with Oxford INCAEnergy energy dispersive analyzer); microhardness measurement (SHIMADZU HMV G21ST and PMT-3); differential thermal analysis (NETZSCH 204 F1 Phoenix); production of magnetoplastics (in the form of cubes with a size of 5 × 5 × 5 mm), including in magnetic fields of different magnitudes (up to 2 T); measurement of the magnetic field on the surface of magnetoplastics (magnetic induction meter ATE - 8702).

MQ-melt quenching GA was performed on the URZhMV-3 unit (UkrNIISpetsstal, Zaporizhzhia) using energy carrier gas (technological argon with the addition of special gas passivating additives) of medium pressure (≤ 1.2 MPa or 12 atm.) And spray nozzles of the "Nozzle Laval". The melt (weighing up to 30 kg) was obtained by induction melting in a rammed (lining based on corundum and oxides of rare-earth elements) overturning crucible (Leibold-Gereus furnace, Germany) with a capacity of up to 45 kg

As charge materials used: technically pure iron grade 005ZhR (TU 14-1-2033-77) with a total impurity content of up to 0.15 wt. %, neodymium grade NM-1 (TU 48-4-271-91); ferroboron FB - 17.6 produced by JSC Zaporizhzhia Abrasive Plant (GOST14021.1-78).

Substances belonging to the "chemically pure" type were used as alloying elements.

The paper considers two main GSP compositions (stoichiometry in at. %): ternary alloy $Fe_{78}Nd_{15}B_7$ and alloy $Fe_{73}Nd_{15}B_8Co_{0,5}Al_{1,0}Mo_{1,5}(Dy, Tb)_{1,0}$.

Fractional composition of GSP was made using measuring sieves with numbers: 0.020; 0.056; 0.100; 0.160; 0.400 and 0.630 mm (GOST 18318–94).

The hysteresis properties of the powders were measured on a VSM-250 vibration magnetometer in fields with an intensity of up to 2 T at room temperature (NUST MISIS). The accuracy of determining the specific magnetization did not exceed 0.003 $A \cdot m^2/kg$, and the coercive force did not exceed 0.005 kA/m.

The hysteresis properties of magnetoplastics and sintered anisotropic powder magnets were measured on a hysteriograf in a closed magnetic flux with two differential (compensated) coils in a field up to 1.4 T in a 5.0 mm gap (Federal State Unitary Enterprise Research and Experimental Institute of Automotive Electronics and electrical equipment).

4. Results and discussion

The paper presents the results of comprehensive studies of the structural-phase state in the material and the functional properties of magnetoplastics from the initial and annealed (argon and / or vacuum) MQ-melt quenching GSP alloys of the Fe - Nd - B system of different chemical and fractional compositions. This choice of the samples under study is due to the following known facts obtained in our earlier works [19, 26–30, 35, 37, 41–42]:

1) The alloying elements used are surface active additives to the alloy and affect the physicochemical properties of the solidifying supercooled melt, including the last solidifying liquid phase component in the MQ GAP alloy (neodymium-based ternary eutectic), which separates (isolates) the grains the main phase and provides predominant mass transfer during their recrystallization under isothermal annealing conditions above the temperature of the eutectic transformation (≥ 510 ° C)

2) Within the framework of the accepted modern mineralogical terminology [44], the processes of grain growth of the $Fe_{14}Nd_2B$ phase can be considered as recrystallization processes in the presence of a liquid-phase component in the alloy.

3) The physicochemical properties of the liquid-phase component are determined by the chemical and fractional compositions of the MQ GAP of the investigated alloys.

4) During MQ by the GA method of alloys of the Fe-Nd-B system, GAP is formed, which has a fractional distribution depending on its chemical composition (content of neodymium, boron and alloying elements) and the parameters of the technological process, first of all: the type of nozzle; type, pressure and consumption of energy carrier gas; temperature and time of draining; heating temperature of the intermediate tank; the size of the injection hole; the shape and speed of the melt jet.

5) GAP with fractional composition from 1 to 800 microns under the conditions of the realized GA process is cooled at a rate from 10^5 to 10 K/s.

6) Smaller GAPs (captured by a cyclone-type cleaning system) can be cooled at a rate of more than 105 K/s. Its amount indicates the fluidity of the melt and the role of the surface energy of the melt in the dispersion of the melt jet by an inert gas flow.

7) Supercooling of the melt in solidifying GAP droplets depends on their size both due to the attainable cooling rate ($V_{cool} \sim d_m^{-3/2}$) and a decrease in the proportion of heterogeneous crystallization in smaller droplets due to a decrease in the likelihood of impurities (crystallization centers) into drops of small volume.

8) The grain structure of GAP in the first approximation is determined by the chemical composition of the alloy, the droplet size and the crystallization mechanism (homogeneous or heterogeneous).

Second-order effects (dendritic shape of grains and interfaces caused by grain dispersion) are associated with heat release at the initial stages of crystallization of the supercooled melt and the forming concentration overcooling at the crystallization front.

The size of the formed particles is determined by the ratio between the flow rate of the melt and the flow rate of the gas, the Weber number, the ratio between the critical viscosity of the melt and the kinetic viscosity of the gas. In this case, there is an equation that characterizes the deviation of the particle size D_l from its average size D_m [45]:

$$\frac{D_m}{D_1} = A * K^B \tag{1}$$

where A, B are constants depending on the process conditions;

$$K = \frac{M_L \nu_L 1}{M_g \nu_g W_e \gamma} \tag{2}$$

where M_L is the melt flow rate (kg/s); M_g – gas consumption (kg/s); {$_L$, {$_g$ – kinetic viscosity of melt and gas (m^2/s); © – specific heat ratio: ©=C_L/C_g; W_e – Weber's number.

The achieved supercooling determines the rate of nucleation and growth of crystals and affects the solidification of the supercooled melt.

Figures 1 and 2 (respectively) show a schematic diagram of the spraying unit URZhMV-3 and the differential curves of the distribution density of the studied hydraulic fracturing by size $n(d) = \frac{d \cdot \Delta M}{M \cdot \Delta d}$ [46].

The appearance of GAP and the shape of an individual powder are shown in figure 3. The characteristic features of their metallographic structure are shown in figure 4.

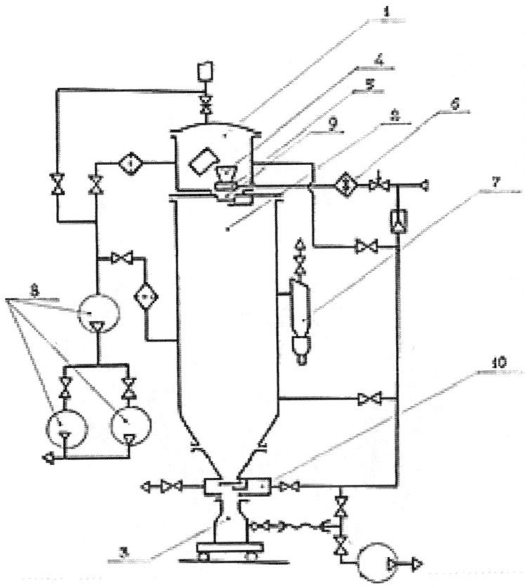

Figure 1. Diagram of the installation for spraying a stream of melt with an inert gas flow (URZHMV-3). 1 - induction furnace; 2 - spray chamber (with a diameter of 1.88 m and a height of 7 m with forced water cooling); 3 (10) - powder receiver with a vacuum seal; 4 - heated metal receptacle; 5 - unit for spraying liquid melt with an inert gas flow under pressure up to 2.5 MPa; 6 - inert gas supply system to the spraying unit; 7 - "Cyclone" - a system for capturing super fine powder from the discharged gas stream; 8 (9) - vacuum system with a vacuum seal.

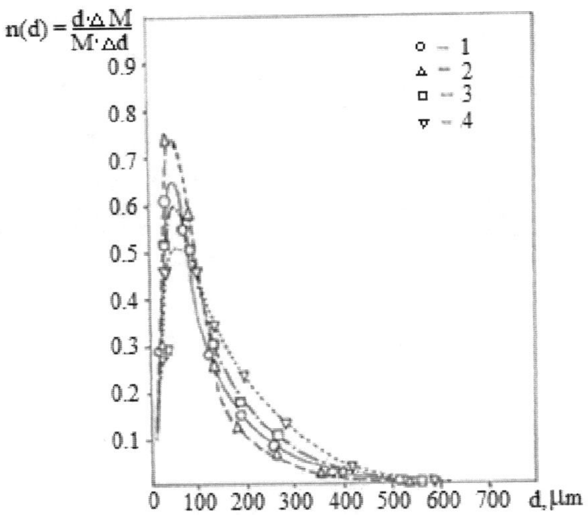

Figure 2. Differential distribution density curve of hydraulic fracturing by size, where 1 (3, 4) - ternary alloy with different charge content of boron; 2 - alloyed alloy.

Figure 3. Appearance of GAP ternary alloy (a) and the shape of its separate powder (b).

The paper investigated the effect of isothermal annealing (1, 2 and 3 hours 600 ° C, vacuum ~ 10^{-3} Pa) on the microstructure of GAP alloy. The obtained results of the digital optical metallography data set of the images of thin sections GAP were statistically processed using the SIAMS Drive System program, summarized depending on the alloying and are shown in figure 5.

There is a clear tendency that for the alloyed alloy (in contrast to the ternary alloy), upon isothermal annealing, an increase in the average maximum grain size of the $Fe_{14}Nd_2B$ phase is observed. As can be seen from figure 4, such a dependence may be due to the presence of multiple contacts of grain boundaries (not separated by a ternary eutectic) of the $Fe_{14}Nd_2B$ phase with each other and their coalescence (growth) by the recrystallization mechanism.

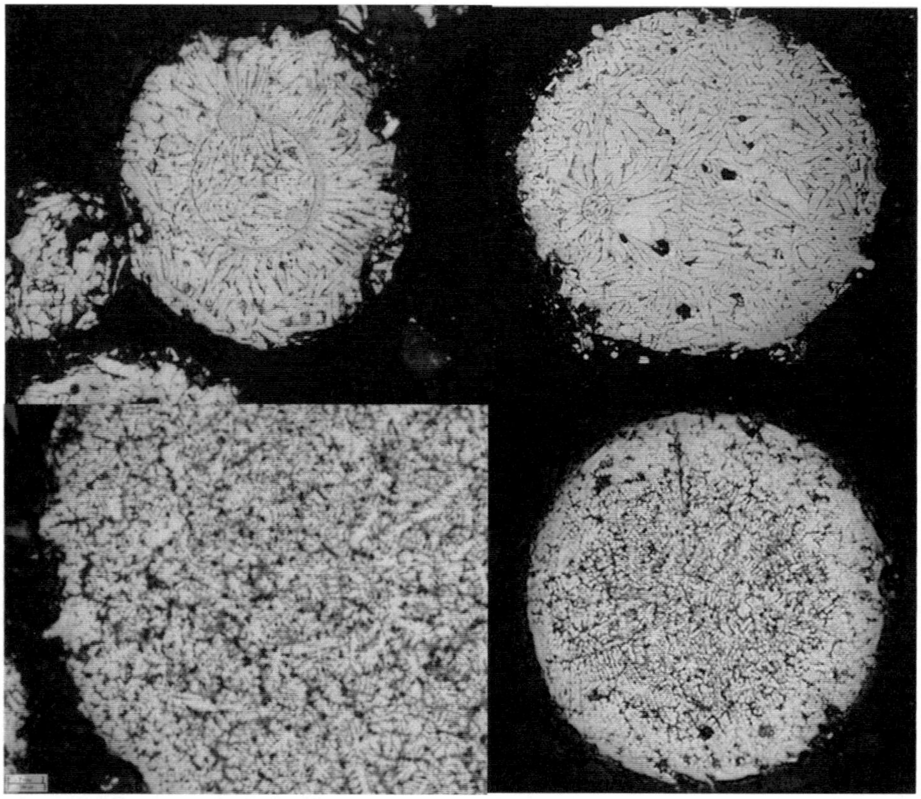

Figure 4. Metallographic structure of GAP of fraction (+ 120–460) μm of ternary (top) and alloyed (bottom) alloy, where the lower left photo is an enlarged image of the particle periphery shown on the right.

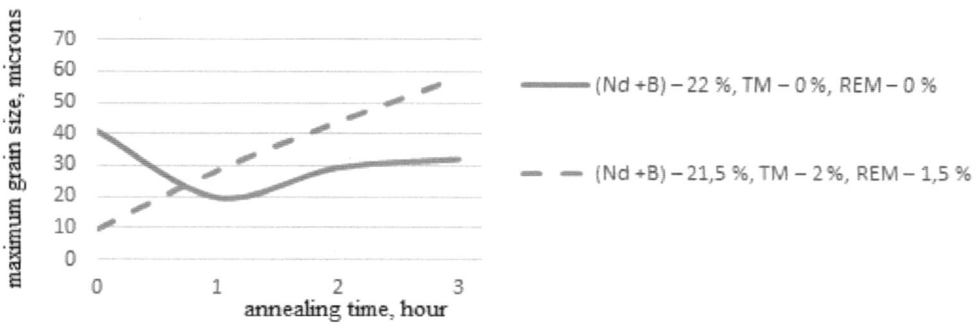

Figure 5. Influence of the time of isothermal annealing (600 °C, vacuum ~ 10^{-3} Pa) on the average maximum grain size of the $Fe_{14}Nd_2B$ phase for the MQ GAP alloy of the fraction (+ 120–460) μm.

This is confirmed by metallographic images (figure 6) of the initial and annealed GSP of ternary and alloyed alloys.

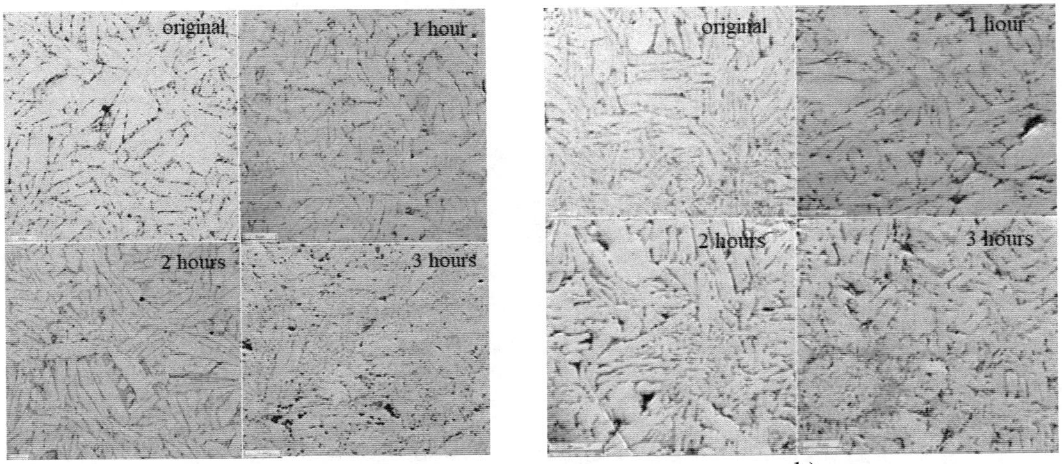

a) b)

Figure 6. Microstructure of the initial and annealed GAP of (a) ternary and (b) alloyed alloys.

Attention is also drawn to the fact that, upon annealing of ternary GAP, the processes of fusion (recrystallization) of $Fe_{14}Nd_2B$ grains also proceed, but are leveled by the processes of grain refinement of the $Fe_{14}Nd_2B$ phase at an early stage of annealing and therefore become noticeable only upon prolonged annealing (more than 2 hours). The grain refinement of the $Fe_{14}Nd_2B$ GAP phase of the ternary alloy at the early stages of annealing is primarily due to magnetostrictive effects [27]. As a consequence, in figure 5, there is a tendency for the average maximum grain size of $Fe_{14}Nd_2B$ to grow during isothermal annealing at 600 ° C for more than 2 hours GAP ternary alloy. In alloyed GAPs of a given fractional composition, already in the initial state, there is a reduced value of the average maximum grain size of the $Fe_{14}Nd_2B$ phase (in comparison with the ternary alloy) and, therefore, its growth is realized immediately from the early stages of annealing. At the same time, there is an increased (in comparison with the unalloyed ternary alloy) tendency of an increase in the number of pores on the metallographic section GAP. It is most likely to associate the observed result with an increased content of additional intermediate phases (intermetallic compounds) based on alloying elements and rare earth elements in the alloy doped with GAP. Such inclusions are always predominantly concentrated in the intergranular space (ternary eutectic enriched in neodymium) [25, 47–49]. As a rule, during standard sample preparation of a GAP metallographic section (mechanical polishing and subsequent chemical etching), selective extraction (precipitation) from the processed surface of the section of solid inclusions will inevitably be present.

For the manufacture of magnetoplastics, we used a standard plasticizer - a two-component epoxy resin. Three schemes have been implemented. According to the first: a previously prepared homogeneous mixture of powder of the fraction (+120 - 240) microns and resin in a ratio of 10: 1 was loaded into a rectangular mold, which was placed on the surface of a magnetic plate (permanent magnet with a field on the surface of up to 0.6 T); the workpiece was compacted using a screw press with a force of up to 1 MPa (0.1 t/cm^2) and held under pressure until the resin was completely solidified; a briquette (magnetoplast) with a size of (5 × 5 × 5) mm was obtained; the sample was magnetized (in the direction of the initial pressing field) in a closed magnetic system of an electromagnet in a gap of 5 mm at an induction of up to 2.2 T (figure 7). For a magnetoplast (using an ATE-8702 device), the field (induction) was measured on different surfaces along (S, N) and across (Sn, Nn) of the pressing field (in the center of the face) of a magnetized (square) sample, and thus its anisotropy was estimated. properties. The surface with the maximum properties (S) corresponds to the surface of the minimum distance with the magnetic plate when making a briquette. Accordingly, the surface opposite to it is (N). The difference in magnetic induction on the S and N - surfaces of such magnetoplastics is associated with the topological (inhomogeneity of the packing density) and approximate (inhomogeneity of rotation in a magnetic field) of the magnet of anisotropic GAP. The difference in magnetic induction on the S and Sn (N and Nn) surfaces is mainly determined by the magnetic anisotropy of the GAP. The results are summarized in table 1.

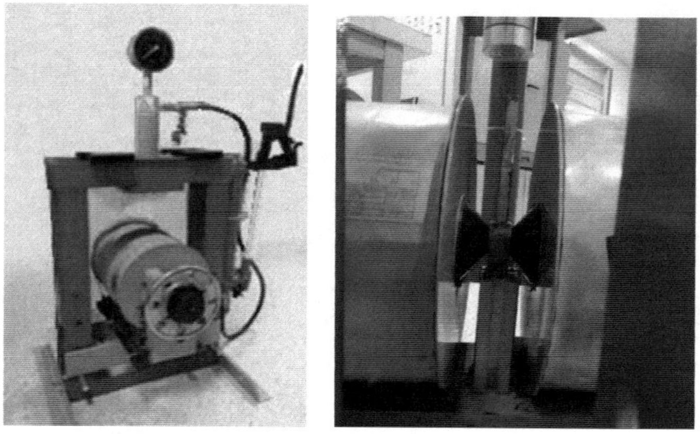

Figure 7. General view and elements (mold, electromagnet tips) of an experimental setup for the manufacture of magnetoplastics in a closed magnetic field (Scheme 2).

The second scheme for the manufacture of magnetoplavs: a pre-prepared sample of a given mass in the form of a homogeneous mixture of GAP fraction (+120 –240) μm and resin in a ratio of 10: 1 was loaded into a rectangular demountable mold made of A316L stainless steel with a wall thickness of 3 mm; the mold was placed in a magnetic system based on an electromagnet with a closed core and special tips made of alloy grade K49F2 ("permendur"); the total gap between the tips during compaction did not exceed 11.0 ± 0.1 mm; induction in the gap during compaction is not less than 1.2 T; compacting of the briquette is performed using a hydraulic press with a force of up to 5 MPa (0.5 t/cm^2); the briquette was kept under pressure until completely solidified (but not less than 3 hours); received a square sample with a size of 5 × 5 × 5 mm; the sample was magnetized (in the direction of the initial pressing field) in a closed magnetic system of the same electromagnet in a 5 mm gap at an induction of 2.4 T (figure 7); then the measurements of the anisotropy of the magnetic properties of the magnetoplast are made, as in Scheme 1. The measurement results are listed in table 1.

Table 1. Values of magnetic induction at the center of the surface of the faces of cubic samples (5 × 5 × 5 mm) of the investigated magnetized magnetoplastics (S, N - faces perpendicular to the field, Sn, Nn - parallel to the compaction field).

GAP condition in magnetoplast		Induction measurement surface	Magnetic induction (± 2 mT) on the surface of the magnetoplast		
			Scheme 1	Scheme 2	
			doped	doped	undoped
Original		S/S_n	37/4	65/7	55/6
		N/N_n	37/4	65/6	55/6
Isothermal annealing in vacuum, hour	1	S/S_n	34/4	61/6	57/6
		N/N_n	33/3	60/6	57/5
	2	S/S_n	39/4	67/5	60/5
		N/N_n	42/4	68/5	60/5
	3	S/S_n	60/6	71/4	62/4
		N/N_n	60/6	72/5	63/4

The first scheme for the manufacture of magnetoplastics (compaction in an open magnetic field) leads to an inhomogeneous distribution of GAP over the volume and insufficient orientation along the direction of the magnetic field during compaction. The orientation of the particles in the GAP magnetic field occurs due to the average magnetic anisotropy existing in them, related to the average inhomogeneity of the grain size of the $Fe_{14}Nd_2B$ phase in a separate powder (see figure 6) and, as a consequence, in the bulk of the magnetoplast as a whole. The values of the magnetic induction on the surface of such a magnetoplast do not exceed 40 ± 5 mT.

The transition to schemes (2) and (3) for the manufacture of magnetoplasts and the use of compaction in a closed magnetic field make the magnetoplast more uniform in the distribution of GAP in the sample volume and make it possible to orient a larger number of powder particles along the direction of the magnetic field, including those with a weakly expressed structural magnetic anisotropy. This makes it possible to achieve a magnetic induction at the center of the GSP on the surface (perpendicular to the compaction field) of the magnetized GAP sample in the initial state of more than 55 ± 5 mT. This corresponds to an increase in this value by more than 30% in comparison with magnetoplastics manufactured according to the scheme (1).

The performed measurements of the anisotropy of the magnetic properties of magnetoplastics (measurement of the magnetic induction on mutually perpendicular faces along and across the compaction field) convincingly prove that in the initial state in the GAP of ternary and doped alloys, a structural-phase state (inhomogeneity of the grain structure of the $Fe_{14}Nd_2B$ phase) on average provides a certain magnetic anisotropy of each powder and anisotropic properties of the magnetoplast as a whole.

During isothermal annealing at 600 °C, structural-phase changes occur in the GAP alloy, increasing their average magnetic anisotropy. As a consequence, the functional properties of the magnetoplast made of such GAP (magnetic field induction on the surface) are improved. Moreover, in the alloyed alloy, the increase in properties with optimal annealing is ~ 15% greater than in the ternary alloy.

The legitimately identified changes are associated with the observed process of fusion (recrystallization) of grains of the $Fe_{14}Nd_2B$ phase, which is more pronounced in the GAP of the alloyed alloy (see figure 6).

In figure 8 shows the results of differential thermal analysis (DTA) of the investigated undoped (a) and doped (b) GAP alloys with fractional composition (+ 120–240) μm. The analysis was carried out in an atmosphere of technical argon at a speed of 20 K / min in a standard Al crucible. The mass of the investigated powder was (2.0 ± 0.1) g. Analysis shows that the calorimetric effects of thermal absorption begin immediately upon the annealing of the GAP and experience additional features in the region of the magnetic phase transition (280 ÷ 350 °C) and melting of the ternary eutectic (510 ÷ 560 °C).

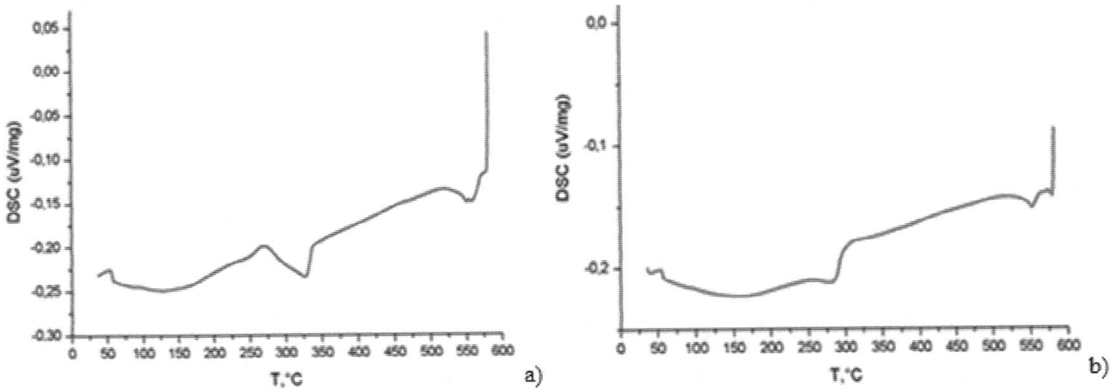

Figure 8. DTA (20 K / min, argon) GAP (weight 2.0 g) fractions (+ 120–240) μm of undoped (a) and doped (b) alloys.

To optimize the properties (to achieve the maximum value of the induction on the surface) of the magnetoplast, the third scheme of their production was implemented in the work: the mass fraction of the plasticizer was reduced to 3 wt%; pressing pressure increased to 10 MPa (1 t / cm^2); the bulk density of GAP in the powder mixture is increased due to the mixing of the powder of three fractions, the sizes of which are close to the requirements of the closest packing of balls, for example, in the fcc lattice, taking into account the presence of voids in the form of octa (R_2) and tetra (R_3) pores. According to the size and ratio of the number of octa (N_{R2}) and tetra (N_{R3}) pores per ball (N_{R1} = 1) of unit size (R_1), such a package has the following ratios: R_1:R_2:R_3=1,00:0,41:0,22; N_{R1}:N_{R2}:N_{R3} = 1:1:2 [52]. A simple recalculation of the number and size of pores per mass fraction in a mixture of a fraction of the appropriate size (for a constant density of the powder material and the existing discrete set of fractional composition) gives the following ratios for a charge charge with a mass of (20.0 ± 1.0) g: m_1 (fraction +400 - 640 microns): m_2 (fractions +120 - 240 microns): m_3 (fractions + 56 - 120 microns) = 5: 5: 10. This mass corresponds to the optimal volume of the mixture in the manufacture of a minimum experimental batch of magnetoplastic samples of a given size (at least 20 pieces with a size of 5 × 5 × 5 mm).

To control the nature of the distribution of GAP particles in the volume of the magnetoplast, metallographic sections were made with a surface along the direction of the magnetic field during compaction (perpendicular to the direction of compaction). A typical example of a metallographic image for a magnetoplast obtained according to the 2-scheme from the initial GAP doped fraction (+ 120–240) μm is shown in figure 9 (a). Alongside are the data of scanning electron microscopy on the morphology of the used GAP (9 b)

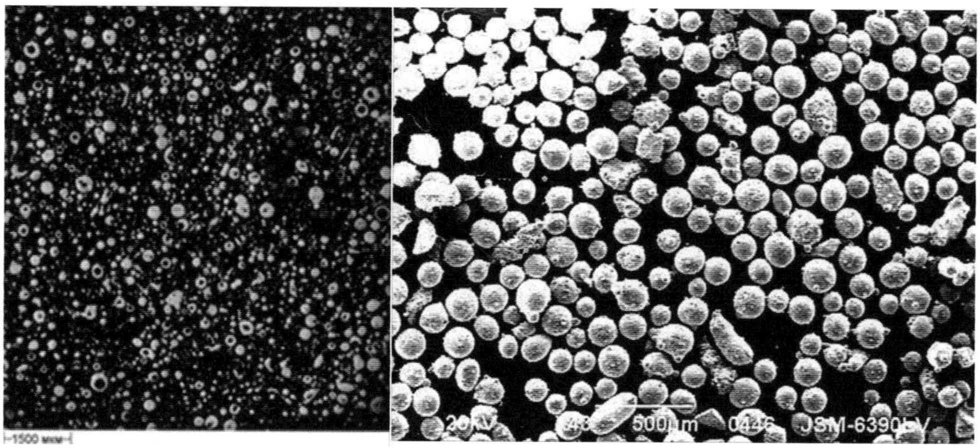

a) b)

Figure 9. Magnetoplastic made according to scheme 2 (a) and the morphology of the used GAP (b).

Metallographic images of other studied samples of magnetoplastics (Scheme 3, from the initial and annealed GAP) are shown in figure 10. The nature and parameters of the distribution of GAP on the cross-sectional area of the magnetoplast section were made by digital metallography (Olympus BX-30 microscope) using the SIAMS Drive System software of the SIAMS-800 solid body image analysis station. The results obtained for all investigated samples are summarized in table 2.

To verify the results on the existence of structural (grain) magnetic anisotropy in hydraulic fracturing, obtained in the study of magnetoplastic samples, the magnetic properties of GAP were measured on a VSM-250 vibration magnetometer in fields with an intensity of up to 2.4 T at room temperature. The magnetometer was calibrated using a reference sample of metallic Pd with a known value of the magnetic moment. The error in determining the specific magnetization (S_s) did not exceed 0.003 A×m^2/kg, and the coercive force (^IH_C) did not exceed 0.005 kA/m.

Table 2. The results of the analysis of the distribution of GAP particles on the surface of the section of the metallographic section of the magnetoplastic samples, depending on the method of their preparation.

Analyzed image parameters of a metallographic section	Magnetoplast Original			After annealing 600 ° C 3 hours	
	doped Scheme 2	Scheme 3	undoped Scheme 3	doped Scheme 3	undoped Scheme 3
Analyzed section area (sq. microns)	82781084	98462648	54687649	57575586	68476800
Number of particles (pcs.)	17824	13047	5690	8418	10416
Particle cross-sectional area (sq. microns)	27905008	44452564	35979964	36134614	39058655
Fraction of particles in the area of a thin section (%)	34.0±1.8	45.0±0.9	62.1±1.4	62.8±1.2	57.0±1.2
Average distance between particles on a thin section (μm)	49.0±2.0	61.0±1.0	60.4±1.6	49.0±1.1	44.6±1.2

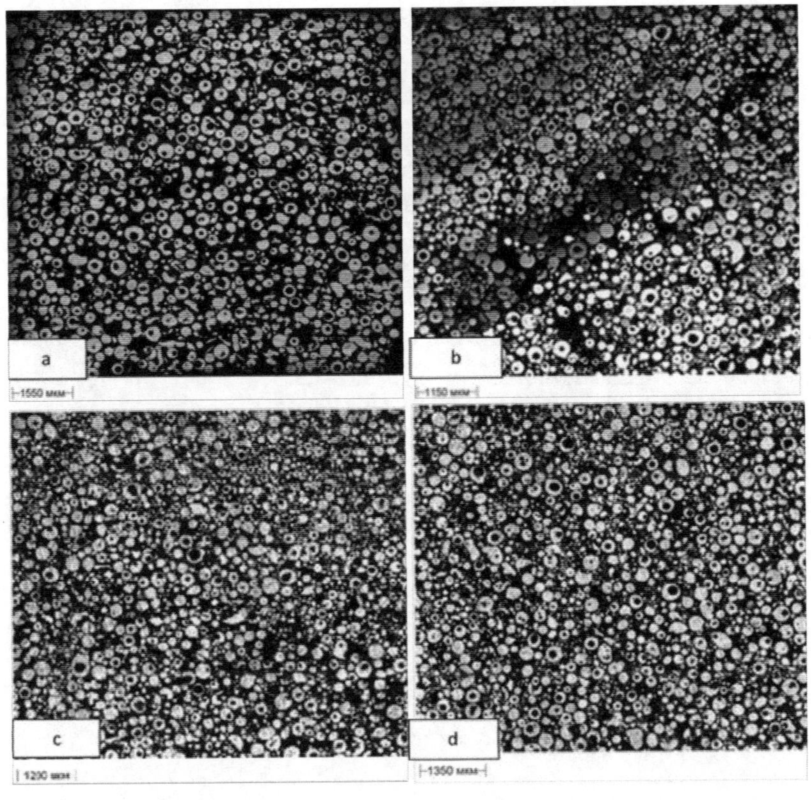

Figure 10. Image of metallographic sections of magnetoplastics obtained according to scheme 3. a - from the original undoped GAP; b - from the original doped GAP; c - from annealed (600 °C, 3 hours) undoped GAP; d - from annealed (600 ° C, 3 hours) alloy doped GAP.

The measurements were carried out on GSP of the fraction (+ 120–240) μm in the initial state. The mass of the powder sample was 500 ± 10 mg. The measurement results are shown in figure 11 for undoped (a) and doped (b) GAP. In both cases, the studied GAP has a magnetic rigidity, which is

IOP Conf. Series: Materials Science and Engineering 1181 (2021) 012014 doi:10.1088/1757-899X/1181/1/012014

determined by remanent magnetization (S_r), coercive force (${}^{I}H_C$) and the shape of the demagnetization loop. These parameters are also shown in figure 10 (a, b). For doped and undoped hydraulic fracturing, the indicated values are respectively equal: S_r = 31.19 and 38.69 A×m²/kg; ${}^{I}H_C$ = 224.1 and 233.4 kA/m or 2.816 and 2.933 kOe.

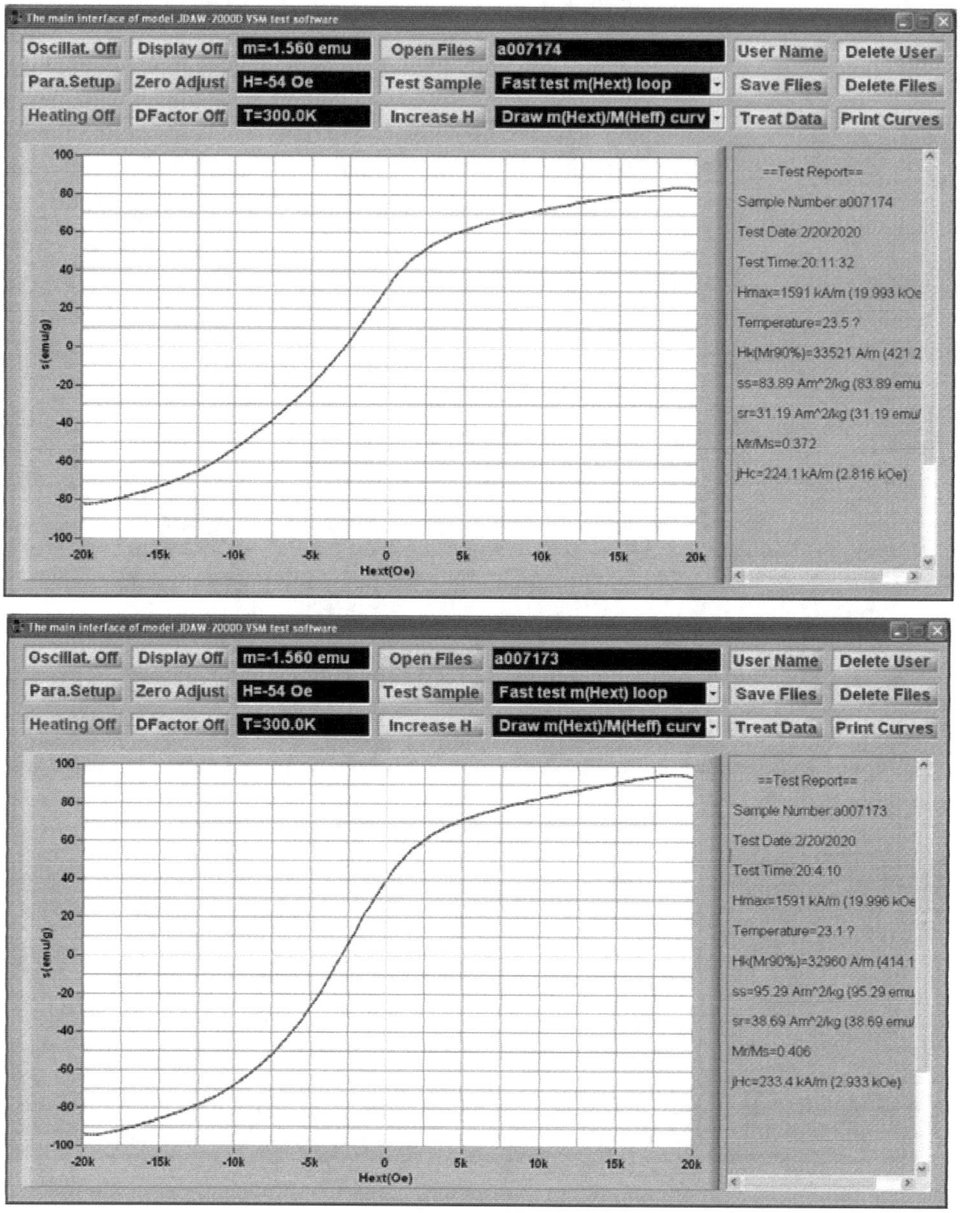

a)

b)

Figure 11. Demagnetizing branch of the hysteresis loop of GAP of the fraction (+ 120-240) μm in the initial state for undoped (a) and doped (b) alloys.

Annealing of GAP at 600 °C does not lead to complete recrystallization of grains of the $Fe_{14}Nd_2B$ phase, but the processes occurring in alloys of the Fe-Nd-B system during the manufacture of powder anisotropic sintered magnets using standard technology [1, 9-16]: vacuum induction melting of an ingot (from 0.5 to 45 kg); crushing to particles with a size of 5 ÷ 50 mm; mechanical grinding in a ball mill in a protective liquid medium to a fineness of 1 ÷ 10 microns; wet pressing (compaction) of samples in a magnetic field with an induction above 1.2 T; sintering in vacuum at ~ 1050 °C for 1 hour; cooling with a furnace up to 600 ÷ 800 °C and accelerated cooling (quenching) by an inert gas flow to room temperature.

To implement the complete recrystallization of the studied GAP and to prove that anisotropic magnets with properties similar to those of a magnet with a similar chemical composition obtained by the standard technology are obtained from GAP (using a technology close to the standard), anisotropic sintered samples were prepared from GAP. The following technological features were implemented: a mixture of different GAP fractions was used; the mixing algorithm was similar to the 3rd scheme - the production of magnetoplastics (the condition for the most dense filling of the space with spherical powders of different sizes); a GAP mixture with a mass of 100 ± 2 g was ground in a PM-100 planetary ball mill to particles with a size of 1 ÷ 10 microns; further, according to the modes of standard powder technology [1, 9-16], taking into account the data obtained in [32, 33, 36, 38, 39], anisotropic sintered magnets (10 × 10 × 10) mm in size were manufactured. From them, by the method of plane-parallel grinding, square specimens with dimensions (5 × 5 × 5) mm were made. The samples were magnetized to saturation in a pulsed magnetization apparatus in a field with an induction of up to 10 T. Their magnetic properties were measured by demagnetization in the gap of an electromagnet in a field with an induction of up to 2.2 T. The hysteriograph used a measurement circuit with two differential compensated coils. The demagnetization rate did not exceed 0.5 T/minute. Standardized $SmCo_5$ reference samples were used to calibrate the measuring system. The results of measurements of demagnetization of anisotropic sintered magnets from undoped (alloy No. 1) and doped (alloy No. 2) GAP of anisotropic sintered magnets magnetized to saturation are shown in figure 12 (a). The same figure (b) shows a typical image of a metallographic structure (in the plane of a microsection parallel to the compaction magnetic field) that forms in the material during sintering (for example, alloy No. 2).

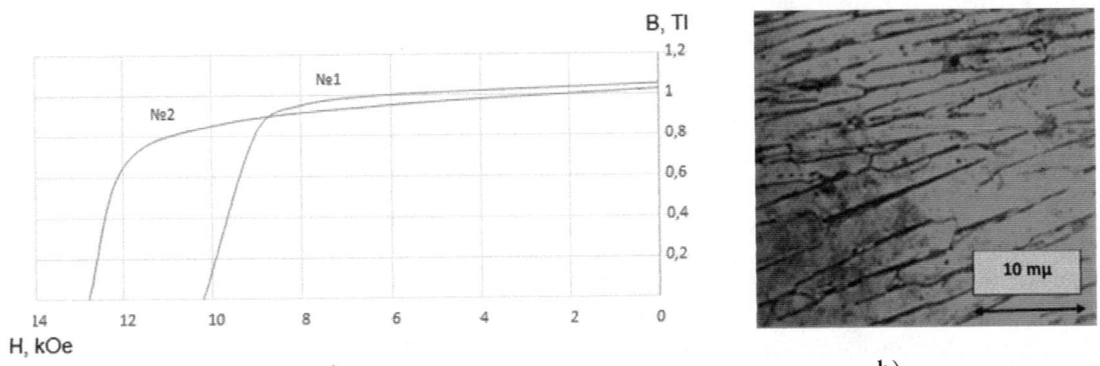

a) b)

Figure 12. Magnetic properties (a) and metallographic structure (b) of anisotropic sintered magnets from unalloyed (alloy No. 1) and doped (alloy No. 2) GAP: demagnetization curves of samples magnetized to saturation (a); image of the metallographic structure formed during sintering by the example of alloy No. 2 (b).

The obtained structural-phase state in the sintered material ensures its good magnetic properties: for alloys No. 1 and No. 2 (respectively) the rectangularity of the demagnetization loop exceeds 85% and 80% (in area); residual induction (Br) reaches 1.050 ± 0.005 and 1.025 ± 0.005 T; coercive force ($^{I}H_{C}$) - 10.2 ± 0.1 and 12.8 ± 0.1 kOe. The results achieved are comparable with similar data for anisotropic sintered magnets of the same composition, manufactured using traditional powder technology [5, 9–12, 17, 33].

5. Conclusions and discussion

The performed studies and the presented results concretize and expand the existing understanding of the formation of magnetic properties in the material of alloys of the Fe-Nd-B system, manufactured by the

MQ method [1, 2, 22, 25]. At least, the results obtained prove that ternary and alloyed alloys of the Fe-Nd-B system in the MQ GAP with a fractional composition of (+120, –640) μm, with their subsequent processing into anisotropic sintered magnets by grinding to particles with dispersion of (+1, -10) μm and sintering (1050 °C, 1 hour), complete recrystallization of the initial grain structure of the $Fe_{14}Nd_2B$ phase occurs and the formation of a relatively homogeneous microstructure with an average size along (parallel to the crystallographic axis "c") and across (parallel to the crystallographic axis "a") grains of the main magnetic phase: 10 ± 5 μm and 5 ± 2 μm, respectively. In this case, the grain boundary has a size approaching 1 μm and is filled with a ternary eutectic (based on neodymium and a mixture of borides $Fe_{14}Nd_2B$ + $Fe_4Nd_{1,1}B_4$) with single inclusions of non-metallic phases (slags) [3] and intermetallic compounds such as Laves phases, μ, σ and others Frank-Kasper related compounds [21-25, 36, 47-51].

It should be especially noted that in the sintered compact obtained by grinding GAP to granules with an average size of ~ 10 μm, there are always grains of the $Fe_{14}Nd_2B$ phase with a size significantly less than the specified size (up to 1.0 μm and possibly less). However, during sintering, a pronounced grain structure of the main magnetic phase is formed with a size comparable to the size of particles in a compact (briquette). And this means that when manufacturing anisotropic sintered magnets, it is necessary and sufficient to form a compact, in which only a certain (minimum sufficient) number of single-grain oriented (by pressing in a magnetic field) granules with a size close to optimal (6 ÷ 10 μm) should be present. In the future, if the necessary thermodynamic and kinetic conditions are created, then by the recrystallization mechanism in all sintered particles of the compact, single-grain structures will be formed.

The boundary (surface) of the compaction particles simultaneously plays two roles. First, it is a barrier to the recrystallization interaction of grains from different granules. Secondly, it is the main volume where the liquid-phase component of the sintered material is concentrated. Thirdly, it is a sink and a source of technological impurities, including non-metallic (oxygen, nitrogen, hydrogen). The first component predominantly determines the optimal (final) grain macrostructure of the sintered material. The second component determines the formation of the grain boundary, the diffusion of the main components and technological impurities between the sintered volumes (granules), the formation of micro- and nano-structures inside the granules and at their boundaries. The third component affects both of the first. Moreover, it can both enhance their course, and still stabilize, for example, by the mechanism of concentration hypothermia (see figure 4, upper pictures). The cumulative result of these three processes occurring simultaneously is the structural-phase state of the sintered material, which determines the entire complex of its magnetic properties. Moreover, the dominant component of this complex process is the process of recrystallization in the MQ GAP alloys at temperatures when the material contains an insignificant amount (up to 6 vol. %) Of the liquid-phase component. The relevance of this issue is multiplying in connection with the introduction of additive technologies in the production of micromagnetic systems based on the alloy. Fe-Nd-B [43].

The presented results prove that when analyzing the magnetic anisotropy of magnetoplastics from GAP, it is legitimate to use the following model [37, 41]: the magnetic anisotropy of GAP is formed by inhomogeneities (in shape and size) of the grain structure of the $Fe_{14}Nd_2B$ phase in the alloy of a powder (micro-ingot). These features are associated with the chemical composition of the alloy (alloying) and the conditions of its solidification (crystallization) during quenching from a liquid state (Melt Quenching) by spraying a melt jet with an inert gas flow (Gas Atomization) of medium pressure (up to 1.2 MPa).

Upon annealing above the melting point of the neodymium-based ternary eutectic (above 510 °C), which is present in a certain amount (up to 5 vol. %) In all industrial magnetically hard alloys of the Fe-Nd-B system, the process of recrystallization of grains of the $Fe_{14}Nd_2B$ phase occurs. The kinetics of this process for a fixed chemical composition depends on the initial grain structure of the alloy and the annealing temperature.

The initial grain structure of the GAP alloy can be controlled by the technological parameters of spraying (drain temperature, metal consumption, nozzle design, type and pressure of energy carrier gas) and fractional composition.

The more dispersed structure of the alloy is more sensitive to annealing. In such GAP, as a rule, the initial stages of alloy recrystallization are realized faster.

The maximum (optimal) properties of the magnetoplast are formed upon recrystallization of grains of the $Fe_{14}Nd_2B$ phase with a size of up to 6 μm (along the crystallographic axis "c"). The temperature and duration of isothermal annealing of hydraulic fracturing in the manufacture of magnetoplastics from them should not lead to significant growth of grains of the base phase (over 10 μm).

The technology of quenching from a liquid state (MQ) by the method of gas atomization (Gas Atomization) of a melt jet by a flow of an inert gas of medium pressure (up to 1.2 MPa) makes it possible to obtain a powder material (GAP), which has functional properties and cost indicators promising for its use in production sintered anisotropic magnets, magnetoplastics and magnetic systems, including additive technologies.

Acknowledgment
The reported study was funded by RFBR and Kaliningrad region according to the research project № 19-48-390002\20 (p_a).

The authors of the article are grateful to the following organizations, specialists and scientists:

- Techno Magnit LLC - for the provided GSP samples.
- Andrey V. Belov - for his help in the manufacture of sintered anisotropic magnets.
- Alexander G. Savchenko - for measuring the magnetic properties of GAP.

References

[1] Sagawa M, Fujimura S, Togawa N, Yamamoto H and Matsuura Y 1984 New material for permanent magnets on a base of Nd and Fe (invited) *J. of Applied Physics* **55(6)** 2083

[2] Nozawa Y, Iwasaki K, Igarashi Y *et al.* 1988 Effect of additive elements of magnetic properties of Die-Upsent Nd-Fe-B-magnets *J. Japan Inst. Metals* **52(9)** 899

[3] Pavlyukov A A , Opanasenko O S , Kravich E V and Korzheva N P 1989 Influence of oxidation of Fe-Nd-B alloy powders on the structure and magnetic characteristics of permanent magnets *Powder metallurgy* **7** 97

[4] Brekhara G P, Konev N N, Nemoshkalenko V V, Savin V V *et al.* 1990 Method of obtaining powder permanent magnets *Auth. Cf.* AC SU1782136AI

[5] Fukada Y, Fujita A and Shimotomai M 1990 Magnetic properties and corrosion characteristics of Nd-(Fe,Co,Ni)-B pseudo-ternary system *Kawasaki Steel Technical Report* **23** 27

[6] Brekhara G P, Vasilyeva E A, Girzhon V V and Savin V V 1995 Method of heat treatment of sintered permanent magnets Fe-Nd-B. Positive decision for invention: NIIGPE N506644 9/02 (039794)

[7] Ternovy Y F, Borkovskykh V A, Savin V V *et al.* 2000 Method of thermomechanical treatment of powder and granular materials Declaration patent 28215A - N95083770 - Ukraine

[8] Ternovy Y F, Borkovskykh V A, Savin V V *et al.* 2000 Method of obtaining metal powders based on rare earth metals Declaration patent 28214A. - N95083764. – Ukraine

[9] Nakamura H, Hirota K, Shimao M and Minowa T 2005 Rare earth permanent magnet US7488393B2 (United States) Priority to JP2005-084087 (2005-03-23) Anticipated expiration (2026-01-27)

[10] Nakamura H, Hirota K and Minowa T 2004 *Preparation of rare earth permanent magnet material* US8211327B2 (United States) Priority to JP2004304543 (2004-10-19) Adjusted expiration (2026-02-16)

[11] Nagata H, Nomura T and Minowa T 2007 «Rare earth permanent magnet and its preparation» US20110090032A1 (United States) Priority to JP2007068823A (2007-03-16) *Application*

filed by Shin Etsu Chemical Co Ltd 2010-10-27
[12] Sagawa M and Fujimoto N 2008 «NdFeB sintered magnet and method for producing the same» US10854380B2 (United States) Priority to JP2008-004845 (2008-01-11) Anticipated expiration (2029-01-09)
[13] Sagaba M (JP) 2006 «Method of making a baked magnet NdFeB» RU2009114155A (Russia) Priority to JP2006-250462 (2006-09-15) Publication of RU2423204C2 (2011-07-10)
[14] Nagata H, Nomura T and Minowa T 2011 «Rare earth permanent magnets and their preparation» US10614952B2 (United States) Priority to JP2011102789 (2011-05-02) Adjusted expiration (2033-06-10)
[15] Popov A G, Vasilenko D Y and Shitov A V 2012 «The method of obtaining high-coercive magnets from base alloys Nd-Fe-B» RU2476947C2 (Russia) Publication of RU2011123182A 2012-12-20
[16] Yamazaki T and Nakajima K 2013 «R-T-B rare earth sintered magnet» European Patent Office: EP2752857B1 Priority to JP2013000445 (2013-01-07)
[17] Schneider G, Goll D, Loffller R and Karim R 2012 High-throughput synthesis and analysis for searching new permanent magnet materials *Proc. Eng. Conf. Int. ECI Digital Archives* (Vail, Colorado, USA http://dc engconfintl org/materials_genome)
[18] Yamamoto M, Inoue A and Masumoto T 1989 Production of Nd-Fe-B alloys powders using high-pressuregas atomization and their hard magnetic properties *Metallurgical Transactions* A **V20A** 5
[19] Brekharya G P, Vasilieva E A, Konev N N, Savin V V *et al.* 1990 Structure and magnetic properties of alloyed Fe-Nd-B alloys quenched from a liquid state *PhMM* **1** 63
[20] Glavaguera mora M T, Baro M D, Surinach S *et al.* 1990 Magnetization Versus heat treatment in rapidly solidified NdFeB alloys *IEEE Transactions on Magnetics* **26(5)** 2613
[21] Zhou G F, Chuang Y C, Sun X K *et al.* 1990 Effect of heat treatment on the microstructure and magnetic properties of niobium-containing (Nd,Dy)-(Fe,Co)-B sintered permanent magnets *J. of the Less-Common Metals* **158** 41
[22] Wronski Z S 1991 Microstructure and magnetic properties of low-neodymium Nd-Fe-B-Si magnets produced from HP gas atomized powder *J. Appl Phys* **69(8)** 5507
[23] Harada T, Ando T and O'Handley R C *et al.* 1991 A microstructural Study of Nd15Fe77B8 magnetic alloy produced by liquid dynamic compaction (LDC) *Materials Science and Engineering* **A133** 780
[24] Szymura S 1991 Modefication of the magnetic properties and corrosion resistance of Nd-Fe-B permanent magnets wiht addition of cobalt *J. of the Less-Common Metals* **175** 185
[25] Savchenko A G 1992 Permanent magnets based on rapidly quenched alloys of the Nd-Fe-B system *Izv. Vuzov Ferrous metallurgy* **1** 59
[26] Savin V V 1993 The methods of diffraction analysis of the structure-phases transformation in magnetic alloys Fe-Nd-B system *Materials Science Forum* **133-136** 437
[27] Brekharya G P, Vasilieva E A, Girzhon V V, Savin V V *et al.* 1993 Influence of thermal cycling treatment on the structure and properties of permanent magnets Fe-Nd-B *PhMM* **76(2)** 129
[28] Savin V V, Ternovoy Yr F, Borkovskih V A and Nedolya A V 1996 Manufacture of gas- atomized of Fe-TM-Nd- REM-B alloys and magnets from them *J. of Magnetism and Magnetic Materials* **157/158** 49
[29] Borkovskikh V A, Kostenko R V, Kostenko E Yr and Savin V V 1999 Using hydrogen for the production of Fe-Nd-B system permanent magnets *International J. of Hydrogen Energy* **24** 263
[30] Brekharya, G P, Savin, V V, Bovda, O M, Shpak A P 2006 Study of influence of copper and carbon on properties of the permanent magnets fabricated on the basis of a $Fe_{76}Nd_{16}B_8$ alloy *Metallofizika i Noveishie Tekhnologii* **28(3)** 383
[31] Zakotnik M, Harris I R, Williams A J 2008 Possible methods of recycling NdFeB-type sintered magnets using the HD/degassing process *J. of Alloys and Compounds* **450** 525

[32] Konev N N and Belov A V 2012 Prospects of the use of methods of super-fast hardening of liquid metal in the production of nanocrystalline magnets based on Fe – Nd – B *Metal Science and Heat Treatment* **54(7-8)** 330

[33] Konev N N and Belov A V 2013 Magnetic properties of Fe – Nd – B sintered permanent magnets obtained by the method of strip casting *Metal Science and Heat Treatment* **55(1-2)** 92

[34] Margarit G, Neil D, Hadjipanayis G and Niarchos D G 2014 Experimental proof of microwave sintering of Nd–Fe–B powders toward fabrication of permanent magnets *IEEE TRANSACTIONS ON MAGNETICS* **50(11)** 2104804

[35] Savin V V, Semin V A and Semina V K 2014 Features of amorphisation alloys of Fe-TM-Nd-REM-B system at melt queching by method of melt spinning and gas atomization *V Int. Conf. Functional nanomaterials and high-pure substances* (Suzdal) 98

[36] Kablov E N, Ospennikova O G, Korolev D V, Piskorsky V P, Valeev R A and Rezchikova I I 2015 Mechanisms of the effect of boron content and heat treatment on the properties of magnets of the Nd – Fe – Al – Ti – B system *Aviation materials and technologies* **S2** 30

[37] Sorokovikov M, Savin V, Suvorov A, Savina L and Chaika V 2019 Investigation of the influence chemical and fractional compositions on the magnetic properties of the gas atomization powder of the Fe-Nd-B system *Power and Material Science IOP Conf. Series: Journal of Physics: Conf Series* **1172** 012025

[38] Xu X D, Sasaki T T, Une Y, Kubo H, Ohkubo T, Sagawa M, Hono K 2018 Origin of the coercivity difference in Nd-Fe-B sintered magnets processed from hydrogenation-disproportionation-desorption-recombination powder and jet-milled powder *Acta Materialia* **151** 293

[39] Ding G, Guo S, Chen L, Di J, Chen K, Chen R, Lee D and Yan A 2018 Effects of the grain size on domain structure and thermal stability of sintered Nd-Fe-B magnets *J. of Alloys and Compounds* **735** 1176

[40] Yina W, Chena R, Tanga X, Jua J and Yana A 2019 Effect of pressless heat treatment on the magnetic performance and microstructure of hot-deformed Nd-Fe-B magnet *J. of Magnetism and Magnetic Materials* **482** 9

[41] Sorokovikov M, Savin V, Savina L, Chaika V, Zherebcov I 2019 Features of Amorphization of the Fe-TM-Nd-REM-B Alloys System at Melt Quenching by Gas Atomization *J. of Metastable and Nanocrystalline Materials* **31** 51

[42] Chaika V, Savin V, Sorokovikov M, Marukhin A and Martyushev A 2020 Effect of annealing on recrystallization of GAP particles of Fe-Nd-B system alloy *IOP Conference Series: Materials Science and Engineering* **744** 012038

[43] Volegovab A S, Andreeva S V, Seleznevaa N V, Ryzhikhina I A, Kudrevatykha N V, Mädlercd L, Okulovacd I V 2020 Additive manufacturing of heavy rare earth free high-coercivity permanent magnets *Acta Materialia* **188** 733

[44] Popov V A 2010 *Metallogeny of ancient and modern oceans* Collection of works (Miass: IMin UB RAS) 249-52

[45] Ternovoi Yu F 1992 *Development and experimental substantiation of the theory and methods of engineering calculations of the parameters of the technology for obtaining powders of alloy steels and alloys by melt spraying*: dissertation for the degree doctorate of technical sciences (UkrNIISpetsstal: Zaporizhzhia)

[46] Savin V V and Kostenko E Yu 2005 *Crystal chemistry of intermediate phases and amorphous alloys based on transition metals: Monograph* (Zaporizhzhia National University, Zaporizhzhia) 346

[47] Matsnura Y, Hirosawa S, Yamamoto H *et al.* 1985 Phase diagram of the Fe-Nd-B ternary System *Jap J Appl Phys* **24(8)** L635

[48] Kuzma Yu B and Chaban N F 1990 *Binary and ternary systems containing boron Handbook* (Moscow) 177-9

[49] Gong W, Li Q, Yin L 1991 Temperature feature of NdDyFeB magnets wiht ultrahigh coercivity *J. Appl. Phys.* **69**(8) 5512

[50] Shiming H, Wei Xu 1990 An isobor-isothermal phase diagram of Nd-Fe-B system at atmospheric pressure and 1000 °C *J. Chin. Rare Earth Soc.* **8(3)** 200
[51] Devi M U, Gupta K P 1990 Phase equilibria in the Re-Fe-B system at 1000 °C *J. of Less-Common Metals* **159** 13
[52] Shaskolskaya M P 1984 *Crystallography Second edition* (M.: Higher School)

Synthesis and properties of high temperature self-curing epoxy resin

Alexei Slobodinyuk[1], Dmitriy Kiselkov, Nadezhda Elchisheva and Daria Slobodinyuk

Institute of Technical Chemistry, Branch of the Perm Federal Research Center, Ural Branch, Russian Academy of Sciences, 614013, Perm, Russia

[1] E-mail: slobodinyuk.aleksey.ktn@mail.ru

Abstract. A new bifunctional epoxy resin was synthesized by the reaction of replacing the hydrogen atom of amino groups from 4,4-diaminodiphenylmethane and epichlorohydrin taken in a double excess in dichloroethane. In order to remove HCl formed during the synthesis, which can interact with the initial diamine, as well as with the oxirane rings of both epichlohydrin and the synthesized product, an excess of potassium carbonate was introduced into the reaction mixture. The structure of the synthesized epoxy resin - 4,4'-methylenebis (N- (oxiran-2-ylmethyl) aniline) was characterized and proved by the methods of FTIR spectroscopy, nuclear magnetic resonance and elemental analysis. The kinetic parameters of the synthesized resin were characterized by differential scanning calorimetry. Using the Kissinger method, the activation energy of the self-curing reaction was calculated, the value of which was 1.139 kJ / mol. Such a low activation energy of epoxy resin allows to cure at low temperatures, but shortens the storage time. Two curing modes were studied: 1 mode - 10 hours at 150 ° C, 2 step mode - 10 hours at 150 ° C + 3 hours of postcuring at 180 ° C.

1. Introduction

Epoxy-based materials are widely used as binders of composite materials, coating, adhesives due to their excellent mechanical, electrical properties, chemical resistance, and easy-to-cut [1,2,3]. Epoxy resins occupy almost 70 % of the whole thermosetting resin market. Technological progress leads to significant increase in requirements for construction material properties in a wide temperature range. Todays demands for compounds based on epoxy resins for integrated circuits require a glass transition temperature above 200 ° C.

Primary and secondary di- and polyamines [4], polybasic acids [5] and their anhydrides [6], thiols [7] are traditionaly used as hardeners for epoxy resins. At the same time, increasing the glass transition temperature of epoxy polymers is a challenging task. Many approaches to increasing the glass transition temperature of epoxy resins, as well as the moisture resistance of epoxy resins by modifying the structure of initial resins, that affects the properties of the final cured epoxy resins. In [8], the authors increase the glass transition temperature of epoxy polymers by including a naphthalene ring in the structure. Kaji and Dndo [9], Pan et al. [10], Wang et al. [11] and Castell et al. [12] synthesized a naphthalene-based epoxy resin containing naphthyl and phenyl structures. Xu et al. [13] and Ren and authors [14] synthesized epoxy resins with a naphthalene-cycloaliphenic group.

Two-component systems "epoxy resin - hardener" consist of microdefects, which give rise to heterogeneity due to incompatibility of components based on different solubility properties of epoxy resins and hardeners [15].

Therefore, in recent years, great interest has been attracted by epoxy resins capable of self-curing, thereby avoiding the above disadvantages and at the same time showing high values of the glass transition temperature.

The aim of this work was to synthesize the structure of a new heat-resistant self-curing epoxy resin. The cure characteristics and physical properties of the cured polymer were investigated using several analytical methods.

2. Materials and Methods

Glass transition temperature T_g^s of epoxy resin was determined using differential scanning calorimeter METTLER TOLLEDO DSC 822^e at scanning rate 5 °C/min.

^{1}H NMR spectra was recorded using Bruker Avance-Neo 400 spectrometer (operating at nominal proton frequency 400 MHz) in CDCl3 using HMDS as an internal standard.

FTIR spectras were recorded using Bruker Vertex80V with 64 scans at 2 cm^{-1} resolution.

2.1. Materials

In this study a commercially available 4,4-diaminodiphenylmethane and epichlorohydrin (Acros Ltd.), magnesium sulfate, potassium carbonate, ethyl acetate, and dichloroethane (Vitakhim Ltd.) were used. Ethyl acetate and dichloroethane were additionally dried by distillation over phosphorus pentoxide. All chemicals used were reagent grade.

2.2. Methods

Glass transition temperature Tgs of epoxy resin was determined using differential scanning calorimeter METTLER TOLLEDO DSC 822e at scanning rate 5 °C/min.

^{1}H NMR spectra was recorded using Bruker Avance-Neo 400 spectrometer (operating at nominal proton frequency 400 MHz) in CDCl3 using HMDS as an internal standard.

FTIR spectras were recorded using Bruker Vertex80V with 64 scans at 2 cm^{-1} resolution.

2.3. Synthesis

In a two-necked flask 5 g of 4,4'-methylenedianiline (1), 7.5 g of K_2CO_3 and 200 ml of dichloroethane were added and kept with constant stirring for 2 hours at room temperature. Then 4.7 ml of epichlorohydrin (2) was added and stirring was continued at 80 ° C for 5 hours. After completion of the reaction, the mixture was cooled to room temperature. The solution was filtered to remove the inorganic residue. The solvent was distilled off on a rotary evaporator. The crude product was treated three times with ethyl acetate and distilled water and separated. $MgSO_4$ was added to the organic layer to remove the remaining water. The organic layer was filtered off, the solvent was evaporated (figure 1).

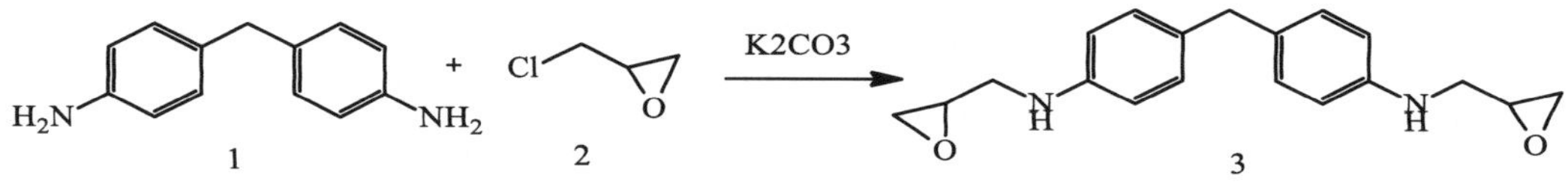

Figure 1. Synthesis of 4,4'-methylenebis (N- (oxiran-2-ylmethyl) aniline).

Compound 3 was a brown viscous substance, readily soluble in chloroform and alcohols, hardly soluble in acetone, and insoluble in alkanes and water.

3. Results

In the ^{1}H NMR spectra of a solution of compound 3 in $CDCl_3$, in addition to the signals of the protons of the aromatic rings, CH and CH_2 groups, there is a singlet of two protons of the NH groups in position 4 of the aromatic fragment in the region (4.08 ppm).

4,4'-methylenebis (N- (oxiran-2-ylmethyl) aniline) (3). The yield is 86%.

^{1}H NMR spectra, δ, ppm: 3.08-3.12 m (4H, CH_2), 3.21 m (4H, CH_2), 3.50 m (1H, CH), 3.57 m (1H, CH), 3.64-3.68 m (2H, CH_2), 4.06 s (2H, NH), 6.50 d (4H, CH_{ar}), 6.90 d (4H, CH_{ar}) (figure 2).

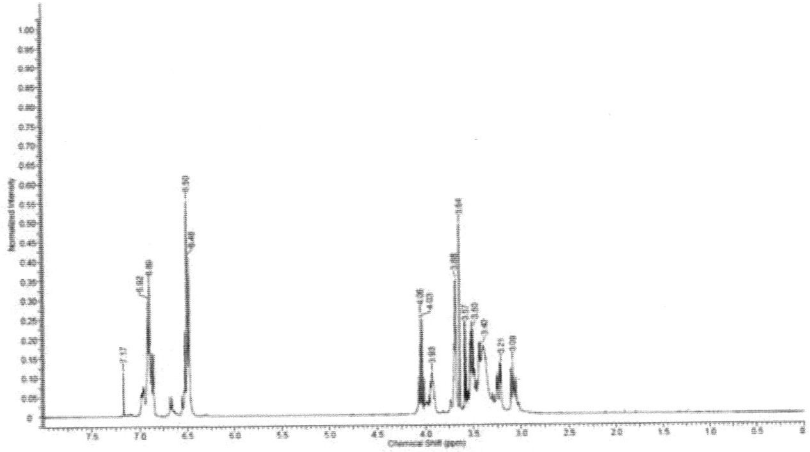

Figure 2. ^{1}H NMR spectra of 4,4'-methylenebis (N- (oxiran-2-ylmethyl) aniline).

The completeness of the reaction (figure 1) was controlled using FTIR spectroscopy by monitoring the absorption band at 3450 cm^{-1} (figure 3), that is attributed to the hydroxyl groups of epichlorohydrin. The absorption band at 3450 cm^{-1} completely disappears during the synthesis.

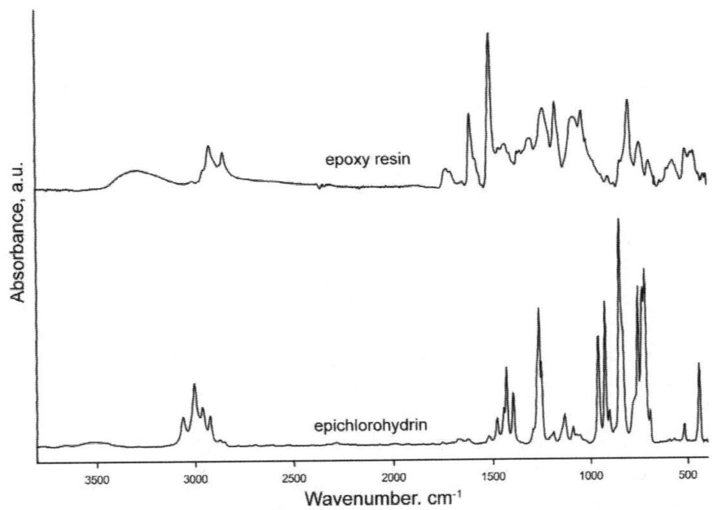

Figure 3. FTIR spectra of epoxy resin and epichlorohydrin.

The quantitative content of free epoxy groups was determined using back titration method ISO 3001:1999. The experimental value - 27.6 % is close to theoretical value - 27.74 %.

Elemental analysis of the synthesized epoxy resin was carried out using Elementar Vario EL cube. Found, %: C 73.70; H 7.25; N 9.20. $C_{19}H_{22}N_2O_2$. Calculated, %: C 73.55; H 7.10; N 9.03.

4. Discussions

Two curing modes of synthesized epoxy resin were studied: 1 mode - 10 hours at 150 ° C, 2 step mode - 10 hours at 150 ° C + 3 hours of postcuring at 180 ° C The completeness of the reaction was monitored using FTIR spectroscopy by the disappearance of the absorption band at 910 cm^{-1} [16]. A schematic diagram of the curing reaction of the synthesized resin is shown in figure 4.

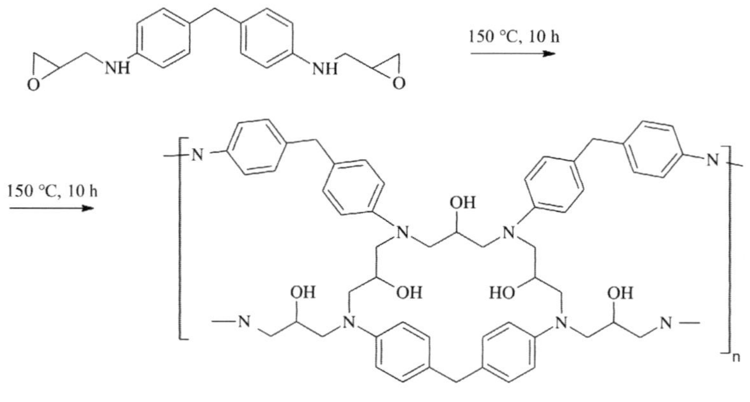

Figure 4. Curing reaction of epoxy resin.

Figure 5 shows the DSC cure curves of the synthesized resin. The parameters of the kinetics were calculated using the Kissinger model [17, 18]. According to the Kissinger model, the activation energy can be calculated using the peak reaction temperatures at different heating rates. The relationship can be expressed by the following equation:

$$In\ (q\ /\ \mathrm{Tm}) = \mathrm{Ea}\ /\ \mathrm{RTm} - \mathrm{In}\ (\mathrm{AR}\ /\ \mathrm{E\ a}) \tag{1}$$

where T_m – maximum temperature of reaction, K; A – pre-exponential factor; q – constant heating rate, °C/min; E_a – activation energy, J/mol; R – universal gas constant 8.314 J/(mol*K).

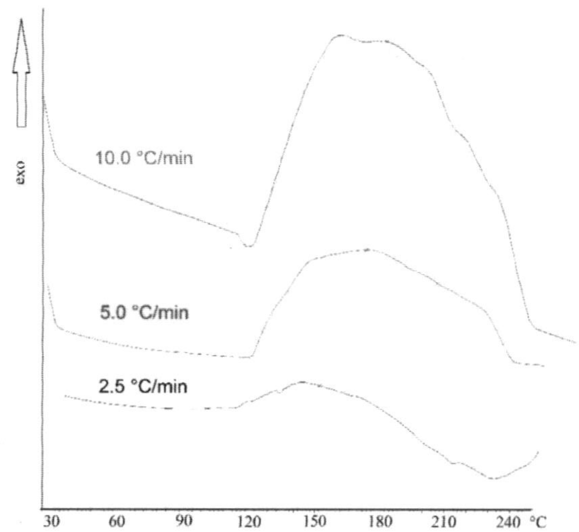

Figure 5. DSC curves of curing reaction of epoxy resin at different heating rates.

Table 1. Calculated kinetic parameters of synthesized epoxy resin.

Heating rate. °C/min	Temperature of reaction peak. °C
2.5	152
5.0	167
10.0	185

The activation energy was calculated using the dependence diagram of $\ln(q/T_m^2)$ from $1000/T_m$ (figure 6), where T_m- peak temperature on DSC curve, K; q – heating rate, K/min.

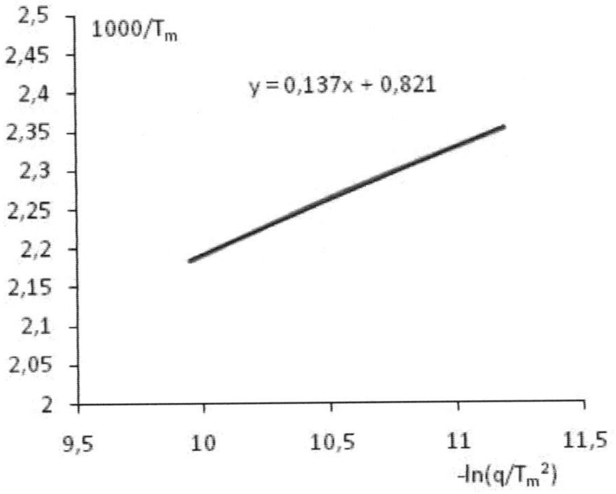

Figure 6. Linear plot of $-\ln(q/T_m^2)$ versus $1000/T_m$ based on Kissinger's equation.

$$E_a = -1.181*8.314 \text{ J/mol*K} * 1000\text{K} = 9.820 \text{ kJ/mol}. \quad (2)$$

The calculated activation energy was 9,82 kJ/mol.

Figure 7 shows the DSC curve of the cured resin using mode 1. The glass transition temperature of self-cured synthesized epoxy resin at scanning rate 5 °C/min is higher than 200 °C.

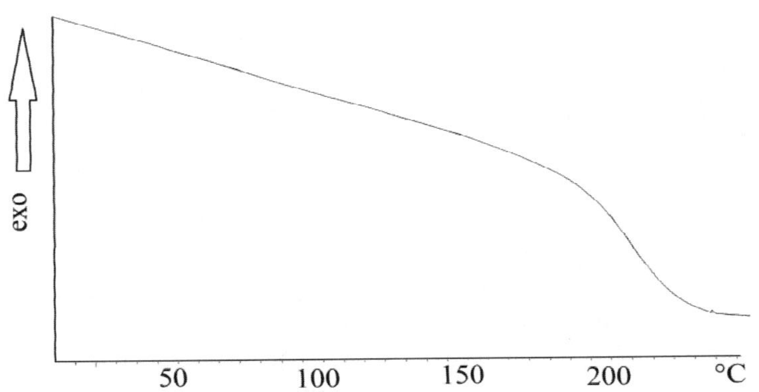

Figure 7. DSC diagram of self-cured epoxy resin using mode 1.

Figure 8 shows the DSC curve for the second curing mode of epoxy resin. The glass transition temperature of the synthesized resin after self-curing is in the range of 210-215 °C.

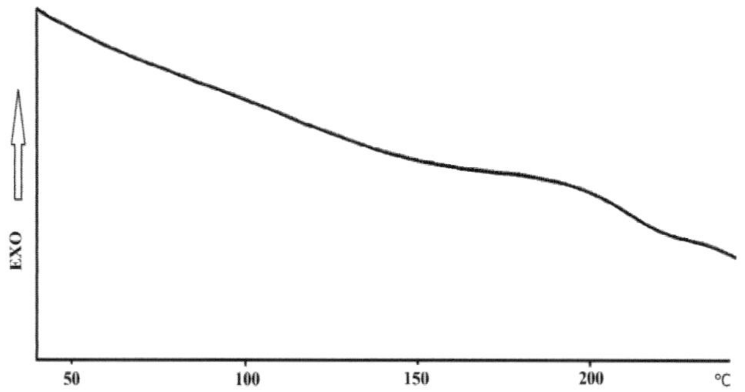

Figure 8. DSC curve of cured epoxy resin using mode 2.

5. Conclusions

Using 4.4-diaminodiphenylmethane and epichlorohydrin taken in a double excess in dichloroethane, a bifunctional epoxy resin was synthesized by the reaction of replacing the hydrogen atom of amino groups. The structure of the synthesized epoxy resin - 4,4'-methylenebis (N- (oxiran-2-yl-methyl) aniline) was characterized and proved by the methods of FTIR spectroscopy, NMR spectroscopy and elemental analysis. Using the Kissinger method, the activation energy of the self-curing reaction was calculated, the value of which was 1.139 kJ / mol. Two curing modes of epoxy resin were studied: mode 1 - 10 hours at 150 ° C, mode 2 - 10 hours at 150 ° C + 3 hours of postcuring at 180 ° C. An epoxy resin was synthesized from 4,4-diaminodiphenylmethane and epichlorohydrin.

The activation energy of this resin was calculated based on DSC results. The high glass transition temperature and low activation energy of synthesized epoxy resin provide opportunity for use as a matrix of modern composite materials.

Acknowledgements

The reported study was funded by RFBR and Perm Territory, project number № 20-43-596010. The study was performed using the equipment of the Center for Shared Use "Studies of Materials and Substance," Perm Federal Research Center, Ural Branch, Russian Academy of Sciences with a support of the state task AAAA-A18–118030790037-7.

References

[1] Lee D, Song S H, Hwang J, Jin S H, Park K H, Kim B H, Hong S H and Jeon S 2013 *Small* **9(15)** 2602-10
[2] Song N, Cui S, Hou X, Ding P and Shi L 2017 *ACS Appl. Mater. Interfaces* **9(46)** 40766-73
[3] Tang L, He M, Na X, Guan X, Zhang J and Gu J 2019 *Composites Communications* **16** 5-10
[4] Spiesschaert Y, Guerre M, De Baere I, Van Paepegem W, Winne J M and Du Prez F E 2020 *Macromolecules* **53(7)** 2485-95
[5] Montarnal D, Capelot M, Tournilhac F and Leibler L 2011 *Science* **334(6058)** 965-8
[6] Giebler M, Sperling C, Kaiser S, Duretek I and Schlogl S 2020 *Polymers* **12(5)** 1148
[7] Hoyle C E, Lowe A B and Bowman C N 2010 *Chem. Soc. Rev.* **39(4)** 1355-87
[8] Duann Y F, Liu T M, Cheng K C and Su W F 2004 *Polym. Degrad. Stab.* **84(2)** 305-10
[9] Kaji M and Endo T 1999 *J. Polym. Sci., Part A: Polym. Chem.* **37(16)** 3063-9
[10] Pan G, Du Z, Zhang C, Li C, Yang X and Li H 2007 *Polymer* **48(13)** 36863693
[11] Wang C S, Leu T S and Hsu K R 1998 *Polymer* **39(13)** 2921-7
[12] Castell P, Serra A and Galia M 2003 *J. Polym. Sci., Part A: Polym. Chem.* **41(10)** 1536-44
[13] Xu K, Chen M, Zhang K and Hu J 2004 *Polymer* **45(4)** 1133-40
[14] Ren H, Sun J, Wu B and Zhou Q 2006 *Polymer* **47(25)** 8309-16
[15] Huang K, Liu Z, Zhang J, Li S, Li M, Xia J and Zhou Y 2015 *European Polymer Journal* **70** 45-54
[16] Strel'nikov V N, Senichev V Yu, Slobodinyuk A I, Savchuk A V, Volkova E R, Makarova M A,

Belov Yu L, Derzhavinskaya L F and Selivanova D G 2018 *Russ. J. Appl. Chem.* **91(12)** 1937-44
[17] Kissinger H E 1957 *Anal. Chem.* **29(11)** 1702-6
[18] Kissinger H E 1956 *J. Res. Natl. Bur. Stand.* **57(4)** 217-221

The clamps of sheet blank movement trajectory simulation under the shell part stretch forming process

V A Mikheev, A F Grechnikova, Roberto de Alvarenga and M M Demidova

Samara National Research University, 34, Moskovskoye shosse, Samara, 443086, Russia

E-mail: vamicheev@rambler.ru

Abstract. The main object for analyzing the results of modeling and optimizing the shaping method by the skin is the graphs of the change in thickness at characteristic or selected points on the surface of the shell of double curvature. The article discusses two variants of the method, which have the same cyclic diagram of step movements of the main and auxiliary working cylinders of the stretch-tightening press RO-630-11. Both options contain kinematic techniques that implement the following two effects. The first is the effect of isometric bending of the shell surface after unloading the sheet blank at the third step motion of the press according to the cycle diagram. The second is the effect of the forced position of a flat and sufficiently long sheet blank in rectilinearly arranged discrete clamping devices of the press. This provided a plastic deformation of the middle part of the sheet blank without localizing the thinning deformation and the alignment of its values in different zones of the sheet blank.

1. Introduction

The essence of the shaping process is that a flat sheet blank is transformed into a spatial shell part under the influence of tensile forces from the moving clamps of the press located at its two narrow opposite ends. The sheet blank is plastically deformed by stretching, but not all at once; it adheres tightly to the surface of the tightening punch and gradually takes the shape of this surface. The main condition for obtaining a shell part of the required geometric shape is the stretching of all sections of the sheet blank to the state of plasticity [1].

Shaping refers to the shaping operations associated with increasing the area of the sheet blank due to the thickness of the sheet. There is a possibility of localization of stretching of the sheet blank in the dangerous section of the part. As a result, the sheet blank is quickly thinned and destroyed, which is the reason for high rejects and large technological waste. The reason for the localization of stretching of the sheet blank is, first of all, the unevenness of the deformation due to the influence of friction forces and the complex shape of the shell. Most often, the localization of stretching during stretching occurs at the section of the sheet blank from the stretching punch in front of the tightening press clamping devices. In this case, the thinning of the sheet blank, which occurs during the covering, does not depend on the covering scheme itself, but depends on the friction conditions, thickness, the nature of anisotropy, and the deformation characteristics of the rolled sheet.

To ensure the monotony of deformation, the shaping of the sheet blank must be carried out under symmetric stretch conditions without its localization in the part of the sheet blank located between the edge of the stretch punch and the clamping jaws of the press. The stretching of a sheet blank during shaping by a wrapping differs not only in the magnitude of the forces, but also in the area of their application along the edge of the blank. In this case, the zone of plastic deformation on the sheet blank is outlined only on one side in the direction of tension, and the second, implicit boundary is the transition between the zones of plastic and elastic deformation, which is established by calculation or experimentally based on the analysis of the physical essence of the process [2,3].

The directions of movement of the boundary of the plastic deformation zone can be changed by moving the clamps of the sheet blank along a certain trajectory. To do this, it is necessary to coordinate the nature of the movement of the working bodies of the tying press by programming. The kinematic grounded schemes for shaping the shell of the considered geometric shape by a wrapper are associated with the fact that the tensile force applied to the discrete clamps is perceived by the workpiece along the clamp width. This feature affects the «picture» of the center of plastic deformation, which will be different depending on the rectilinear or curvilinear position of the discrete clamps on the clamping plate of the press [4].

To maintain the stability of the plastic state of the shell, it is necessary to prevent the appearance of bending deformation of the middle surface due to the application of external forces in the tangential direction at the section of the sheet blank coming off the surface of the tightening punch. The middle surface, reduced to the vertical planes of symmetry due to the orientation of the surface relative to the curvature lines, will ensure the constant position of the shaping contour of the surface of the shaping punch in the vertical plane of symmetry of the shaping press, as well as the perpendicularity of the plane of the shaping plate of the shaping press to the horizontal plane of the sheet blank at the vanishing point of the shaping contour of the shaping punch.

Stable control of the shaping processes by the skin is based on the initial position of the plastic deformation zone in the central zone of the sheet blank. The shaping of the sheet blank according to one or another kinematic scheme is accompanied by the oncoming movement of the boundaries of the plastic deformation zone while maintaining the location of the greatest deformations in the central region of the sheet blank. This allows you to increase the degree of deformation with a minimum probability of sheet breakage in one of its free sections between the edge of the punch and the press clamps.

Geometric study of the shaping punch to implement the conditions of symmetric shaping will make it possible to quickly perform complex calculations of the parameters of shaping processes even without the use of the traditional finite element method, which is characterized by the duration of calculations and a relatively high probability of error [5]. However, the achievements of modern information technologies, which raise the level of production and non-production technologies based on the digital twin - the digital representation of an object, have already led to the need to use existing developments in the field of storing information, operating with it and applying it in various fields.

The development of new methods of shaping by a stretch is to a certain extent restrained by the poor knowledge of the rheology of the behavior of the aluminum alloy in the annealed, and especially in the freshly quenched state. Due to the sheet texture formed during rolling, aluminum alloys are more prone to the predominant development of deformation along the sheet thickness or to thinning [6]. This tendency can be minimized to the recommended limits by creating a favorable texture of the sheet during the rolling process. This is a kind of guarantee for the stable control of the wrapping forming processes. For this, the correspondence of the texture axes of symmetry of the properties of the sheet blank and the directions of the lines of curvatures of the shell of double curvature intersecting at the pole of the surface of the covering punch was provided [7].

Under conditions of symmetrical wrapping, it is possible to reduce the likelihood of corrugations and breaks, which creates a solid basis for controlling the shaping processes by the wrapping, and increases the degree of shaping of the sheet blank, which is achieved practically in one pass without localizing the thinning deformation and leveling their values in different zones of the sheet blank. To do this, you only need to correctly position the pulling punch on the press table and know the trajectory of the clamps relative to its shaping contour and its location in the vertical plane of symmetry of the pulling press. The above principles of fulfilling the conditions of symmetric wrapping were used by us in modeling the shaping of a wrapped shell part by moving the clamps of the sheet blank along a certain trajectory.

2. Results of modeling the process of shaping by a wrap

The initial stage in the modeling of the process is the preparation of geometric models of the elements involved in the shaping of the tight-fitting: sheet blank; cover punch and double curvature shell in the LS-DYNA software package. The initial data for these elements was obtained from the enterprise. Sheet billet made of aluminum alloy 1163RDMV, 1.5 mm thick and dimensions in terms of 1880x10500 mm. The mechanical properties of this alloy were obtained during testing of five samples and statistical processing of the results: ultimate strength - 174 MPa; yield point - 78 MPa; uniform elongation - 18%; elastic modulus - 241.4 MPa and anisotropy coefficients μ_{21} = 0,428 and μ_{12} = 0,40.

Initially, it should be noted that the preparation of geometric models of elements for the calculation in LS-DYNA is carried out in the LS-PREPOST program. All geometry is exported to the program in * STEP format for further processing, which is presented in the form of creating a mesh, specifying the main parameters, generating an output calculation file, and so on. We omit the details of the preparation of the initial data here and show only the results of the geometric study of the covering punch and for visualizing the picture of the simulation results in kinematic steps due to the trajectory of movement of the clamps of the sheet blank. The main object for analysis and optimization of the method will be graphs of thickness changes at characteristic or selected points on the surface of the sheet blank. The tightening punch after geometric processing with symmetrical positions of the shaping contour, the central cross-section and the «pole»of the surface relative to the longitudinal axis of tension during tightening is shown in figure 1.

When describing the method of shaping a sheathing part by means of moving the clamps of a sheet blank along a certain trajectory, it is necessary to provide the results of a change in thickness during shaping by a shaping at the points we have designated on the surface of the sheet blank. The sheet blank is located in the initial position relative to the covering punch, starting from the edges A, B and C along the left contour of the part; the middle between the left contour of the part and the central section D, E and F; central section G, H and I; the middle between the center section and the right contour of the part J, K and L; the edge of M, N and O along the right contour of the part (figure 2). These points are located on three lines located on the sheet front front (blue), center (red) and back (green) in relation to the position of the press operator. The points of the «blue» and «green» lines correspond to the elements of the modeling grid located next to the line of the side contour of the part, the so-called trimming of the part by the contour template.

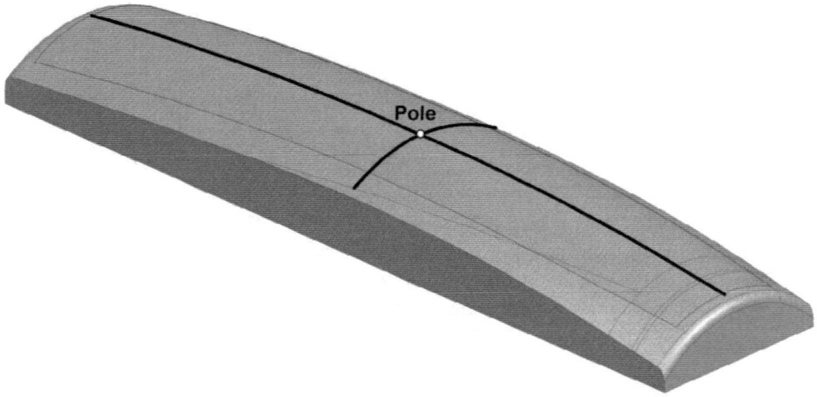

Figure 1. Tightening punch after geometric processing with symmetrical positions of the shaping contour, central cross-section and «pole» of the surface relative to the longitudinal axis of tension during stretching.

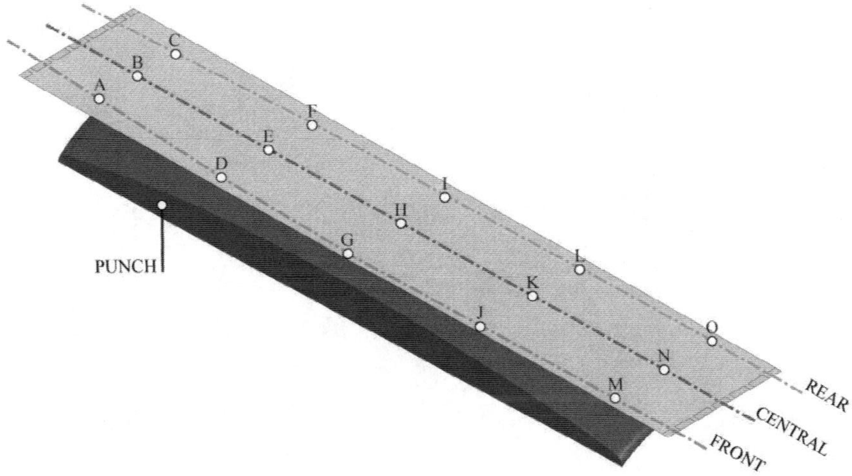

Figure 2. The points we marked on the surface of the sheet blank to identify the results of the thickness values.

The diagram in figure 3 indicates the initial position of the flat sheet when aligning the longitudinal axis of tension relative to the shaping contour of the tightening punch with the rectilinear position of the discrete clamping devices of the press. The longitudinal axis of tension coincides with the texture axis of symmetry of the sheet stock properties.

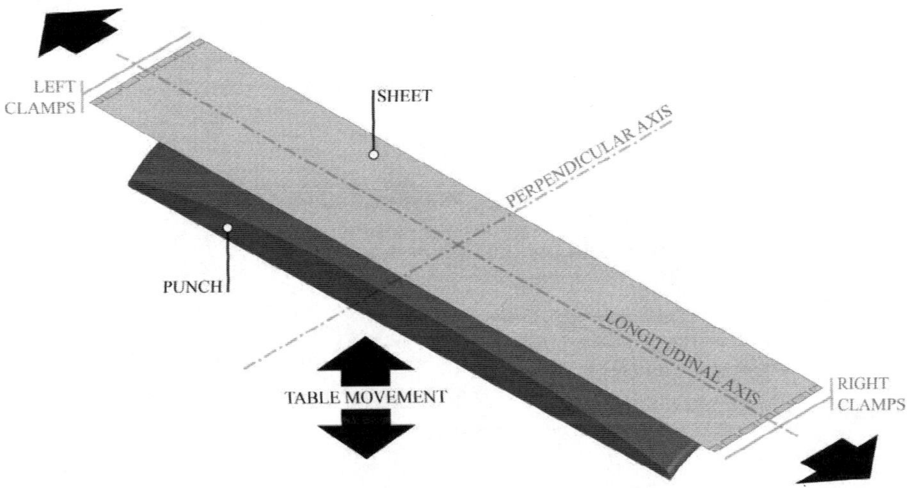

Figure 3. Alignment of the longitudinal axis of tension relative to the shaping contour of the tightening punch in the rectilinear position of the discrete clamping devices of the press.

The «black» arrows highlighted in the figure correspond to the main shaping movements of the stretching press (extension cylinders and table cylinders). The left and right clamps are mounted on clamping plates, which are pre-tilted by means of pivot cylinders, ensuring that the sheet blank is wrapped around the forming contour of the drawing punch when the press table moves upward by a

IOP Conf. Series: Materials Science and Engineering 1181 (2021) 012016 doi:10.1088/1757-899X/1181/1/012016

certain amount. Then stretching cylinders are turned on, which occupy a stationary horizontal position in the left and right base carriages of the RO-630 11 press (figure 4).

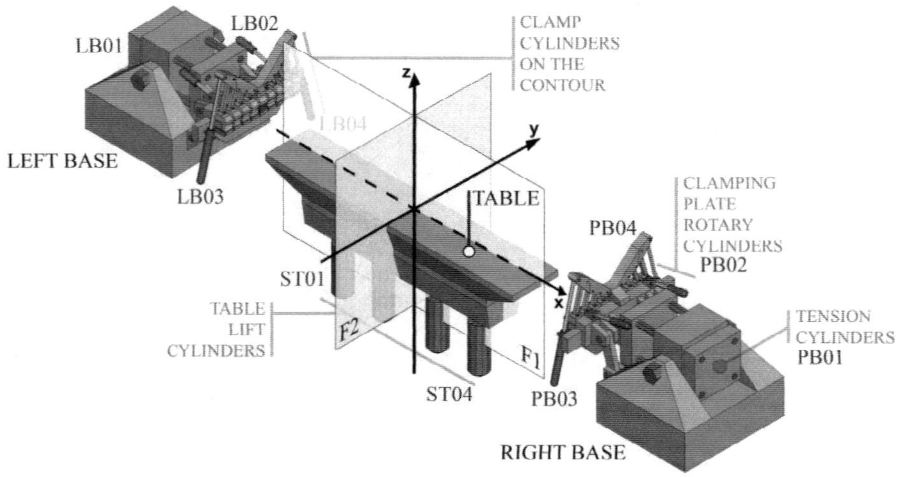

Figure 4. Main working cylinders of stretching and press table, left and right base carriages, left and right rotary clamping plates with installed discrete clamping devices, cylinders for setting the clamps along the contour and cylinders of rotation of the clamping plate of the RO-630 11 press.

Two variants of the method of shaping a wrapped part by means of moving the line of clamps of the sheet blank along a certain trajectory are proposed for discussion, having the same cyclic diagram of step movements of the main and auxiliary working cylinders (figure 5), where the indices of the kinematic state of the press and the conditional time of calculation are shown in circles.

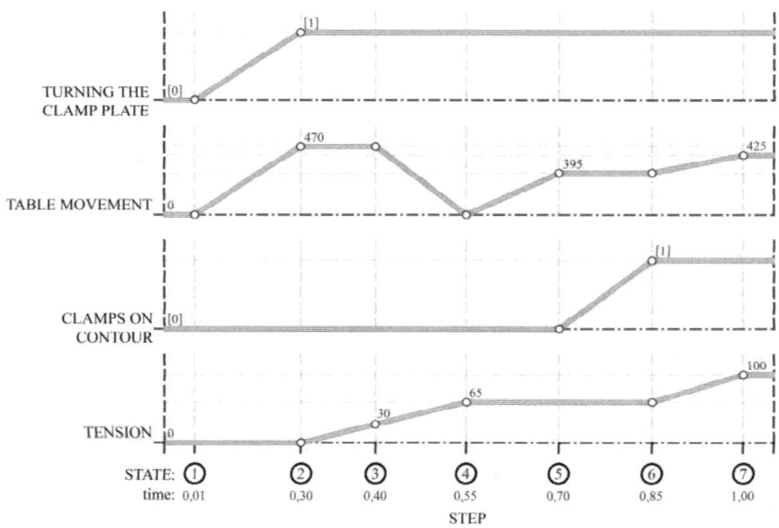

Figure 5. Cyclic diagram of the step movements of the main and auxiliary working cylinders of the press.

The cycle diagram does not show the initial installation of the sheet stock into straight-line discrete clamping devices. Other default settings are present in the cycle diagram and are indicated. For example, for a clamping plate: [0] - the initial position is vertical; [1] - inclined installation position; for a press table with a tightening punch, 0 - zero position of the press table, when the "pole" of the tightening punch has a «tight» contact with a horizontally positioned sheet in clamping devices, for discrete clamps: [0] - initial rectilinear arrangement; [1] - curvilinear installation position and for stretching cylinders: 0 - horizontal position and extended plunger at a distance that ensures the initial installation of the sheet blank into straight-line discrete clamping devices.

The first variant of the method relates to the case when all the clamps are triggered, ensuring full gripping of the left and right edges of the sheet. The second variant of the method refers to the case when, for each base carriage, the central and two adjacent clamps are not turned on, leaving a part of the left and right edges of the sheet free. This variant of the method is able to reduce the difference in thickness values between the central part and the edge of the sheet that occurs during the shaping process. The cyclic diagram of the step movements of the main and auxiliary working cylinders of the press for both variants of the method is the same (figure 6).

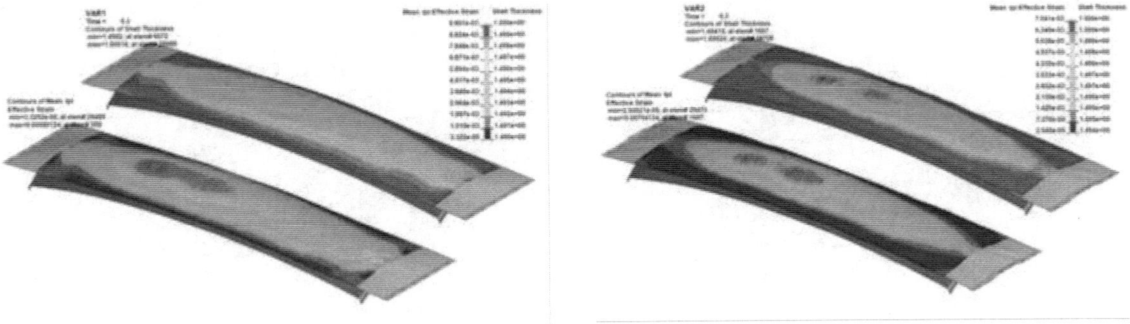

(option 1) STATE 2 (option 2)

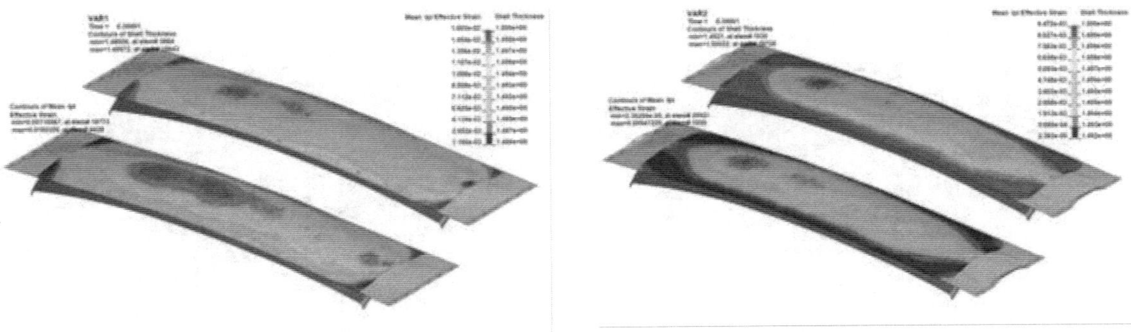

(option 1) STATE 3 (option 2)

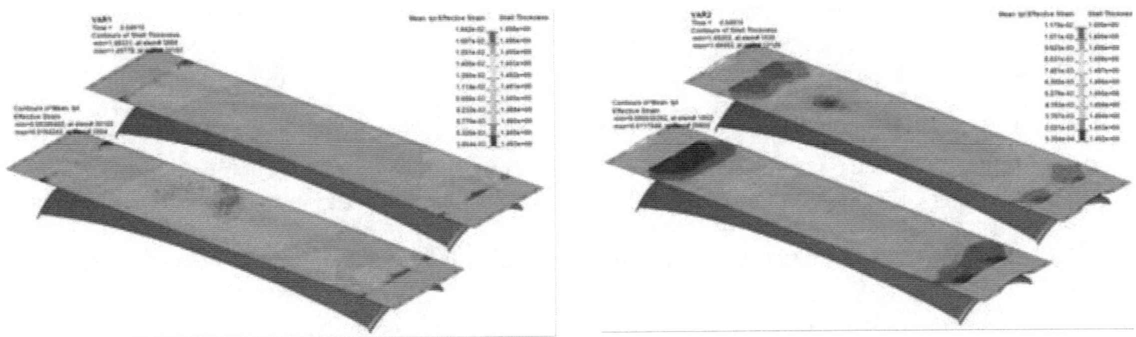

(option 1) STATE 4 (option 2)

IOP Conf. Series: Materials Science and Engineering 1181 (2021) 012016 doi:10.1088/1757-899X/1181/1/012016

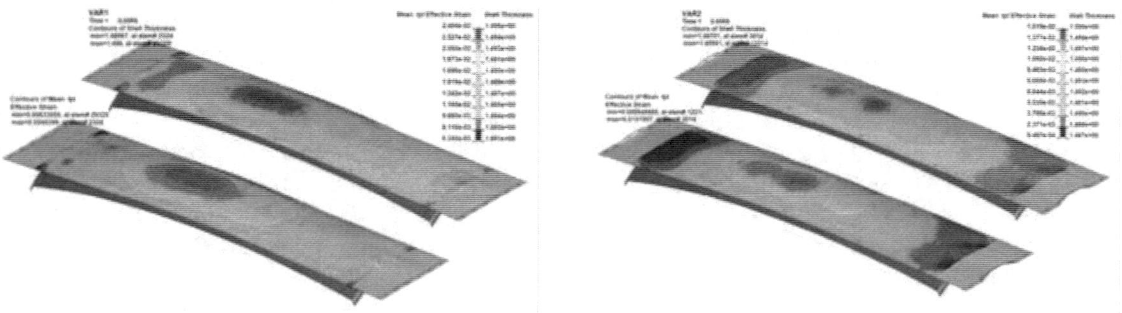

(option 1) STATE 5 (option 2)

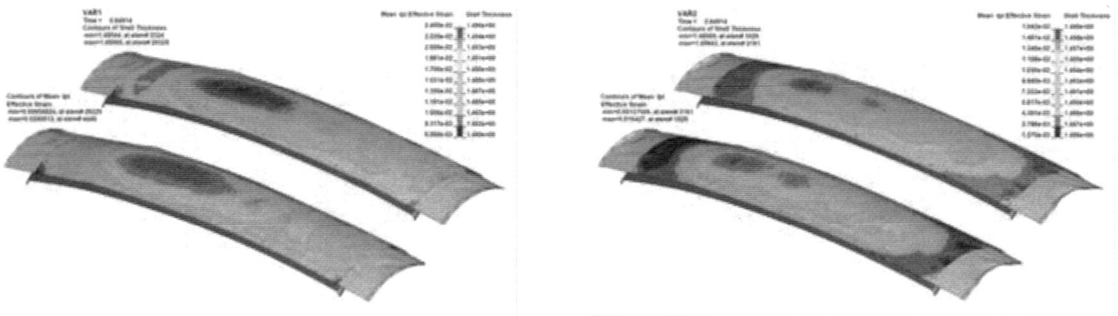

(option 1) STATE 6 (option 2)

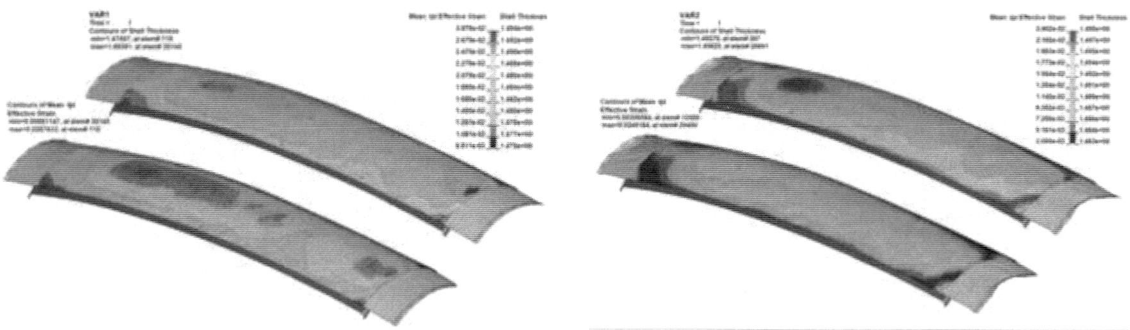

(option 1) STATE 7 (option 2)

Figure 6. The results of visualization of the equivalent deformation and thickness of the sheet blank in the process of shaping by the tight-fitting according to the steps of the press movement for option 1 (left) and option 2 (right) with the corresponding kinematic state of the press.

The first step movement (from state 1 to state 2) - turns the left and right clamping plates into the setting state and the table movement up to a value of 470 mm to ensure the wrapping of the sheet blank along the forming contour of the tightening punch. During wrapping, the pre-stretched blank remained forced flat with discrete clamping devices in a straight line. This state of the sheet blank provides its initial configuration of a sufficiently narrow central region in accordance with the amount of curvature at the pole of the surface of the covering punch.

The second step movement (from state 2 to state 3) - the extension cylinders are switched on at the moment of the end of the rotation of the clamping plate and the movement of the table upwards. Plastic stretching of the wrapped sheet blank along the shaping contour of the tightening punch fixes the longitudinal curvature and the subsequent plastic configuration of the sheet blank, increasing the plastic deformation zone of the central part towards the clamps and practically covering its width, remained forcibly flat with straight-line discrete clamping devices.

The third step movement (from state 3 to state 4) - the table is lowered simultaneously with stretching of the sheet blank due to the continuation of the stretching cylinders to a value of 65 mm. At this step, due to isometric bending of the sheet blank, which leads to an increase in the transverse curvature in its central part, the lateral sections of the sheet blank (frontal and rear lines) are significantly stretched, while the center line is not stretched. As a result of this effect, the transverse curvature is plastically configured and the convergence of the values of the thickness of the sheet blank in the transverse direction is observed.

The fourth step movement (from state 4 to state 5) - the table moves up to a value of 395 mm, the stretching cylinders do not work. The gradient of stretching of the sheet stock along the center line increases, while maintaining the boundaries of the plastic deformation zone without sliding towards the clamps, reducing the likelihood of breaking the sheet blank at one of its free sections between the edge of the punch and the press clamps. The values of the equivalent plastic deformation in the middle part of the sheet blank confirm the obtaining of the geometric shape of the shell with the corresponding curvature.

The fifth step movement (from state 5 to state 6) - the cylinders are turned on to set the left and right discrete clamps along the corresponding circuits to the setting state. The diagram shows a normalized value from 0 to 1, with a total clamping angle of 63 degrees on the left press base and 78 degrees on the right press base. Provides a tight clamping of the corner parts of the sheet blank to the surface of the tightening punch.

Sixth step movement (from state 6 to state 7) - stretching cylinders are switched on to a value of 100 mm and the table rises up to a value of 425 mm. As a result, the sheet blank, which received a full plastic configuration at the previous steps, undergoes plastic stretching of the corner parts pressed at the fifth step and a calibration adjustment of the conformity of the surface of the molded shell to the surface of the tightening press, i.e. completely lay down on the tightening punch.

When simulating both variants of the shaping method by the wrapping, we adhered to the specified sequence of step movements by moving the line of clamps of the sheet blank and preserving the parameters of the state of the equipment, determined by the sequence and values of the displacements of the working bodies of the wrapping press PO-630 11. The choice of two variants of the method was planned to test their effectiveness in order to achieve the main goal in understanding the achievement of the geometric shape and the minimum thickness difference of the double curvature shell.

When the RO-630 11 press is in operation, the displacements of the plungers of the left and right stretching cylinder must be monitored. When shaping the wrapping of the shells that are symmetrical with respect to the «pole» of the wrapping punch, the movements of the plungers must be the same so that the "pole" remains stationary. In our case, the lengths of the left and right sections of the shaping contour of the tightening punch relative to its "pole" are not the same. Therefore, the movement of one of the plungers is ahead of the movement of the other plunger by a certain amount. In this case, the movements are synchronized so that in the process of stretching the «pole»on the workpiece remains necessarily motionless.

In addition, when stretching the sheet blank using a horizontally located tension cylinder, the movement of its plunger with an inclined clamping plate in the installed state changes the angle of descent of the sheet blank relative to the tangent to the trailing edges of the tightening punch. To restore the position of the angle of departure of the sheet blank relative to the tangent, it is necessary to re-enable the table lift.

If the condition of synchronous operation of the left and right stretching cylinders to ensure the position of the «pole» stationary during modeling can be fulfilled, then to restore the position of the angle of departure of the workpiece relative to the tangent, we can implement the simulation of the process directly on the virtual press after installing the tightening punch on its table.

To analyze the results of the equivalent deformation and thickness in the process of shaping by a skinny, obtained in the process of modeling the first (left) and second (right) variants of the method, we will place their pictures in an amount corresponding to the kinematic states of the press indicated on the cycle diagram, and the graphs of changes in thickness values in selected points of the sheet surface in figure 7.

(option 1)

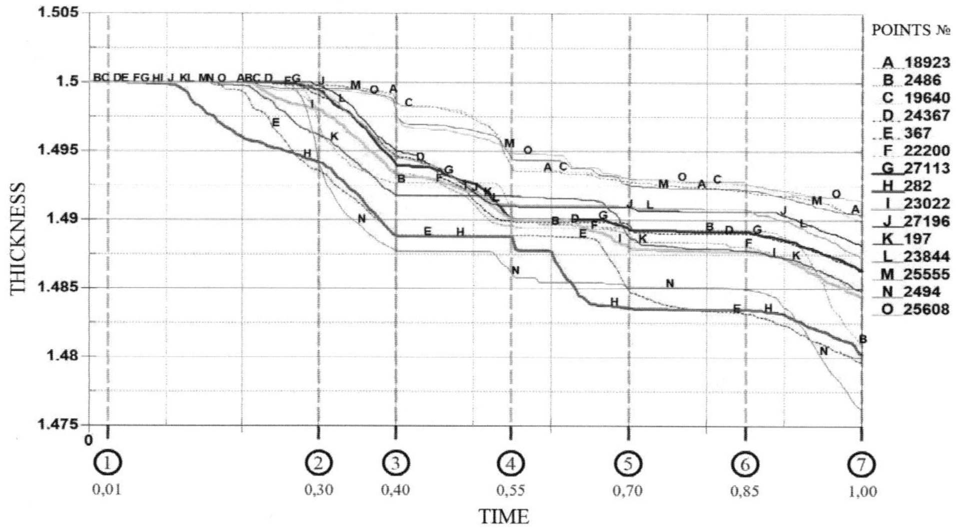

(option 2)

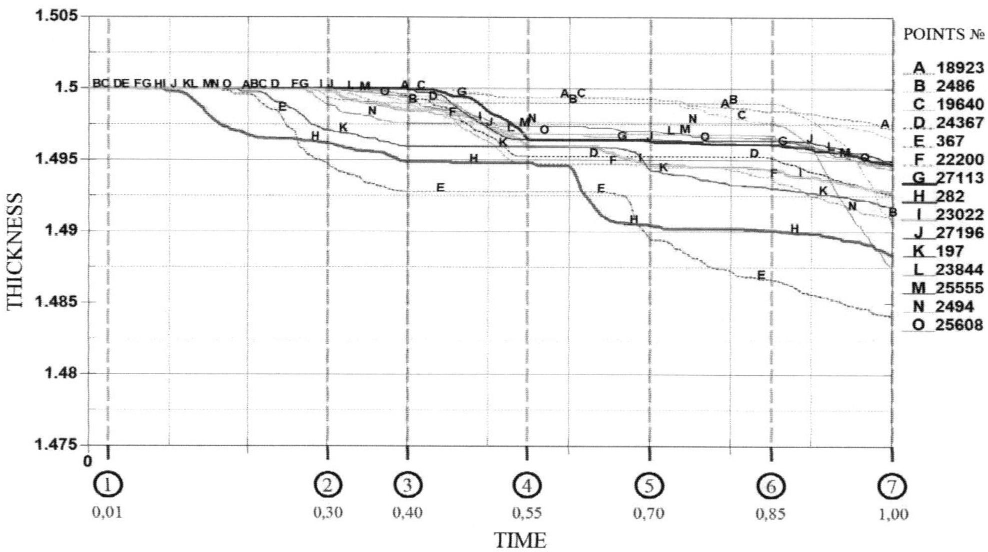

Figure 7. Graphs of the distribution of the thickness of the sheet blank in the process of shaping by a tight-fitting for option 1 of the method (upper) and option 2 of the method (lower).

Let us transfer all visual attention to the graphs of the thickness distribution in the process of shaping by the skin, obtained in the process of modeling the first and second variants of the method (figure 7). According to figure 2, which shows the center line (red, including points B, E, H, K, N), the front line (blue, including points A, D, G, J, M) and the back line (green, including points C, F, I, L, O). If in figure 7 the blue, red and green lines are bold, then these are points G, H and I. If the blue, red and green lines are of average thickness, then these are points D, E, F, J, K and L. If blue, red and the green lines are thin, drawn intermittently, then these are points A, B, C, M, N and O.

If you look closely at the graphs of the thickness distribution over the area of the finished part, where the division value between the digits of the ordinate is 0.005 mm, then the absolute difference in thickness of the part of the first version of the method is 0.016 mm and, accordingly, the second version is 0.013 mm, which is 18.5% better than the first option. This is due to the fact that deformations at the points on the central line B and N located at the edge of the end of the tightening punch in the first variant of the method could be localized, which increases the probability of sheet breaking. In the second variant of the method, plastic deformations are more stable, their greatest values are located in the middle part of the sheet blank. However, the method of implementing partial gripping of the edge part of the sheet blank, only by the extreme gripping devices, has never been used in practice.

The manifestation of the effect of stretching the outermost fibers of the workpiece (blue and green lines) indicates the presence of the effect of isometric bending of the shell surface after unloading the sheet blank at the third step motion, when the table was lowered, and the sheet continues to stretch up to 65 mm. The manifestation of the effect, when the sheet blank remained forcibly flat with rectilinearly arranged discrete clamping devices, provided a plastic configuration of the middle part of the sheet blank without localizing the thinning deformation and equalizing their values in different areas of the sheet.

3. Summary

In conclusion, we note that high requirements are imposed on the skins, especially in the quality of the surface of the skin parts, in the accuracy and shape of their external contours, obtained after shaping operations. At the same time, high requirements are imposed on the accuracy of the dimensions of the skins themselves and on the spread of the wall thickness of the parts. In addition, the designation of processing modes or the number of transitions during shaping by wrapping on the basis of reference data, even for well-known alloys, led to the appearance of a large number of rejects, which It was found that due to the significant non-uniformity of deformation when covering complex shells, the likelihood of defects associated with exceeding the permissible thinning of the sheet blank and the permissible grain size of the metal increased.

This led to an unjustified increase in the number of transitions due to unacceptable thinning and critical deformation of the recrystallization of the sheet material. "Critical" recrystallization is the main reason for the appearance of a coarse crystalline structure. This process is accompanied by intense grain growth as a result of heat treatment after cold deformation with critical degrees. For the annealed alloy D16, according to the reference data, the critical deformations are in the range of 10...12%, and for the material of the cladding layer of annealed cold-rolled aluminum AD0 - 4...6%. The number of transitions during the wrapping of the clad sheet material was determined from the last values.

The efforts of our research were aimed at the development of methods of shaping with a wrap, which reduce the degree of uneven deformation. It was necessary to minimize the thickness difference to the recommended limits, but for this a new technology was created to increase the degree of deformation during tightening, which is achieved practically in one transition without localizing the thinning deformation and equalizing their values in different areas of the sheet blank.

As a result of the studies performed, the conditions for symmetric wrapping were created and the trajectories of movement of the clamps of the sheet blank were modeled during the shaping of the blanket shells close in geometric shape to the skin, with a uniform change in thickness in different areas of the sheet blank without folds and breaks. This is only possible with programmed tying presses. For the transition to the level of adaptive program control, we propose to control the degree of deformation at characteristic points of the shell surface of double curvature due to the contactless measurement of deformations by the method of correlation of digital images.

The modern level of information technology makes it possible to create a digital twin of the press. You can create simulation models, promote them, create software that automates the process of building simulation models, i.e. develop this side of modeling.

References

[1] Bratukhin A G and Ivanov Yu L 1999 *Modern technologies of aircraft construction* (Moscow: Mechanical Engineering) pp 49-63
[2] Mikheev V A, Klochkov Yu S, Kuzina A A, Grechnikova A F and Savin D V 2012 The choice of the kinematic scheme of shaping by covering bypass-forming shells of complex spatial shape *Vestnik of the Samara State Aerospace University* **5(36)** 239-45
[3] Mikheev V A, Grechnikov F V, Samokhvalov V P, Savin D V, Dementyev S G and Surudin S V 2014 Investigation of the kinematic diagram of sequential wrapping of biconvex shells on the FEKD wrapping press *Vestnik of the Samara State Aerospace University* **16** 180-6
[4] Mikheev V A and Surudin S V 2017 Fundamentals of the calculation of the process of forming tight thin shells of double curvature *Izvestia of Samara Scientific Center of the Russian Academy of Sciences* **19(1-3)** 555-62
[5] Mikheev V A, Dement'yev S G, Samokhvalov V P, Savin D V and Surudin S V 2013 Isometric conditions for the formation of a tight shell of double curvature with minimal thickness difference *Izvestia of Samara Scientific Center of the Russian Academy of Sciences* **15(6-1)** 161-6
[6] Grechnikov F V 1998 *Deformation of anisotropic materials (intensification reserves)* (Moscow: Mashinostroenie)
[7] Grechnikov F V 2004 Optimal deformations during the shaping of double curvature shells with a covering *Blank production in mechanical engineering* (Moscow: Mechanical Engineering) pp 23-7

Structure and properties of elastomers with hard segments and crystallizing soft segments suitable for the Arctic and the Far North

Alexei Slobodinyuk[1]**, Dmitriy Kiselkov and Daria Slobodinyuk**

Institute of Technical Chemistry, Branch of the Perm Federal Research Center, Ural Branch, Russian Academy of Sciences, 614013, Perm, Russia

[1]E-mail: slobodinyuk.aleksey.ktn@mail.ru

Abstract. The physicomechanical and structural characteristics of several elastomers based on terminal epoxy oligomers synthesized from oligotetramethylene oxide diol, various diisocyanates and oligotetramethylene oxide diamine were studied. Additionally, two model compositions of urethane containing elastomers with hard segments were synthesized to reveal the peculiarities of the organization of the supramolecular structure. It was found that the use of 2,4-toluene diisocyanate allows to obtain elastomers with a higher degree of microphase separation and, accordingly, a lower glass transition temperature. It was shown that regardless of the type of diisocyanate used in the synthesis, the elastomer contains crystalline phase. New materials can be used as compounds, adhesives and sealants for various purposes with the possibility of using them under extreme conditions of the Arctic and the Far North.

1. Introduction

Elastic materials with urethane hydroxyl hard segments, synthesized from epoxyurethane oligomers, are used in various fields of technology: in sealing of electrical products, industrial equipment for load-bearing and enclosing structures of buildings, as inserts to reduce vibration from ventilation systems, for bonding items suitable for high mechanical, shock and vibration loads [1-6].

The most famous synthesis of epoxyurethane oligomers is two-stage, with the formation of an intermediate product - oligodiisocyanate. At the second stage, the oligodiisocyanate is chemically modified by a compound with functional epoxy and hydroxyl groups - glycidol. Compounds with terminal amine, carboxyl, anhydride groups are used as hardeners for epoxyurethane oligomers [7-8].

Elastomers from epoxyurethane oligomers consist of alternating soft and hard segments, the difference in polarity of which causes phase separation with the formation of a separate phase – hard segment domains that play the role of a reinforcing filler and nodes of a specific physical network [9]. Hydrogen bonds stabilize the domain structure.

The use of various polyesters in the synthesis of oligodiisocyanates makes it possible to regulate the structure of the soft segment. The rigid segment structure allows the selection of isocyanate, epoxy alcohol and chain extender used in the synthesis of the polymer to be regulated.

For reliable operation in the Arctic and the Far North, materials are required that retain elastic properties at extreme cold temperatures. Such materials include compositions based on oligodientetraurethane epoxides [10], the glass transition temperature of which is -78 °C. Moreover, this type of materials is characterized by a sufficiently low strength - no more than 5 MPa and consequently low adhesive strength [10].

In work [11] a new method to increase the strength characteristics was introduces. It was proposed to use oligotetramethylene oxide tetraurethane diepoxide as an epoxyurethane oligomer, when cured by low molecular weight hardeners (isophorone diamine, aminoethylpiperazine). It is proposed to use an oligomeric linear hardener with a molecular weight of more than 1000 amu. in order to create conditions for the crystallization of the polymer.

2. Materials and methods

The epoxyurethane oligomer was synthesized using two step method through an intermediate product - oligodiisocyanate. Oligodiisocyanate was synthesized from polyether - oligotetramethylene oxide diol (polyfurite) with a mass fraction of hydroxyl groups of 1.69 % and 2 diisocyanates: 2,4-toluene diisocyanate (TDI) or isophorone diisocyanate (figure 1). The molar ratio between the NCO groups of the diisocyanate and the OH groups of the polyether was 2.05.

Synthesis epoxyurethane oligomers produced according to the method [12] of the scheme:

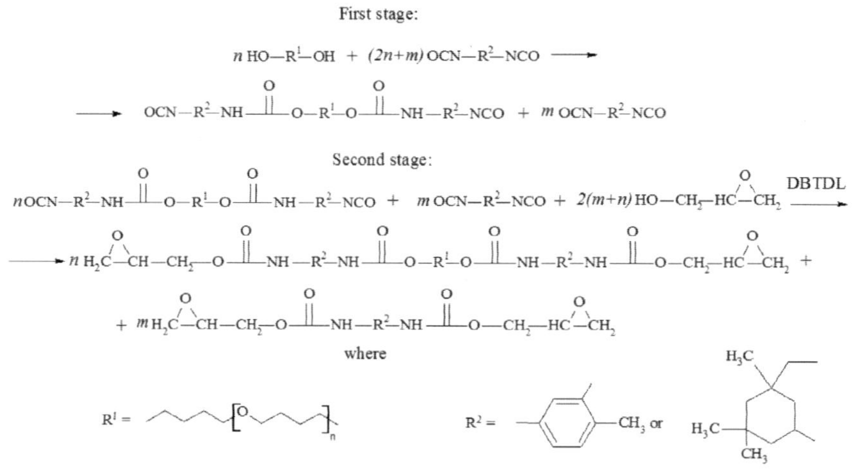

Figure 1. Synthesis of epoxyurethane oligomers.

The above oligomers were cured oligotetramethylene oxide diamine (OTMODA), with a mass fraction of amine groups of 3.18 wt% was used as a hardener.

. The completeness of the conversion of epoxy groups was determined by FTIR spectroscopy by the disappearance of the absorption band at 910 cm^{-1} [13].

The glass transition temperature Tg of the cured samples was determined by differential scanning calorimetry (DSC) using METTLER TOLEDO DSC 822e calorimeter at a scan rate of 5 °C/min.

The FTIR spectra of the cured samples were recorded on a Bruker Vertex 80V FTIR spectrometer with 64 scans at 2 cm^{-1} resolution. For convenience of comparison, the spectral curves were normalized to the band at 1600 cm^{-1}, that corresponds to the stretching vibrations of the C - C aryl ring.

3. Results

Table 1 shows the theoretical and experimentally determined values of free isocyanate groups in the oligodiisocyanate and free epoxy groups in the epoxyurethane oligomer.

Table 1. Characteristics of epoxyurethane oligomers and intermediate synthesis products.

Used ether	Used	Content of free functional groups in oligomer, %

	diisocyanate	isocyanateinoligodiisocyanate		epoxyinepoxyurethaneprepolimer	
		calculated	Measured	calculated	measured
Polyfurite 2000	TDI	3.55	3.62	3.48	3.32
Polyfurite 2000	IPDI	3.41	3.58	3.45	3.41

4. Discussions

An important feature of urethane-containing elastomers is the microphase separation factor due to the difference between hard and soft segments with the formation of nanodispersed filler - domains of hard segments, which plays the role of a reinforcing filler. With an increase in the content of such a filler, the deformation-strength characteristics and wear resistance of the cured polymers increase. The degree of microphase separation of soft and hard segments in elastomers based on epoxyurethane oligomers is lower than in polyurethanes / polyurethane ureas. The principal molecular structure of elastomers based on epoxyurethane oligomers is shown in figure 2.

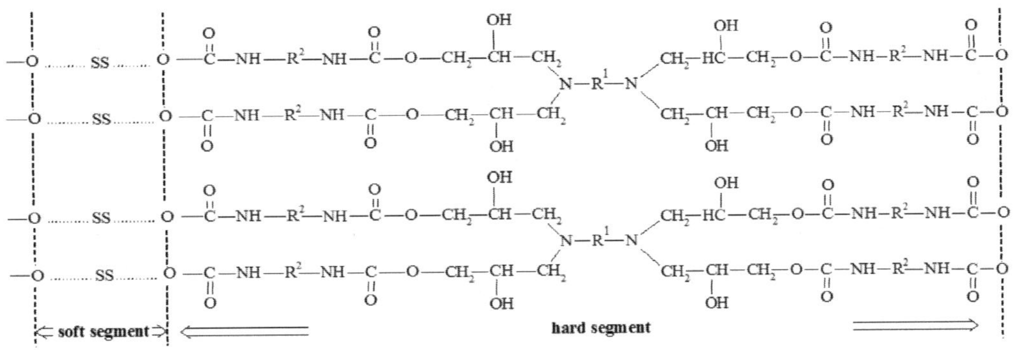

Figure 2. Molecular structure of elastomers from epoxyurethane oligomers.

Meanwhile the use of a low-molecular-weight chain extender (R1 in figure 2) causes a high segmental mobility of the hard segment. When an oligomeric diamine is used, the molecular weight of the hard segment increases and thus steric hindrances appear for the phase organization of the segment structure. On the other hand, the linear structure of the diamine should favor the appearance of the crystalline phase of the polymer. Thus, two opposite factors are superimposed on the value of the strength characteristics: a decrease in the degree of microphase separation should lower the strength value, and the appearance of a crystalline phase should in contrast increase the strength value.

Different methods are used to determine the degree of microphase separation, including FTIR spectroscopy. The degree of microphase separation in polyurethane-ureas synthesized from 2,4-toluene diisocyanate and 3,3'-dichloro-4,4'-diaminodiphenylmethane is determined by the absorption band at 1640 cm^{-1}. Polyurethanes synthesized from 1,4-butanediol and diphenylmethane diisocyanate is determined by the absorption band at 1702 cm^{-1}. Some shift of absorption band depends on the type of initial component that build the hard segments. For polyurethane-ureas, the absorption band of self-associates of had segments made of isophorone diisocyanate shifts by 20 cm^{-1} up to 1660 cm^{-1}.

To analyze the FTIR spectra of the obtained samples, model compounds DA-1 and DA-2 were additionally prepared that contained two types of urethane hydroxyl segments based on TDI, glycidol and OTMODA, as well as IPDI, glycidol and OTMODA, respectively. According to the synthesis method [14] at the first step the ratio of amine to isocyanate was taken 2 to 1, and at the second stage the synthesized compound reacted with a twofold excess of glycidol.

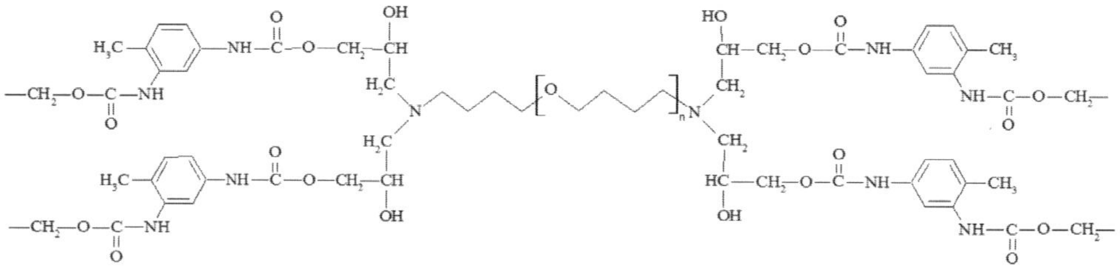

Figure 3. ModelDA-1.

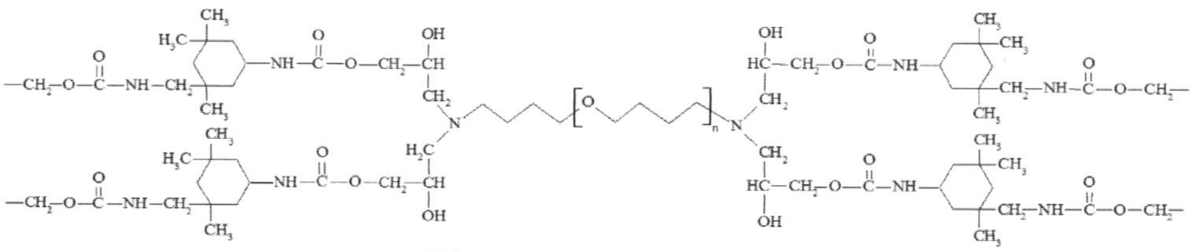

Figure 4. Model DA-2.

When using low molecular weight chain extenders in the synthesis of models for determining the wave number by FTIR spectroscopy of hydrogen bonds between NH and C = O groups of hard segments and thus phase separation in the elastomer, there is one band in the region of carbonyl stretching vibrations absorption (when 2,4-toluene diisocyanate is used, two more bands appear at 1600 cm^{-1}, 1612 cm^{-1}). In this study, the chain extender is an oligomeric diamine. Therefore, when analyzing the FTIR spectra of the synthesized model compounds, it is necessary to take into account the fact that due to the polyester component of the diamine used in the synthesis, the FTIR spectrum in the region of carbonyl stretching vibrations should show several absorption bands characterizing the hydrogen bonds between the oxygen group of the PTMO polyester and the NH group 1695 - 1700 cm^{-1}, as well as free carbonyl groups 1730-1732 cm^{-1} in the case of 2,4-toluene diisocyanate, 1721 cm^{-1} in the case of isophorone diisocyanate. The types of hydrogen bonds in the synthesized elastomers are schematically shown in figure 5.

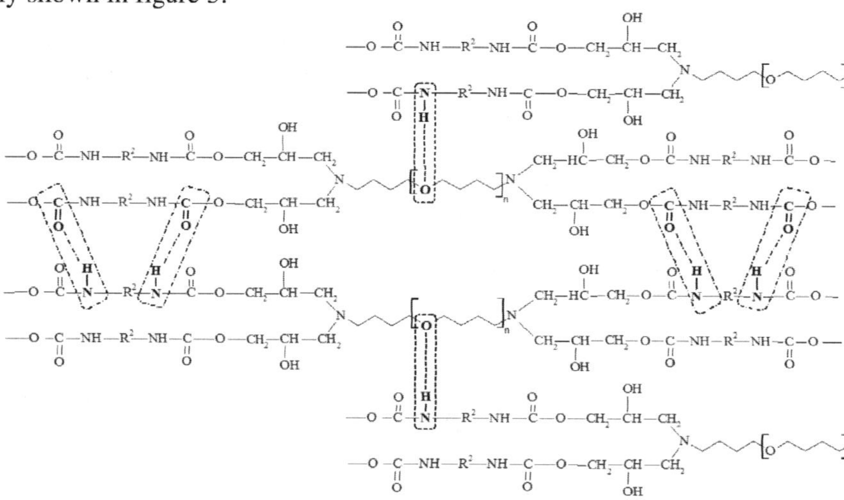

Figure 5. Types of hydrogen bonds in synthesized elastomers.

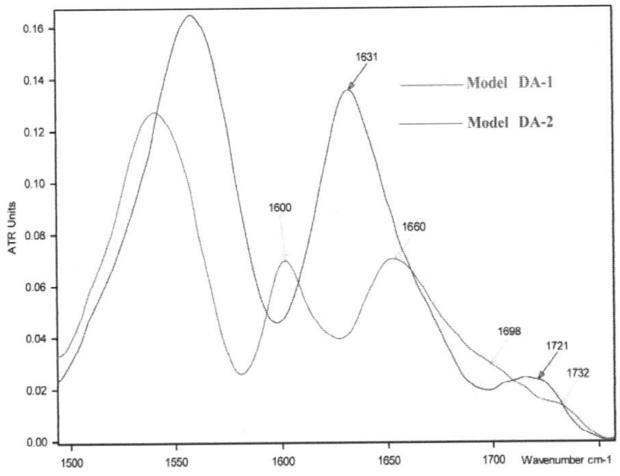

Figure 6. FTIR spectra of synthesized model compositions.

In FTIR spectra of the model synthesized from 2,4-toluene diisocyanate absorption bands appear: at 1600 cm^{-1}, characteristic of the stretching vibrations of the C - C aryl ring, 1732 cm^{-1}, as well as a peak at 1660 cm^{-1} characterizing phase separation (figure 6). For the model from isophorone diisocyanate, two absorption bands appear - at 1721 cm^{-1}, and at 1631 cm^{-1} characterizing phase separation. It should be noted that the FTIR spectrum of model DA-2 lacks an absorption band at 1698-1700 cm^{-1}, which is associated with steric hindrances of hydrogen bonding due to the bulky structure of isophorone diisocyanate.

Analysis of the FTIR spectra of the synthesized elastomers shows that microphase separation appears only on the sample synthesized from 2,4-toluene diisocyanate - an absorption band at 1660 cm^{-1} (figure 7).

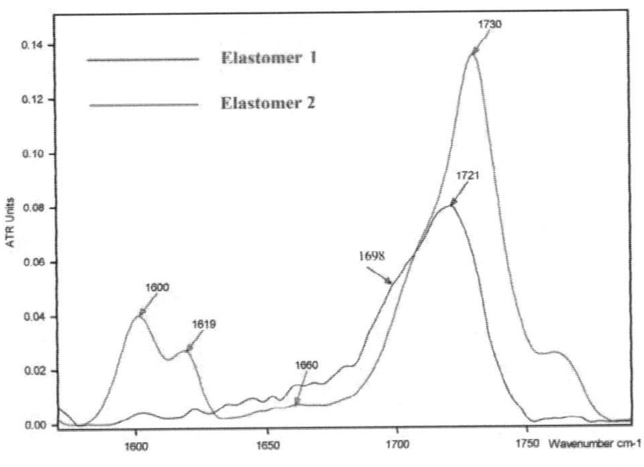

Figure 7. FTIR spectra of synthesized elastomers.

The absorption band at 1612 cm^{-1} attributed to the vibrations of the benzene ring with some substituents is typical for urethane-containing elastomers synthesized from TDI [15]. For the elastomer synthesized from isophorone diisocyanate there are only two absorption bands at 1721 cm^{-1} and 1698 cm^{-1} which can be attributed to interchain interaction (figure 7).

Figure 8 shows DSC curve of synthesized elastomers. The elastomer from isophorone diisocyanate has a higher glass transition temperature due to absence of microphase separation and the presence of interchain interaction. Both elastomers contain crystalline phase.

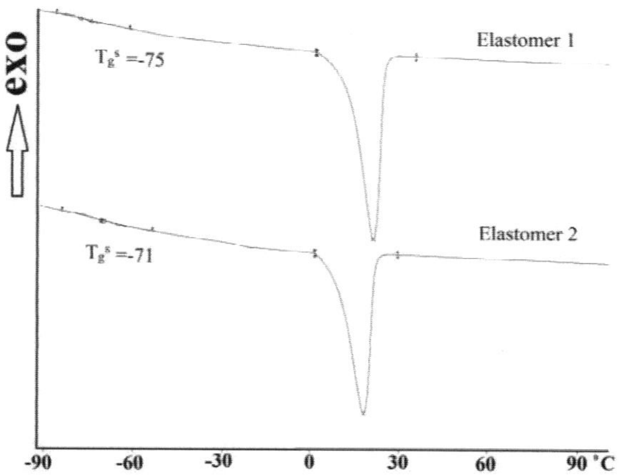

Figure 8. DSC curve of synthesized elastomers.

5. Conclusions

Two epoxyurethane elastomers from oligotetramethyldeneoxidiol with a molecular weight of 2000, 2 types of diisocyanate, glycidol, and an aliphatic diamine - oligotetramethylene oxide diamine were synthesized. It was found that, regardless of the type of diisocyanate used in the synthesis of the epoxyurethane oligomer, elastomers contain a crystalline phase. It has been proven that the use of 2,4-toluene diisocyanate, in contrast to isophorone diisocyanate, promotes phase organization with the formation of a separate hard segments, which leads to a lower glass transition temperature of the elastomer which makes it possible to use these elastomers in conditions of the Arctic and the Far North.

Acknowledgements

The reported study was funded by RFBR and Perm Territory, project number № 20-43-596010. The study was performed using the equipment of the Center for Shared Use "Studies of Materials and Substance," Perm Federal Research Center, Ural Branch, Russian Academy of Sciences with a support of the state task AAAA-A18–118030790037-7.

References

[1] Chen D S, Ma C C M, Hsia H C, Wang W N and Lin S R 1994 *J. Appl. Polym. Sci.* **51(7)** 1199-206
[2] Hsia H C, Ma C C M, Li M S, Li Y S and Chen D S 1994 *J. Appl. Polym. Sci.* **52(8)** 1137-51
[3] Yeganeh H, Mehdipour-Ataei S and Ghaffari M 2008 *High Perform. Polym.* **20(2)** 126-45
[4] Pavlova M, Draganova M and Kabaivanov V 1988 *Polym. Commun.* **29(3)** 80-2
[5] Chuvilina L F2007Patent 2291176 RU
[6] Sidorov O I, Milekhin Yu M, Matveev A A, Poisova T P, Bykova K A and Sadchikov N V 2012*Klei. Germet. Tekhnol.* **9** 15-21
[7] Chen W H, Chen P C, Wang S C, Yeh J T, Huang C Y and Chen K N 2009 *J. Polym. Res.* **16** 601-30
[8] Madhavan K and Reddy B S R 2006*J. Polym. Scie.Part A: Polym. Chem.***44(9)** 2980-9
[9] Hsia H C, Ma C C M and Chen D S1994 *Macromol. Mater. Eng.* **220(1)** 133-149
[10] Fedoseev M S, Tereshatov V V and Derzhavinskaya L F 2010 *Russ. J. Appl. Chem.* **83(8)** 1367-71
[11] Senichev V Y, Slobodinyuk, A I, SlobodinyukD G, Savchuk A V, KulakovaM V, Oshchepkova T E and Dolinskaya R M2020 *Russ. J. Appl. Chem.***93(8)** 1172-8
[12] Slobodinyuk A, Strelnikov V, Kiselkov D and Slobodinyuk D 2021 *Zeitschrift für Naturforschung B.* 1-5

[13] Socrates G 2001 *Infrared and Raman characteristic group frequencies: tables and charts*(ChichesterNew York : John Wiley & Sons) 349
[14] Zhang S, Ren Z, He S, Zhu Y and Zhu C 2007 *Spectrochim. Acta, Part A***66**(**1**)188-93
[15] Kothandaraman H and Nasar A S 1993 *Polymer***34**(**3**)610-5

Stress-strain condition of flat manhole covers under internal pressure in presence of mechanical defect with different dislocation on surface

A V Rubtsov[1], V V Gracheva, P A Kulakov and Z R Mukhametzyanov

Ufa State Petroleum Technological University, Ufa, 450000, Russia

E-mail: sunset202@mail.ru

Abstract. One of the most common damage to the structural elements of vessels and apparatus of hydrocarbon processing plants is mechanical damage, which mainly occurs during repair and installation operations. One of the prerequisites for the occurrence of this type of defects is the use of a locksmith tool and other devices, tools and devices, upon contact of which with the surface of the elements to be repaired and mounted, risks, scratches, balls, nicks and other damage are formed. The vast majority in terms of the degree of occurrence on the surface of structural elements of all defects of a mechanical nature occupy scratches. There is practically no technical device operating in a hazardous production facility and does not have at least one of the elements of this type on the external surface. With respect to the surface on which the scratches are arranged, these defects can be both small geometrical dimensions and sufficiently extended. Such defects generally do not limit the life cycle of the object, but can have a significant effect on the redistribution of the stress-strain state in the element. The redistribution of the stress-strain state can serve as an initial stage for the generation of potential fracture sites from the concentration of increased stresses in these zones. Also, not only the presence of a scratch, but also the length, as well as the nature of the location on the surface of the element, can affect the stressed-deformed state of the element. Therefore, the actual work is a study of the change in the stress-deformed state of the flat manhole cover with a scratch, the arrangement of which on the surface has various options. In this work, the stress-strain state of the flat manhole cover is simulated depending on the change in the angle of inclination and the location of the scratch on it relative to the central zone.

1. Introduction

The stress-strain state (SDS) of vessels and apparatus of the oil and gas industry depends on many factors. These include design features, operating conditions, various external factors, etc [1-5]. A special place among these factors is occupied by the presence and level of defects in the material of structural elements, which is a determining parameter in predicting the further service life and assessing the actual technical condition. A permanent tool for detecting defects is non-destructive testing, which allows you to detect defects and evaluate the standards for the admissibility of detected defects in accordance with regulatory and technical documents in this area. These documents define strict criteria for "admissibility" or "inadmissibility" of defects of structural elements of vessels and devices operating under excessive internal pressure. For example, cracks and fistula that occur during welding and during operation are considered unacceptable defects. Such defects unambiguously put the object in an inoperable state, before they are eliminated. However, there are damages, the presence of which does not imply the transition of the object to the limit state, and there are no risks of the development of such defects during operation. Such defects include scratches that occur during repair or installation cycles. The presence of such damage does not endanger the transition of the object to an inoperable state, but can make significant adjustments to the formation of SDS of the elements on which this damage is located. The use of various application programs for modeling 3D objects with existing damage allows you to evaluate SDS and develop certain recommendations for ensuring reliability [6-8]. Therefore, the actual work is the assessment of SDS of the flat manhole cover depending on the location of the scratch on it from the impact of loads established by the technological regulations.

2. Methodology of the research

To examine the SDS of the flat manhole cover of the apparatus, if there is a scratch on it with different dislocation, the manhole of the apparatus of one of the process units is selected. The device is operated at operating pressure P = 2.0 MPa and operating temperature t = 20 °C. The material of the manhole and its cover is high-quality carbon steel 10. The hatch has the following technical characteristics: size of a branch pipe of the union 472×18 mm; flange of connector is flat welded with thickness s = 18 mm; thickness of flat cover s = 10 mm; outer diameter of flat cover D = 510 mm; cover is equipped with 20 holes ∅ 18 mm for bolted connections.In the process of modeling, the licensed software complex "KOMPAS-3D" was used with the APM FEM system integrated into it for solving engineering and research problems [9-11]. At the first stage, a hatch-eye model was built in the COMPAS-3D program. Then, using the APM FEM strength analysis system, which is part of the COMPASS-3D program package, loads such as: pressure and temperature were applied; fasteners are installed [12-15]. Further, a scratch with different dislocation on the surface of the flat cover with respect to its central zone was simulated. The scratch is an extended linear plane defect with sharp vertices. The length of the scratch was L = 50 mm, width B = 1 mm, depth H = 1 mm. The shape of the scratch with the geometric dimensions is shown in figure 1. The angles of inclination were taken as follows: 15°, 30°, 45°, 60° and 75°. The various options for dislocating the scratch on the flat cover of the manhole relative to the central part are sketched in the figure 2.V the first of the considered options, one sharp tip of the scratch comes from the center of the flat cover (figure 2a). In the second version, the center of the flat cover divides the scratch into two equal halves (figure 2b).

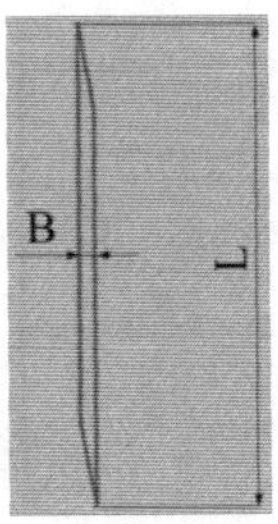

Figure 1. Scratch shape with geometric dimensions.

3. Research results and discussion

At the first stage of research, a SDS assessment of a flat cover without scratching was carried out. Results of calculation of SDS of flat manhole cover without scratch are shown in figure 3.

Analyzing the result shown in figure 3, we can conclude that the maximum stresses are 76.59 MPa and are concentrated in the area close to the bolted circle of the flat manhole cover. Moreover, the distribution zone of maximum stresses is uneven.

Next, the SDS simulation of the flat cover with different dislocation on the surface of the flat cover of the manhole-eye with respect to the center of the cover was carried out according to figures 2a-b. Results of SDS calculation are given in figures 4-8.

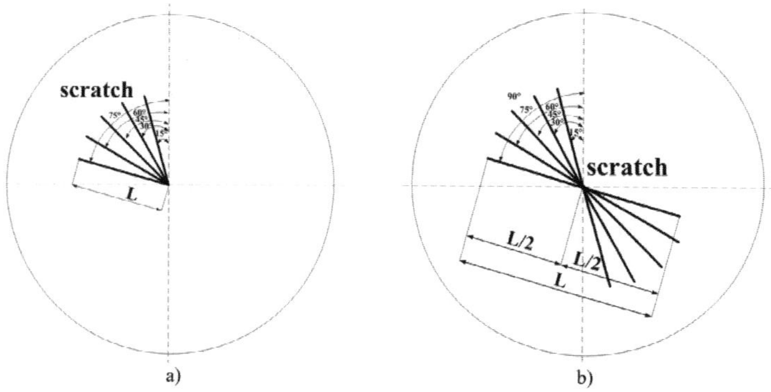

Figure 2. Various options for dislocating the scratch on the flat manhole cover relative to the center.

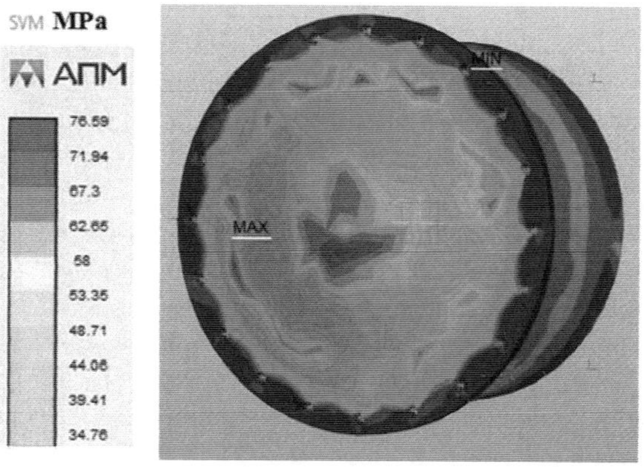

Figure 3. Results of calculation of SDS of flat manhole cover without scratch.

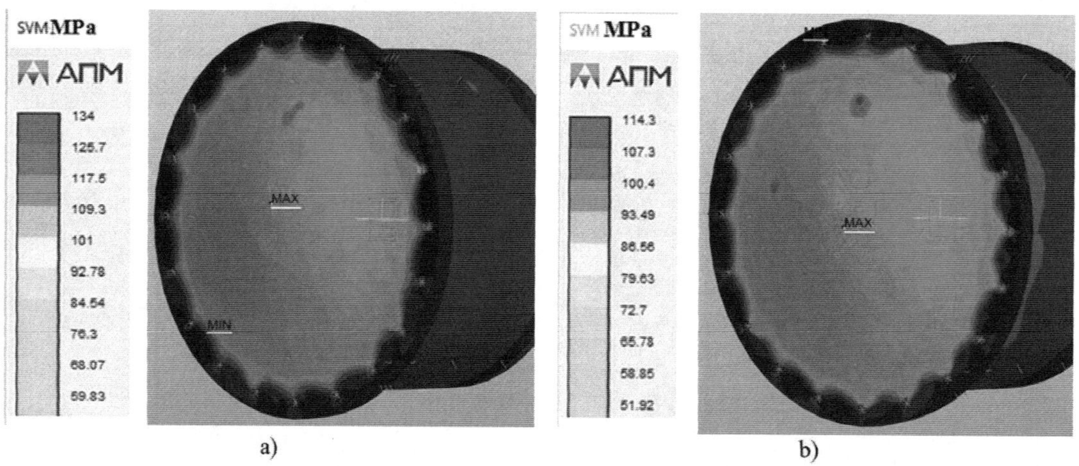

Figure 4. Results of calculation of SDS of flat manhole cover in case of scratch on the surface with inclination angle 15°. a) results of calculation of SDS of flat manhole cover with scratch dislocation relative to the cover center in accordance with figure 2a; b) results of calculation of SDS of flat manhole cover with scratch dislocation in relation to the cover center in accordance with figure 2a.

From figure 4, it can be seen that not only the presence of a scratch, but also its dislocation has an effect on the SDS of the flat manhole cover. Thus, the maximum stresses are concentrated in the scratch area and are 134 MPa in the embodiment where one sharp tip of the scratch comes from the center of the flat cover and the maximum stresses are 114.3 MPa in the embodiment where the center of the flat cover divides the scratch into two equal halves. In the first version, maximum active voltages are 1.2 times higher than in the second. Compared to the maximum stresses of the flat cover without scratching, the increase in stresses occurred by 1.7 and 1.5 times, respectively.

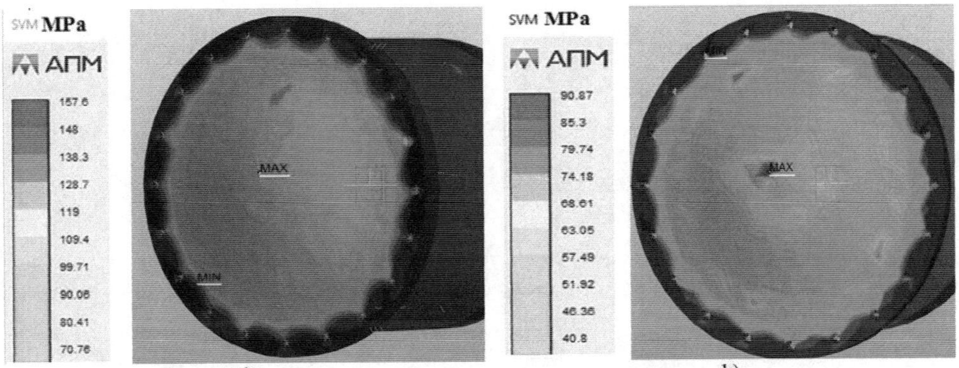

a) b)

Figure 5. Results of calculation of SDS of flat manhole cover in case of scratch on the surface with inclination angle 30°. a) results of calculation of SDS of flat manhole cover with scratch dislocation relative to the cover center in accordance with figure 2a; b) results of calculation of SDS of flat manhole cover with scratch dislocation in relation to the cover center in accordance with figure 2a.

By analyzing figure 5, we can conclude that when the angle of the scratch inclination increases to 30° maximum, the stress is already 157.6 MPa in the case where one sharp tip of the scratch comes from the center of the flat cover. This is 1.2 times higher than with the identical scratch arrangement, but at an angle of 15°. For the variant, when the center of the flat cover divides the scratch into two equal halves, the maximum stresses decreased by about 1.3 times.

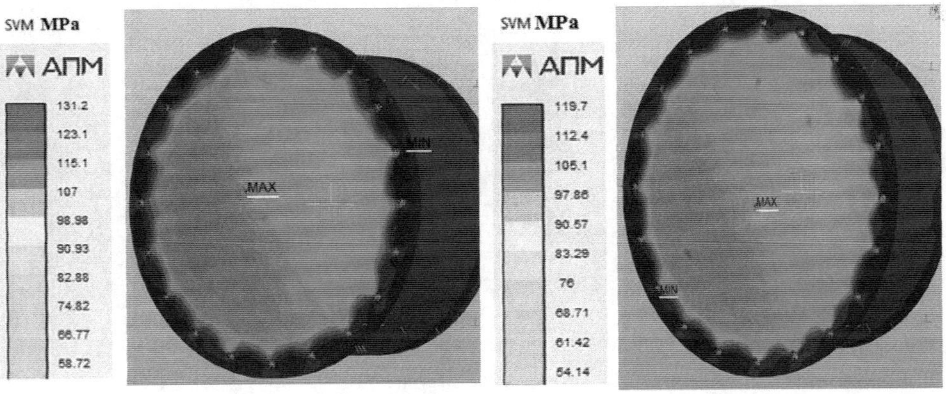

a) b)

Figure 6. Results of calculation of SDS of flat manhole cover in case of scratch on the surface with inclination angle 45°. a) results of calculation of SDS of flat manhole cover with scratch dislocation relative to the cover center in accordance with figure 2a; b) results of calculation of SDS of flat manhole cover with scratch dislocation in relation to the cover center in accordance with figure 2a.

By performing a comparative analysis of the results of figure 6 and the results of figure 5, it can be seen that by increasing the angle of the scratch to 45°, the maximum stresses are reduced to 131.2 MPa

IOP Conf. Series: Materials Science and Engineering 1181 (2021) 012018 doi:10.1088/1757-899X/1181/1/012018

(1.2 times) in the case where one sharp tip of the scratch comes from the center of the flat cover. There was an increase in maximum stresses to 119.7 MPa (1.3 times) when the center of the flat cover divides the scratch into two equal halves.

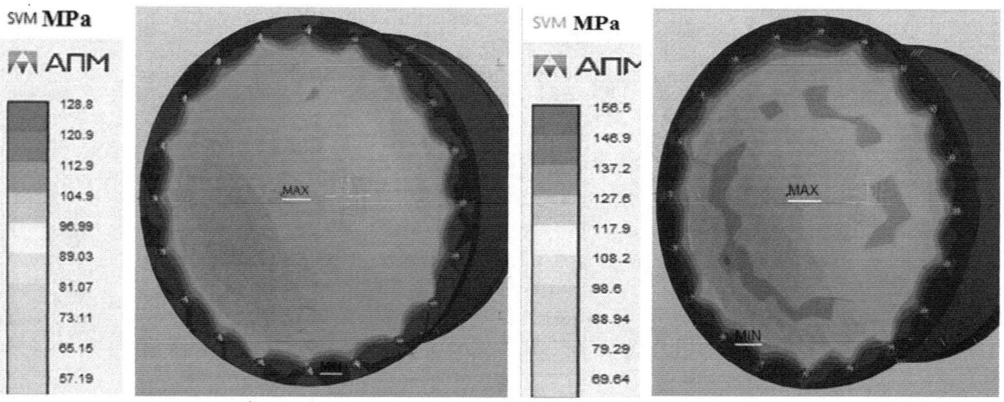

a) b)

Figure 7. Results of calculation of VAT of flat manhole cover in case of scratch on the surface with inclination angle 60°. a) results of calculation of SDS of flat manhole cover with scratch dislocation relative to the cover center in accordance with figure 2a; b) results of calculation of SDS of flat manhole cover with scratch dislocation in relation to the cover center in accordance with figure 2a.

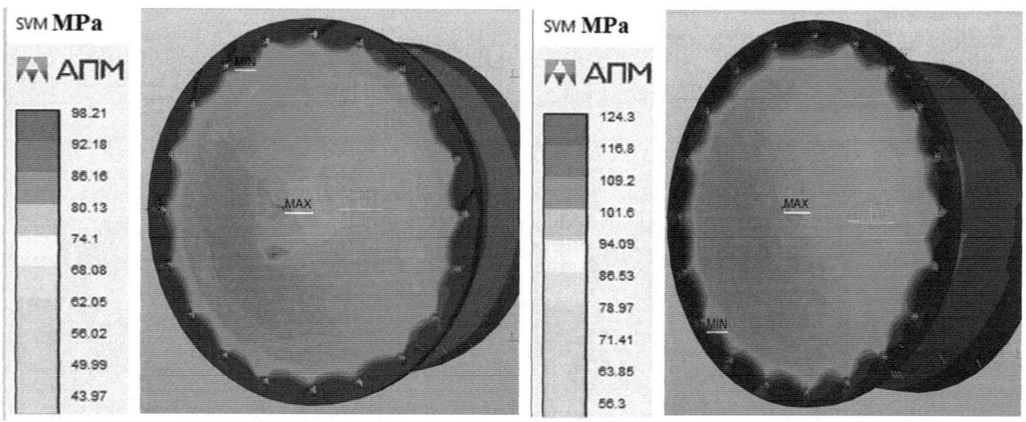

a) b)

Figure 8. Results of calculation of SDS of flat manhole cover in case of scratch on the surface with inclination angle 75°. a) results of calculation of SDS of flat manhole cover with scratch dislocation relative to the cover center in accordance with figure 2a; b) results of calculation of SDS of flat manhole cover with scratch dislocation in relation to the cover center in accordance with figure 2a.

Results which are given in figures 7-8 show that at increase in a tilt angle of a scratch to 60° and 75° the maximum tension is 128.8 MPa and 98.21 MPa respectively at option when one sharp top of a scratch comes from the center of a flat cover. The maximum stress values increased to a maximum value of 156.5 MPa in the case where the center of the flat cover divides the scratch into two equal halves with the angle of the scratch 60°. As the angle of inclination increased to 75°, the maximum stresses decreased to 124.3 MPa.

4. Conclusions

Based on the results of the SDS study of the flat manhole cover with scratch length L = 50 mm depth H = 1 mm and width B = 2 mm and with different dislocation relative to the center of the cover when exposed to operating pressure P = 2MPa and operating temperature t = 20 °C, the following conclusions can be drawn:

- it has been established that the SDS of the flat cover is affected by the dislocation of the scratch relative to the central part of the cover. The maximum stress values are significantly influenced by both the angle of the scratch and its location relative to the center of the cover for two versions:

  1) when one sharp tip of the scratch comes from the center of the flat cover;
  2) when the center of the flat cover divides the scratch into two equal halves.

During technical diagnostics, in order to assess the actual technical condition, it is necessary to simulate SDS of flat covers of nozzles with existing scratches to detect maximum active stresses.

- it is shown that with the scratch inclination angle values in question, which are equal to 15°, 30°, 45°, 60° and 75°, the maximum stresses occur in the flat cover when the scratch is arranged at an angle 30° and amount to 157.6 MPa in the embodiment where one sharp tip of the scratch comes from the center of the flat cover. The maximum stresses in the flat cover are 156.5 MPa in the case where the center of the flat cover divides the scratch into two equal halves with the angle of the scratch 60°.

References

[1] Mishchenko A 2020 *IOP Conference Series: Materials Science and Engineering* **953(1)** 012004
[2] Sładkowski A, Proydak Y and Ruban V 2020 *Transport Problems* **15(3)** 139-51
[3] Vasylyshyn V, Taras I, Bekish I, Kornuta O and Kornuta V 2020 *Management Systems in Production Engineering* **28(2)** 97-103
[4] Aistov I P and Vansovich K A 2020 *Journal of Physics: Conference Series* **1441(1)** 012083
[5] Bambura A, Mel'nyk I, Bilozir V, Prystavskyi T and Partuta V 2020 *Eastern-European Journal of Enterprise Technologies* **1(7-103)** 34-42
[6] Harionovskij V V 2017 *Gas Industry Magazine* **752(5)** 56-61
[7] Tyusenkov A S 2017 *Journal of Chemical Technology and Metallurgy* **4(52)** 766-72
[8] Kulakov P A, Rubtsov A V, Afanasenko V G and Gracheva V 2018 *MATEC Web of Conferences* **224** 02090
[9] Bakhtizin R N, Mustafin F M, Bykov L I, Khasanov R R and Kunafin R N 2016 *SOCAR Proceedings* **3** 52-8
[10] Kulakov P A, Rubtsov A V, Afanasenko V G, Zubkova O E, Sharipova R R and Gudnikova A A 2019 *Journal of Physics: Conference Series* **1399(5)** 055052
[11] Abhilash J, Apeksha Acharjee B 2018 *IOP Conference Series: Materials Science and Engineering* **455(1)** 012113
[12] Turbinsky S, Urbanovich V and Antonovich V 2009 *Reviews on Advanced Materials Science* **20(2)** 136-42
[13] Faritov A T, Rozhdestvenskii Y G, Yamshchikova S A, Minnikhanova E R and Tyusenkov A S *Russian Metallurgy (Metally)* **11** 1035-41
[14] Kozlov I 2020 *Lecture Notes in Civil Engineering* **49** 287-93
[15] Ovchinnikov N 2019 *AIP Conference Proceedings* **2188** 050036

Model of intelligent control system of road-building material compaction by pavers

A P Prokopev, Zh I Nabizhanov, V I Ivanchura and V L Sabinin

Siberian Federal University, prospect Svobodny, 79, Krasnoyarsk, Russia, 660041

E-mail: prok1@yandex.ru

Abstract. The article considers the task of designing intelligent controllers of automatic control systems by working modes of pavers during compaction of road-building materials. Intelligent controller is submitted in a form of fuzzy PID-controller. Mathematical model which takes account of metallic structures of compacting working body and road-building materials is provided. Simulation model of working process under the study in terms of MATLAB/Simulink programme is obtained. Numerical example of comparison between system's transient characteristic and continuous, digital and fuzzy controllers is provided.

1. Introduction

Low-quality asphalt road pavements are the topical issue nowadays. The quality of road pavements strongly depends on road materials, framework of asphalt pavers and road rollers, mode parameters of road building machines. A significant improvement in the quality of asphalt road pavements can be obtained by automatically controlling the mode parameters of road materials' compaction by asphalt pavers. Taking into account the multifactorial nature of the technological process of compacting road materials by pavers, the use of intelligent (fuzzy logic, artificial neural networks, expert system) regulators is the effective tool.

Modern pavers with efficient working bodies can ensure the compaction of asphalt concrete mixture to the standard (set) values of the compacting factor. The regulating effect is provided by changing the frequency of impacts of the ramming bar in the range of 15-30 Hz and the frequency of vibration of the plate in the range of 20-60 Hz. Controlled value is a compaction coefficient, which varies in the range of 0.92-0.99, at a constant speed of the paver - 0.033-0.083 m/s.

High-order models are obtained in the mathematical formulation of objects of automatic control systems (ACS) [1]. The problem of designing continuous controllers for high-order systems is characterized as relevant. In the general theory of automatic control, the controller's structure is selected based on the model of the control object. At the same time, more complex control objects correspond to controllers that are more complex [2].

The idea of controlling the working process of compaction by asphalt pavers using fuzzy controllers is proposed. The efficiency of using fuzzy controllers in automatic control systems by technological processes is the subject of scientific research and discussion [3-7].

The aim of this paper is to investigate the effectiveness of using a fuzzy PID controller in the control system of the working process of road materials' compaction by asphalt pavers in real time.

The following tasks were solved to achieve this goal:

- mathematical and simulation models of the interaction process between the compacting working body of the asphalt paver and road materials are obtained;
- a method for designing PID controllers for high-order systems is proposed; a parametric synthesis of the controller for a specific example is performed;
- the design of a fuzzy PID controller for a system of a fourth-order control object is performed;
- a comparison between the quality of regulation by the transient characteristics of control system models and continuous, digital and fuzzy controllers is carried out.

2. Methods

As a result of the implementation of the method in the state space the transfer function (TF) of the fourth-order control object is obtained

$$W(s) = \frac{0,001115 \cdot s^2 + 0,02509 \cdot s + 153,3}{s^4 + 29,41 \cdot s^3 + 1,553 \cdot 10^5 \cdot s^2 + 7,963 \cdot 10^5 \cdot s + 7,436 \cdot 10^8}.$$

The design of a continuous controller for a fourth-order system is carried out according to the method proposed by the authors [8]. The simulation model in the MATLAB/Simulink program is shown in figure 1.

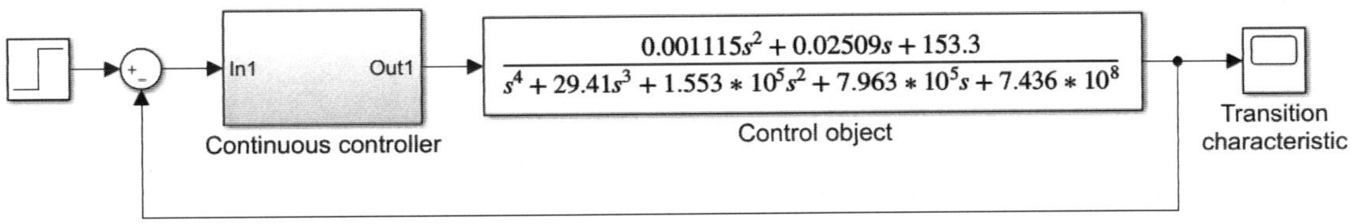

Figure 1. Simulation model of the control system in MATLAB / Simulink.

The model of the subsystem of the continuous PID controller is shown in figure 2.

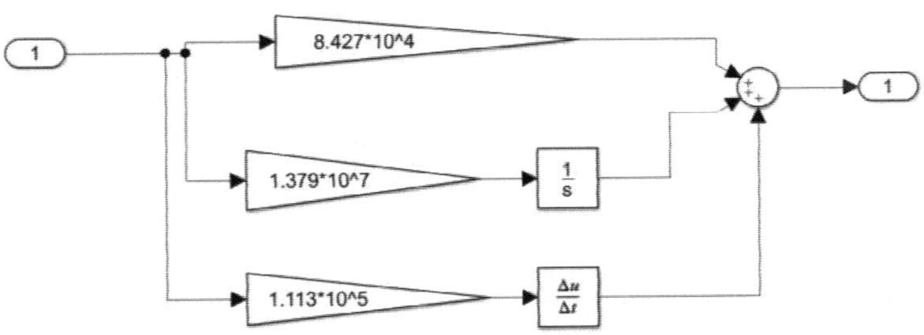

Figure 2. The model of the subsystem of the continuous PID controller.

According to the simulation results, the process is aperiodic, the transition time is 1 s, there is no overshoot.

A simulation model of a control system with a digital controller in terms of the MATLAB/Simulink program is obtained, figure 3.

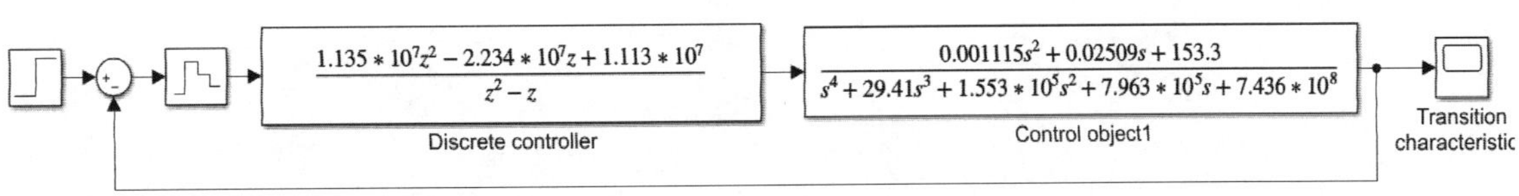

Figure 3. Simulation model of a control system with a digital controller.

IOP Conf. Series: Materials Science and Engineering 1181 (2021) 012019 doi:10.1088/1757-899X/1181/1/012019

The study of the model with a digital PID controller was carried out with a sampling period of T = 0.01 s. According to the simulation results, the process is aperiodic, the transition time is 1.2 s, there is no overshoot. This result satisfies the task for designing an automatic control system.

The development of a simulation model of a fuzzy ACS in the MATLAB/Simulink program has been completed. Based on the selected structure the PID controller equation is written as the following expression [5]:

$$u(t)=\left(\frac{1}{2}\cdot K_p\cdot e(t)+K_d\cdot\frac{de}{dt}\right)+\left(\frac{1}{2}\cdot K_p\cdot e(t)+K_i\cdot\int e(t)\cdot dt\right).$$

Here, the first bracket describes the PD controller, the second bracket describes the PI controller (figure 4).

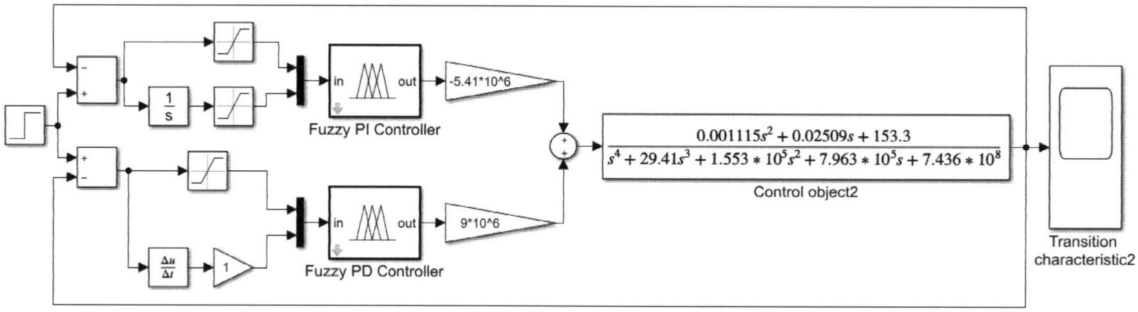

Figure 4. Simulation model of a control system with a digital controller.

The fuzzy logical inference is based on the Mamdani algorithm. In order to minimize the number of rules, we will use 7 terms for the input variables (which gives 49 monitor rules). In this case, each variable corresponds to its own range of the base scale, presented in table1 (figure 5, figure 6).

Table 1. Law of control of FLC_PD and PI.

	NB	NM	NS	Z	PS	PM	PB
e	-1	-0.66x2	-0.33x1	0	0.33x1	0.66x2	1
u	-1	-0.66	-0.33	0	0.33	0.66	1

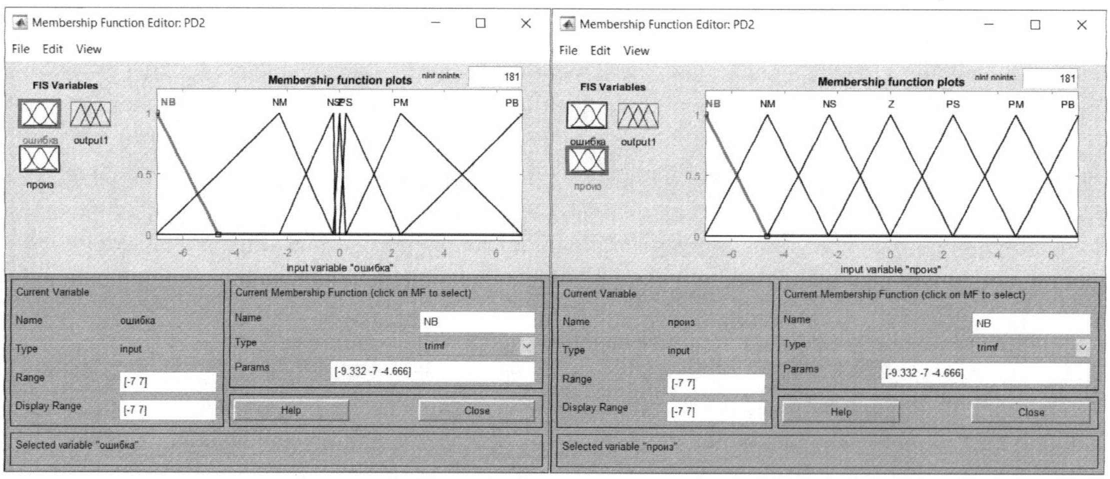

Figure 5. Linguistic description of input variables of the PD type.

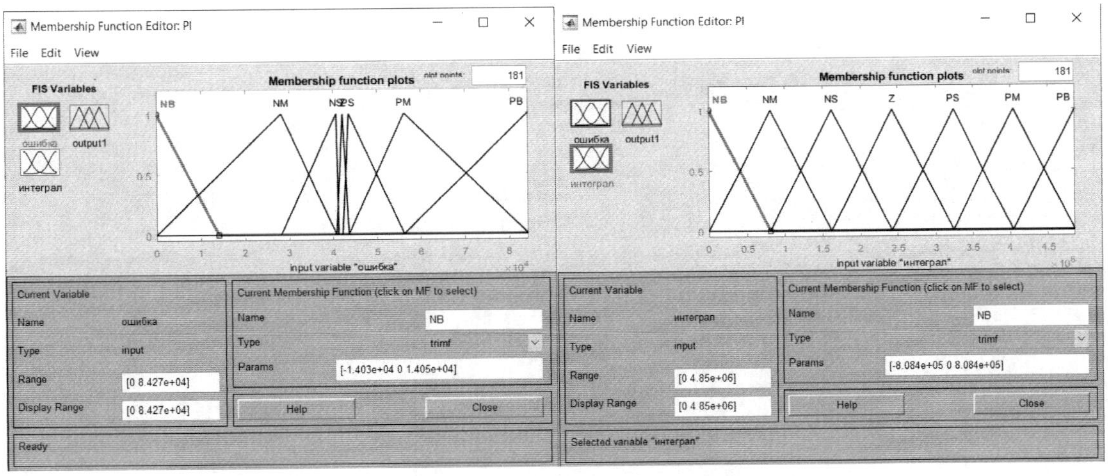

Figure 6. Linguistic description of input variables of the PI type.

7 terms are used to describe the control signal (figure 7).
The linguistic control law takes the form shown in table 2.

Table 2. Rules of the FLR_PD-and PI regulator

Table of rules		e^*						
		NB	NM	NS	Z	PS	PM	PB
Δe^*	NB	NB	NB	NB	NB	NM	Z	PB
	NM	NB	NB	NB	NM	NS	PS	PB
	NS	NB	NB	NM	NS	Z	PM	PB
	Z	NB	NM	NS	Z	PS	PM	PB
	PS	NB	NM	Z	PS	PM	PB	PB
	PM	NB	NS	PS	PM	PB	PB	PB
	PB	NB	Z	PM	PB	PB	PB	PB

The general rule table takes the following form: it contains 49 rules, and each rule has 7 meanings (figure 8).

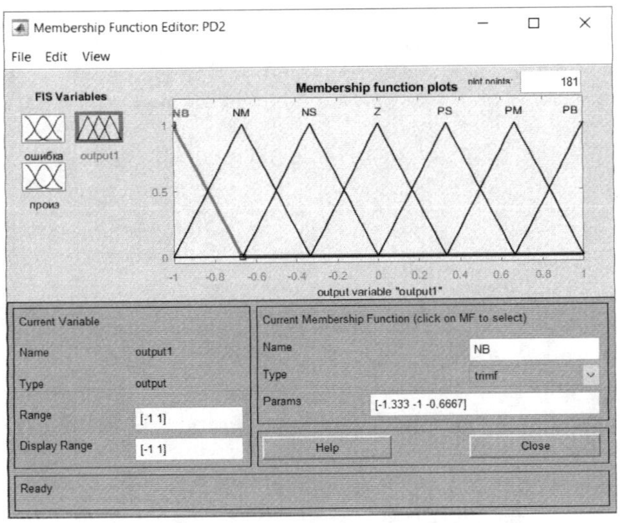

Figure 7. Linguistic description of the output variable of the PD and PI type.

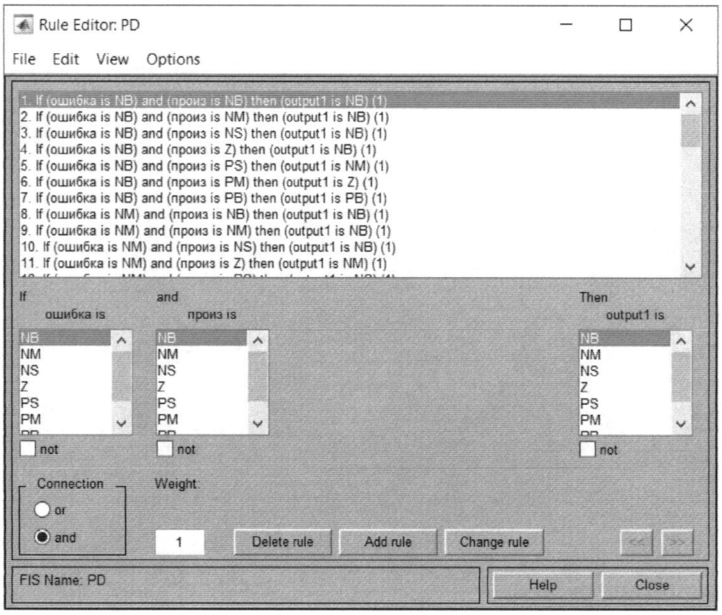

Figure 8. Rule Base Editor.

3. Research results

The transient characteristics of the systems with synthesized regulators are compared. Modeling is necessary to analyze the quality of ACS by dynamic characteristics and the nature of the dynamic process.

The transient response of the fuzzy controller shows better dynamic properties than the classical and digital PID controllers: the transition time is 0.1 s, so the practical effect of replacing the controller is quite obvious (figure 9).

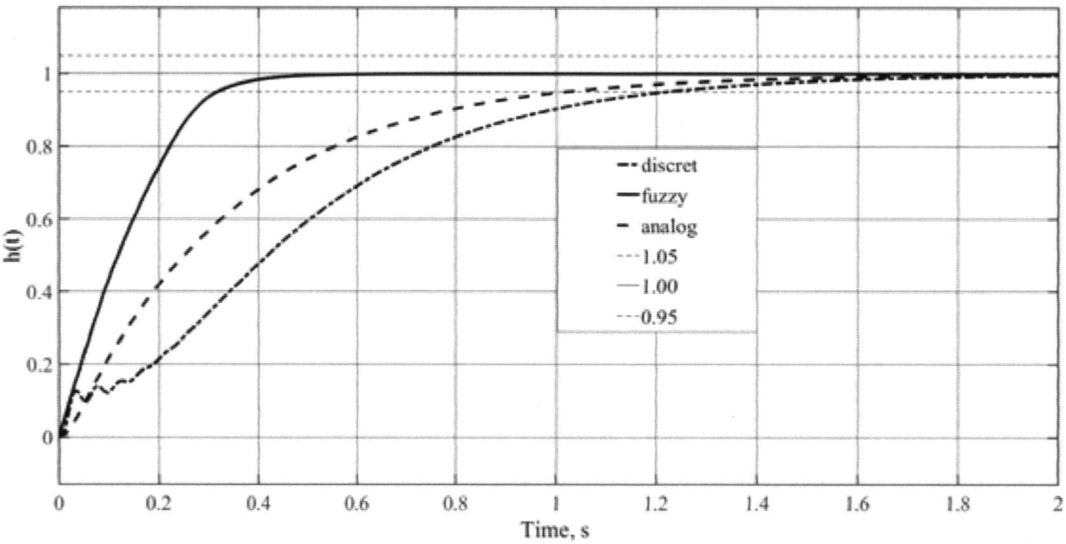

Figure 9. Transient characteristic of the control system with continuous, digital and fuzzy PID controllers.

It can be noted that overshoot is not observed in any of the control systems according to the simulation results. The control system with a fuzzy controller shows the shortest transition time.

References

[1] Prokopev A P, Nabizhanov Zh I, Ivanchura V I and Emelyanov R T 2019 Modeling the non-destructive control of road surfaces compaction *J. Phys.: Conf. Ser.* **1399** 044094

[2] Denisenko V V 2008 PID-regulators: questions of modification implementation *Modern autom. technolgies* **1** 86-99

[3] Mamdani E H 1974 Applications of Fuzzy Algorithm for Control a Simple Dynamic Plant Proc. *IEEE* **121(12)** 1585-8

[4] Mamdani E H and Assilian S 1975 An Experiment in Linguistic Synthesis with Fuzzy Logic Controller *Int. J. Man-Machine Studies* **7(1)** 1-13

[5] Filimonov A B and Filimonov N B 2018 To the question of the construction of fuzzy PID controllers *Mekhatronika, avtomatika i robototekhnika* **2** 112-6

[6] Filimonov A B and Filimonov N B 2018 Sertain Problematic Aspects of Fuzzy PID Regulation *Mekhatronika Avtomatizatsiya Upravlenie* **12** 762-9

[7] Burakov M V and Konovalov A S 2011 Synthesis of fuzzy logic regulators *Information and control systems* **1** 22-7

[8] Prokopev A P, Nabizhanov Zh I, Ivanchura V I and Emelyanov R T 2020 The method of PID controllers synthesis for sixth-order systems *J. Phys.: Conf. Ser.* **1515** 022080

Modeling the stress-strain state of plates in a form of Reuleaux Triangle profile during grinding of plastic materials

D A Krivov

Krasnoyarsk State Agrarian University, 90 Mira Avenue, Krasnoyarsk, 660049, Russia

E-mail: krivovdm@yandex.ru

Abstract. The article deals with the mechanical characteristics that arise in the crushing zone during grinding plastic material in a crusher with type-setting plates in a form of Reuleaux Triangle Profile. It is described the methodology for calculating internal and external energy and the load in the crushing zone. Much attention is given to the modeling of the stress-strain state of the plates during grinding plastic materials. The analysis of deformations made it possible to determine the critical zones of the impact of loads on the plates.

1. Introduction

Flow plastic chips are formed in industrial production during the mechanical processing of plastic materials (low-carbon steel, aluminum, copper, brass). The processing of such chips is difficult due to the plastic properties of the material. One of the possible solutions to the problem of processing flow chips is grinding materials in a roller crusher with type-setting plates in the form of a Reuleaux Triangle Profile [1-3]. Constructive solutions in the manufacture of a crusher should take into account deformation loads and forces acting on the plates. It is important to calculate the loads arising in the crushing zone and make a model of the stress-strain state of the plates to find appropriate solutions.

2. Methodology for calculating loads in the crushing zone

A literature review of mechanics of plastic deformation does not make it possible to determine the optimal ways to solve the problems of calculating the forces and deformations between plates in the form of Reuleaux Triangle Profile during grinding plastic materials. The objectives of determining deformations with significant material broadening are quite complex, as the calculations of stresses are usually made on the basis of empirical formulas [4-6]. In the process of crushing plastic materials with a width to thickness ratio of more than 10, the broadening of the material occurs insignificantly and the task is reduced to the analysis of the plane-deformed state. The deformation process of the material occurs in a tapering channel formed by the arcs of the RK-profile plates and their lateral surfaces (figure 1).

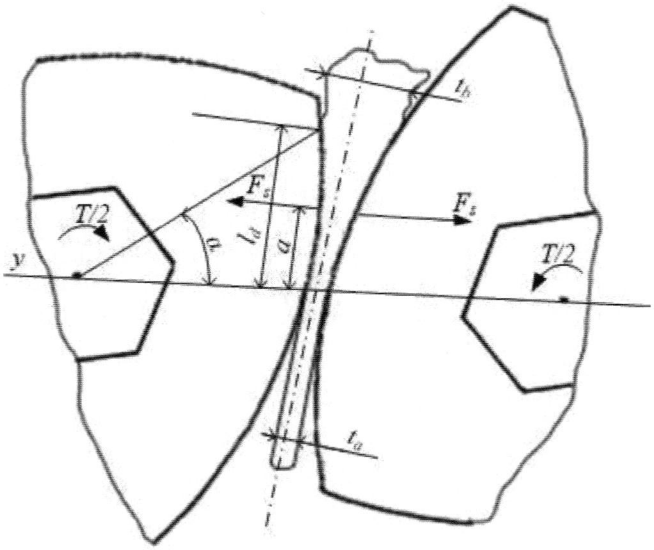

Figure 1. Deformation scheme of a piece of material.

In the process of rotation of the plates, the friction forces on the surfaces of the rolls change direction due to the advance and backward movement of the material relative to the neutral section of the roll. As a result, a pressure peak arises in the region of the neutral cross section. The magnitude of this peak depends on the deformation resistance of the crushed material and the coefficient of friction between the material and the rolls [7, 8].

The process of grinding plastic materials can be described with help of the energy balance method. However, the deformation work balance method gives approximate values of the deformation energy and the amount of force on the roll. This method does not allow to determine the distribution of stresses on the contact surface of the deformable material. Therefore, some assumptions were made [9]:

- there is no broadening (the outflow of material is limited by the side surfaces of the plates in the form of Reuleaux Triangle Profile);
- the deformation is uniform;
- there is no deformation of the rolls.

Then, equating the work of deformation during grinding specific internal energy, can be written:

$$u_i = \int_0^{\bar{\varepsilon}} \bar{\sigma} d\bar{\varepsilon} \quad (1)$$

where $\bar{\sigma}$ – constant stress causing plastic deformation; $d\bar{\varepsilon}$ - infinitesimal effective deformation under plane stress conditions, which, is calculated with accepted assumptions by the formula:

$$d\bar{\varepsilon} = \frac{2}{\sqrt{3}} d\varepsilon_y \quad (2)$$

For the accepted condition of uniform deformation

$$d\varepsilon_y = \frac{dt}{t} \quad (3)$$

With a constant value of stress $\bar{\sigma}$ and in the absence of mechanical hardening, equation 1, after substituting the values of quantities from expressions 2 and 3 into it, can be integrated, as a result:

$$u_i = \frac{2}{\sqrt{3}}\bar{\sigma}\ln\frac{t_b}{t} \tag{4}$$

Equation 4 can also be used to mechanically hardening material if it is assumed that $\bar{\sigma}$ is the average stress causing plastic deformation. It is possible to obtain a value of the specific internal strain energy that is closer to the real value: By introducing the correction factor C into equation 6, which takes into consideration additional shifts, it is possible to obtain a value of the specific internal strain energy that is closer to the real value:

$$u_i = \frac{2}{\sqrt{3}}C\bar{\sigma}\ln\frac{t_b}{t} \tag{5}$$

External deformation energy. The roll torque T shown in figure 1 determines the required external energy, which can be expressed as:

$$u_b = \frac{2T\alpha_0}{(t_b + t_a)l_d} \tag{6}$$

where α_0 is the angle of rotation of the rolls, during which the volume under the rolls is displaced, and can be approximately taken equal to α_b.

Equating the external specific energy to the internal specific energy, provided there are no losses, we get:

$$T = \frac{C(t_b + t_a)l_d\bar{\sigma}\ln(t_b / t_a)}{\sqrt{3}\alpha_b} \tag{7}$$

The force acting on the roll is applied at a distance a from the centers of rotation of the rolls. If we assume that the torque of the rolls depends on this force, then it can be calculated by a simple equation:

$$F_s = \frac{T}{2\alpha} \tag{8}$$

For a small reduction, it can be assumed that

$$\alpha = \frac{l_d}{2} \tag{9}$$

The force acting on the rolls in this way may not fully correspond to the actual force, since the torque of the roll also includes the horizontal component of the force acting on the rolls. The aluminum D16 alloy was chosen to calculate the load according to the presented method. The calculation result:

$$T = \frac{1(0.02 + 0.008)0.032 \cdot 440 \cdot \ln(0.02 / 0.008)}{\sqrt{3} \cdot 0.016} = 13.09[MPa] \tag{10}$$

The force acting on the roll will be as follows:

$$F_S = \frac{13.09}{2 \cdot 0.016} = 409.6[MPa] \tag{11}$$

3. Modeling the stress-strain state during grinding of plastic materials

During the grinding of plastic materials, shear stresses and compressive stresses arise. If the size of a piece of material is greater than the width of the plate, then shear deformations will be limited by the side surfaces of adjacent plates (excess material is cut off, the guillotine effect is realized) and directed tangentially to the arc of the roll profile (figure 2) [10-12].

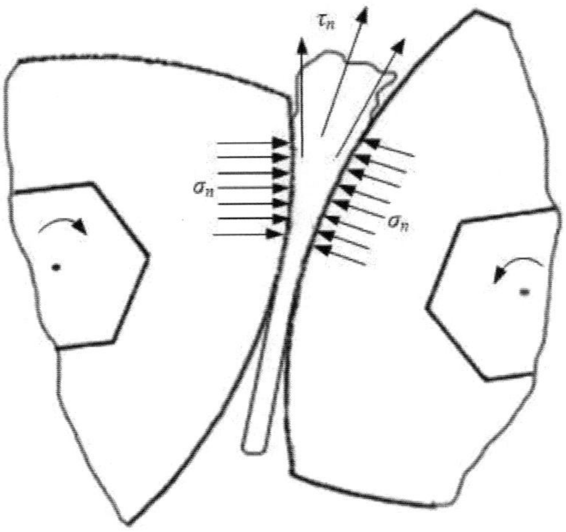

Figure 2. Distribution of stresses during grinding of plastic materials.

The distribution of normal and tangential loads is uneven due to different angular velocities in different parts of the roll. The analysis of the stress-strain state of the plate in the form of Reuleaux Triangle Profile is carried out by the finite element method in SolidWorks Simulation. The analysis was realized taking into account the nature of the stresses arising in the slotted gap. The numerical values of the loads are determined by formulas for calculating the loads arising from the grinding of plastic materials. Steel X12M was chosen as the material of the plates, and the ultimate strength of the aluminum alloy D16 was chosen as the limiting load. The result of modeling the stress state during deformation of plastic materials with a constant slotted gap is shown in figure 3.

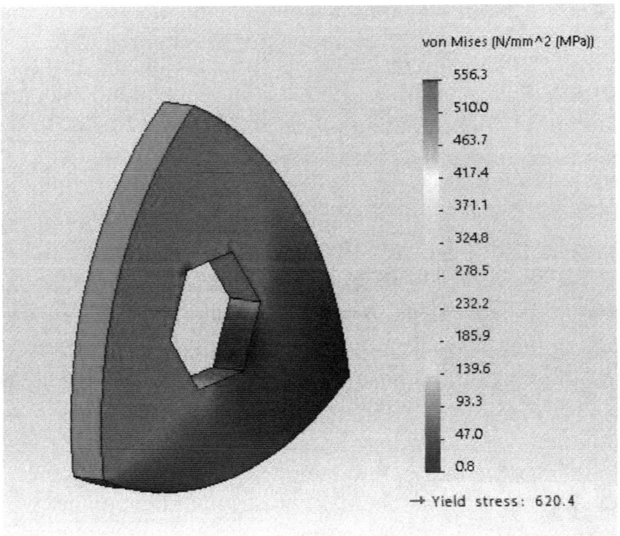

Figure 3. Diagram of stresses during grinding of the aluminum alloy D16.

A load corresponding to the stress distribution scheme was selected to analyze the stress-strain state (figure 2). The diagram shows that stresses are distributed over the volume of the plate in the range from

0.8 to 556.3 MPa, which does not exceed the threshold values of the strength of the plate material. The greatest stresses arise on the edges of the hexagonal bore as a result of the occurrence of tangential stresses directed in the opposite direction of plates rotation and amount to 556.3 MPa. A similar stress concentration was noted in the process of modeling the stress-strain state during grinding of brittle materials [3]. The design of the crusher is provided with rounded interfacing surfaces of the plate along the radius R0.5 mm to reduce the probability of occurrence and propagation of the main crack from the corner of the window. Similar radius roundings are made on the opposite seating surfaces of the hexagon shaft. Average values of stresses in the working zone of the plates are 180-320 MPa, which are much lower than the ultimate strength of the plate material. However, the loads are of a short-term nature, and the stress vectors are displaced along the plate surface.

The deformations arising under the influence of the load (figure 4) are insignificant and are compensated by the elastic properties of the plate material.

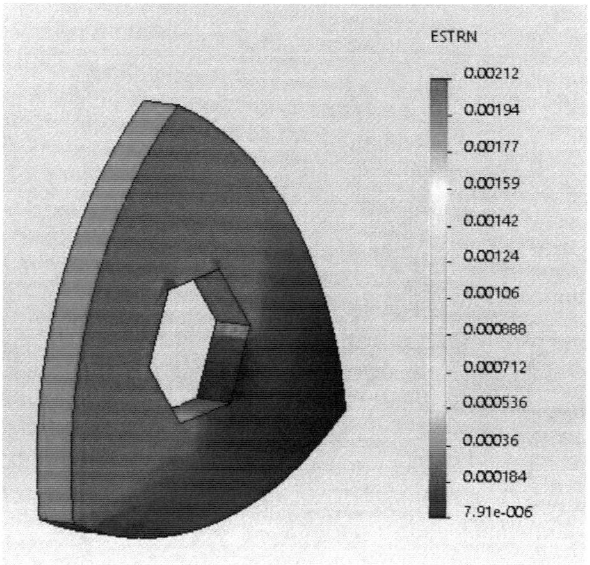

Figure 4. Diagram of deformations during grinding of the aluminum alloy D16.

Based on the results of modeling the stress-strain state, it can be concluded that during grinding plastic materials in the plates in the form of Reuleaux Triangle Profile, there are no critical stresses and deformations that can affect the reliability of the crusher. All calculations on the deformation of plastic materials during grinding with a constant slot gap were confirmed by the simulation.

4. Conclusion

The analysis of the work and stresses in the grinding zone showed that an effective combination of various destruction mechanisms (abrasion, crushing, cutting, alternating loads) increases the intensity of deformation processes and specific loads on the material. The analysis of the stress-strain state of the plates in the form of Reuleaux Triangle Profile showed that the maximum stress values do not exceed the safety margin of the plate material. It is identified zones of maximum stress concentration values. The necessary Constructive solutions have been taken to reduce the risk of defects in these zones.

References

[1] Krivov D A and Gordeev Y I 2017 The development of the schematic diagram of the crusher with rolls in the form of PK-profile *Bulletin of KSAU* **7(130)** 78-83

[2] Krivov D A 2020 Modeling of the stress-strain state of the set plates in the form of the RC-profile of a roller crusher *Proc. Int. Conf. on Science and Education: Experience, Problems, Development Prospects* (Krasnoyarsk: KSAU) pp 101-3

[3] Krivov D A, Gordeev Y I, Krivova D D and Chzhan E A 2019 Simulation of the stress-strain state of the combined rolls plates in the form of a Reuleaux Triangle Profile roller grinder *J. of Phys.: Conf. Ser.* **1353** 012076
[4] McClinton F and Argon A 2000 *Deformation and destruction of materials* ed E M Morozova (Moscow: Mir)
[5] Ecobori T 1971 *Physics and Mechanics of Fracture and Strength of Solids* (Moscow: Metallurgy)
[6] Balovnev V I 1988 Determination of resistance and energy when grinding materials *Construction and road machines* **1** 24-5
[7] Parton V Z 1974 *Mechanics of Elastoplastic Fracture* (Moscow: Nauka)
[8] Grigoriev D P, Khokhlov N I and Petrov I B 2017 Calculation of dynamic fracture in solid deformable bodies *Math modeling* **29(4)** 45-58
[9] Thomsen E, Chng Ch and Kobayashi Sh 1969 *Mechanics of plastic deformations in metal processing* (Moscow: Mashinostroyeniye)

IOP Conf. Series: Materials Science and Engineering 1181 (2021) 012021 doi:10.1088/1757-899X/1181/1/012021

Real-time neural network system for non-destructive control of asphalt mixtures compaction

Zh I Nabizhanov, A P Prokopev, V I Ivanchura and R T Emelyanov

Siberian Federal University, 79, Svobodny ave., Krasnoyarsk, 660041, Russia

E-mail: prok1@yandex.ru

Abstract. The problem is considered of designing a neural network system for control of road materials compaction in real time. The system is designed to ensure the functioning of non-destructive technology in road construction. A model of an artificial neural network (ANN) for monitoring the quality of compaction of road materials during the operation of vibrating rollers in real time is obtained. The system provides normalization of the input data of the ANN model. For the design of the ANN, the following methods were used: the least mean square deviation method; Levenberg-Marquardt algorithm; backpropagation. The structure of a non-destructive neural network system for monitoring compaction in real time is obtained. The results of numerical simulation of a system with an ANN for predicting the compaction coefficient of road materials are presented

1. Introduction

In the Russian Federation the most widespread are highways with non-rigid asphalt concrete pavements. When building layers of road surfaces, there are problems with the use of non-destructive technologies for quality control of asphalt concrete mixtures. There is a significant lag in application of automated system of continuous compaction control of road materials by vibrating rollers. This situation is related to the problem of import substitution because foreign-made devices are very expensive.

Taking into account innovative technologies for the construction of road surfaces, requirements for the quality of pavements, the tasks of implementing digital BIM technologies, an important scientific and technical direction is the use of automated road building machines equipped with modern technical means, including "smart" devices, for functioning as elements of cyber-physical systems.

One of the most important technical processes in road construction is the compaction of road materials: soils; asphalt concrete mixtures. High-quality compaction can solve many problems that lead to defects in the road's operation. These facts have been confirmed in many scientific studies, experience in the construction and operation of road surfaces of highways [1].

There are known achievements in automatic systems of compaction control that implement non-destructive technologies for road rollers. These include equipment installed on vibrating road rollers, equipment for intelligent compaction (IC) systems [2, 3], continuous compaction control (CCC) [4, 5]. Global manufacturers Trimble, Topcon, MOBA, etc. produce ready-made sets of automation equipment for installation on foreign-made rollers. The systems provide measurement of compaction indicators that have a correlation with the compaction coefficient. A significant disadvantage of CCC/IC systems is a poor accuracy in measuring of compaction indicators.

One of the modern directions of development of CCC/IC systems is the use of artificial neural networks. For many years the group of Professor S. Commuri (USA) worked in this direction [6-9]. To train the ANN the authors used the signals of accelerometer sensors installed on the frame of a vibrating roller as input data. Then, the spectral transformation method was used. A similar approach is used for almost all compaction measurement systems of CCC/IC indicators. The disadvantages of the intelligent system developed by the S. Commuri group are large errors due to interference from accelerometer sensors and constant changes in the properties of the asphalt concrete mixture. In the available information sources, there is no data on the implementation of developments based on ANN for real models of vibration rollers.

The long-term development of CCC and IC technologies has created a theoretical and experimental basis for the current stage in the development based on the use of artificial intelligence, intelligent optimization methods. The development of artificial intelligence models for the IC system is the results of scientific research by Xu G. and Chang G. K., who proposed a system including an artificial neural network with a genetic training algorithm [10].

The article is devoted to the creation of a model of a neural network system for quality control of compaction of road materials for installation on vibrating rollers. The paper offers a justification of the input data vector for training the ANN and continuous predicting of the mixture compaction coefficient. The software implementation of the ANN model is provided in the MATLAB program.

2. Methods

Compaction resistance in the asphalt mixture occurs due to the pressure of the rollers on the mixture during the operation of vibration rollers. A complex indicator that characterizes the compressive resistance of an asphalt concrete mixture is the modulus of deformation of the mixture, which has a significant impact on the compaction process at all stages of rolling the pavement by rollers.

The dynamic modulus of deformation E has a strong correlation with the temperature of the mixture, it can be determined by the regression model [1]:

$$E = 691{,}13 \cdot e^{-0{,}03 \cdot t},$$

where t – mixture temperature, °C.

The program code for calculating E in the MATLAB program is defined by the function

```
«m_deformation:»
«function E = m_deformation(t)»
«E = 691.13*(exp(t.*(-0.03)));»
«return»
```

The neural network system consists of the following functional elements (figure 1): a microcontroller with a monitor for entering initial data – machine parameters, process and properties of the compacted material, as well as a system for continuous data collection in real time.

The initial stage of the algorithm includes preliminary processing of input data, which determines the performance of the training process of the ANN, the value of the training error and other properties of the neural network.

Normalization of the input data vector is performed to reduce the negative impact on neurons. In this case, the data is reduced to the interval [0, 1] or [-1, 1]. To equate the quantitative values of the input data, the method of linear shift of the interval value of the characteristic value is usually used.

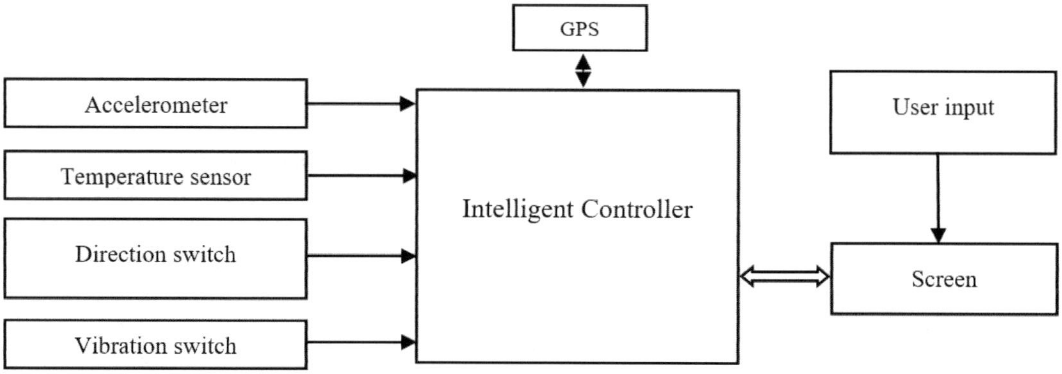

Figure 1. Functional scheme of intelligent information technology.

Determination of the characteristic value x for the i-th example of sample using the formula for representing data in the interval [a, b]:

$$y = \frac{(x_i - x_{min}) \cdot (b - a)}{(x_{max} - x_{min}) + a},$$

where x – the parameter value to be preprocessed; x_{max}, x_{min} – the maximum and minimum sample value of the parameter x.

All input data for training the ANN are preprocessed in order to bring it to the interval from 0 to 1. Data normalization function on the example of the mixture deformation module in the MATLAB program language:

```
«function e0_n=m_def_norm(e0,a,b,e0min,e0max)»
«for j = 1:length(e0)»
«e0_n(j)=((e0(j)-e0min)*(b-a))/((e0max-e0min)+a);»
«end»
«return»
```

Designing a model of a neural network system. The software environment creates the information for the machine learning model. The learning element uses received information to upgrade knowledge database. The functional element from the knowledge base is then used to solve the task. The signals of the measuring transducers transmitted from the external environment are random, so the learning element cannot fill in the gaps correctly or ignore insignificant details. The system functions in an arbitrary way, then it receives an information feedback signal from the functional element. The presence of feedback allows the machine learning system to test operational hypotheses and change them if necessary.

Artificial intelligence consists of two processing stages – the training process and testing. Preliminary studies and analysis of the results led to the conclusions that a multi-layer neural network gives a better result compared to such networks as a generalized regression network, a single-layer Perceptron, an Elman network, a cascade network with direct signal propagation, and others. The first layer of the network is the input data vector of the network, which corresponds to the necessary data for training the neural network-the parameters of the machine, process and asphalt concrete mixture. The second layer is a hidden network layer with ten artificial neurons between the input and output layers. The third layer – the output layer of the network, performs the prediction of the degree of compaction of the asphalt concrete mixture.

Supervised learning is shown in figure 2. This assumes that there is a database of training examples for the network. Each sample of the training set collected from the results of laboratory and field tests is given to the input layer of the network. Then it is processed inside the ANN structure.

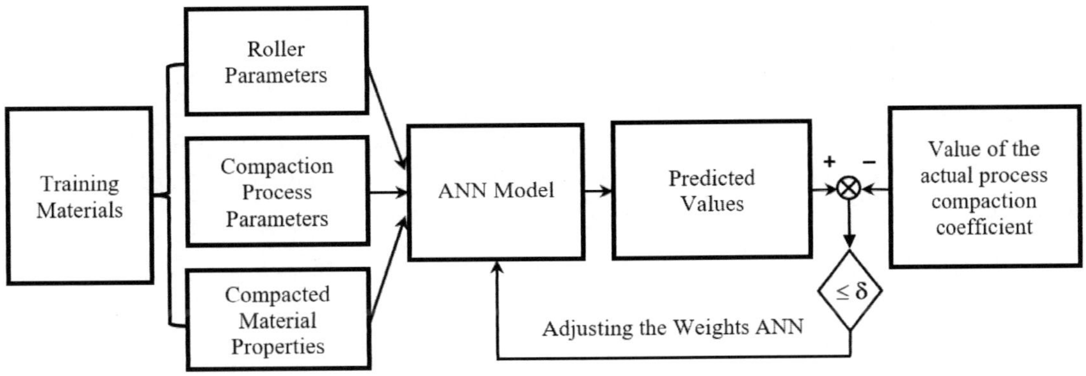

Figure 2. Functional scheme of training an artificial neural network model.

The coefficient of compaction of the asphalt concrete mixture is one of the most important indicators that determine the quality of the compacted pavement of the road. The obtained value of the compaction coefficient, figure 2, is compared in real time with the corresponding parameters from the prepared samples. Next, the training error of an artificial neural network is calculated based on the mean square error method. If the specified error value is exceeded, the weight coefficients of the connections between the layers (internal and input) of the artificial neural network are changed. The weight coefficients are changed based on the algorithm of the error back propagation method, which is suitable for the network structure. The weights change at each iteration until the network training error reaches a low level.

The sigmoidal function is used as an activation function in the synthesis of ANN, which provides an equalization between linear and nonlinear behavior:

$$y(j)=\frac{1}{1+\exp(\upsilon_j)},$$

where υ_j – weighted sum of all synaptic inputs with a threshold value; $y(j)$ – the output of an artificial neuron.

All hidden neurons have 8 local connections, where the input values of hidden neurons υ_{input} are calculated using the following formula:

$$\upsilon_{input}=\sum_{i=1}^{8} w_{ij}x_i,\ j=1,2,3...n,$$

where w_{ij} – a set of weights that are iterated over for all neurons; x_i – data of the vector of input neurons.

After calculating the vector of input data of hidden neurons, the output data of hidden neurons is determined by substituting the input values of each neuron into the activation function.

The input value of the output neuron is calculated according to the following dependence:

$$y_{input}=\upsilon_{output}\cdot w_h,$$

where w_h – matrix of weights of the layer of hidden neurons.

To determine the output neuron, the sum of the products of the weighting coefficients and the output values of the hidden neurons is activated using the activation function:

$$y_{output} = f_{act}(y_{input}).$$

Based on the method of back propagation of the error, gradients are determined that are applied when changing the weight coefficients of the ANN. The concept of this approach is to transmit information error signals from the outputs of the neural network to its inputs towards the reverse of the direct propagation of information signals.

As a result of analytical and experimental studies of the pilot scientific project, an ANN model was built for a system of continuous quality control of compaction of asphalt concrete mixture for use on vibrating rollers. A simulation model of the system is obtained in the MATLAB program. The simulation of the ANN is performed on a test example. The results of the graphical interpretation of the change in the compaction coefficient are shown in figure 3.

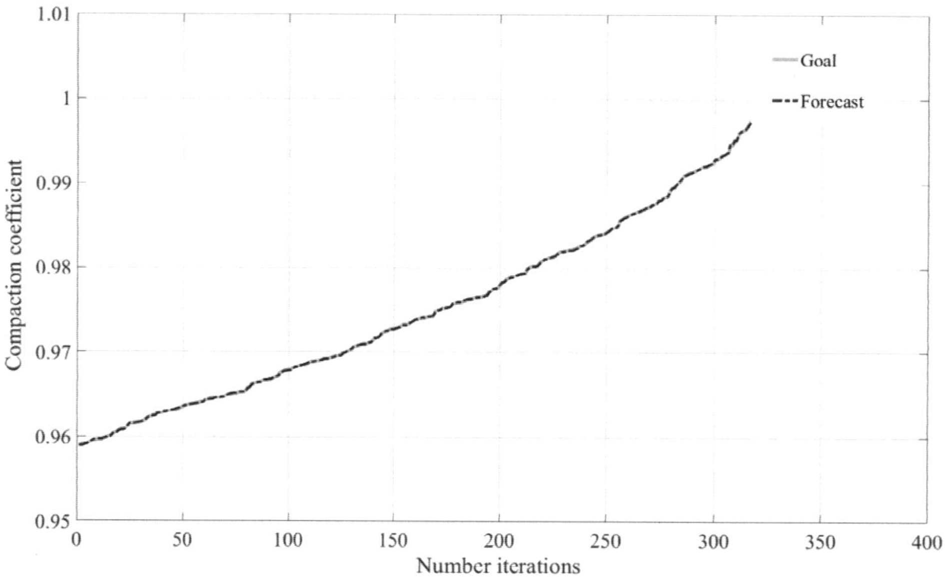

Figure 3. Results of testing an artificial neural network.

The average error in predicting the compaction coefficient of the asphalt concrete mixture is less than 10%, which indicates a good efficiency of using the compaction control method based on the ANN method.

3. Conclusion

As a result of the work performed, a neural network model of the compaction control system was obtained, which determines the value of the compaction coefficient of the asphalt concrete mixture in real time. Checking the trained ANN on a test sample showed that the maximum deviation of the data of the compaction coefficient of the mixture obtained from the neural network and the training set does not exceed 10%, which is an acceptable result for researchers.

Acknowledgments

The research was performed in Siberian Federal University and supported by Russian Foundation for Basic Research (RFBR grant no. 19-37-90052).

References

[1] Kostelov I P and Parkhomenko D V 2007 How to seal asphalt concrete in coatings when changing its type, state and thickness of the layer *Road equipment and technology* 70-85

[2] Xu Q and Chang G K 2015 Adaptive quality control and acceptance of pavement material density for intelligent road construction *Autom. Constr.* **62** 78-88

[3] Chang G K, Mohanraj K, Stone W A, Oesch D J and Gallivan V (Lee) 2018 Leveraging Intelligent Compaction and Thermal Profiling Technologies to Improve Asphalt Pavement Construction Quality: A Case Study *J. Transp. Res Board* **2672(26)** 48-56

[4] Pistrol J, Villwock S, Völkel W, Kopf F and Adam D 2016 Continuous Compaction Control (CCC) with Oscillating Rollers *Proc. Eng.* **143** 514-21

[5] Hu W, Shu X, Huang B and Woods M 2017 Field investigation of intelligent compaction for hot mix asphalt resurfacing *Front. Struct. and Civ. Eng.* **11(1)** 47-55

[6] Commuri S and Zaman M 2008 A novel neural network-based asphalt compaction analyzer *Int. J. Pavem. Eng.* **9(3)** 177-88

[7] Commuri S, Mai A T and Zaman M 2011 Neural Network-Based Intelligent Compaction Analyzer for Estimating Compaction Quality of Hot Asphalt Mixes *J. Constr. Eng. and Manag.* **137(9)** 634-44

[8] Imran S 2016 *Modeling and analysis of the interaction between roller drum and pavement material during compaction* Doctoral dissertation Univ. of Oklahoma Norman OK

[9] Asif Imran S, Barman M, Commuri S, Zaman M and Nazari M 2018 Artificial Neural Network–Based Intelligent Compaction Analyzer for Real-Time Estimation of Subgrade Quality *International Journal of Geomechanics* **18(6)** 04018048

[10] Chang G K 2017 *Intelligent Compaction Measurement Value (ICMV) – A Road Map* US FHWA Tech Brief FHWA-HIF-17-046 9

IOP Conf. Series: Materials Science and Engineering 1181 (2021) 012022 doi:10.1088/1757-899X/1181/1/012022

Modeling of the stress-deformed state of a capacitive apparatus with geometric deviations when mounting on a foundation

A V Rubtsov[1], K S Arkhipova, P A Kulakov, V G Afanasenko, O E Zubkova and Z R Mukhametzyanov[1]

Ufa State Petroleum Technological University, Ufa, 450000, Russia

[1] E-mail: sunset202@mail.ru

Abstract. The reliable and trouble-free operation of the capacitive equipment of hazardous production facilities is significantly influenced by the quality of the installation performed before commissioning. So, during the installation of capacitive equipment, various geometric inconsistencies and deviations may occur, which are established by the current regulatory and technical documents for the installation of technological equipment. If unacceptable deviations and inconsistencies established by regulatory and technical documents are found, they must be fully eliminated. However, in certain cases, the regulatory and technical documentation allows operation with various insignificant geometric deviations and inconsistencies that do not go beyond the established limits. This may relate to deviations from the design horizontal position of the apparatus when installed on a foundation. It should be noted that during operation, under the influence of increased pressure, temperature and other loads, an apparatus with deviations of this type can undergo a completely different stress-deformed state compared to an apparatus without geometric deviations, which is not taken into account in expert diagnostics, strength analysis and assessment of residual resource. The solution to this issue can be approached using software systems that allow building 3D models of the objects under consideration with the introduction of all the necessary parameters into these models and getting a real picture of the distribution of zones of increased stress. In this work, the stress-deformed state of the process vessel is simulated due to the impact of working loads and different directions of deviation from the horizontal design position of the body when the apparatus is mounted on the foundation.

1. Introduction

One of the most important stages before putting technological devices into operation is installation and installation on a foundation. The quality of installation work has a significant impact on the safe and trouble-free operation of the equipment. Various deviations and inconsistencies that arise during the installation process can make a significant contribution to the redistribution of places where the maximum acting stresses are concentrated [1-5]. Such a redistribution can worsen the forecast of the residual resource of the apparatus. One of such deviations may be a violation of the horizontal position of the apparatus on the foundation, for example, due to non-compliance with the horizontal foundation under the supports. Due to such unevenness, a skew of one end of the apparatus is formed relative to the other. The slightest bias can affect both the values of the arising maximum stresses and the areas where they are concentrated. The current regulatory and technical documents establish the requirements for the methodology for the installation of devices and its quality control. However, insufficient attention is paid to minor deviations that do not exceed the permissible deviations during diagnosis. However, sometimes even insignificant values of deviations can lead to the fact that the distribution of stresses in the elements of devices will not be uniformly distributed but concentrated at a certain point. This can be a harbinger of premature destruction of the most loaded element. In the question of studying the nature of stress distributions, software systems can help that allow analysis of the stress-deformed state with various existing deviations and defects [6-8]. Therefore, the actual work is the assessment of the stress- deformed state of the capacity with different directions of deviation from the horizontal design position of the body when the apparatus is mounted on the foundation and the simultaneous action of operational loads.

2. Methodology of the research

To carry out studies on the impact on of different directions of deviation from the horizontal design position during installation, the capacity of the process oil refinery was selected. The structural material of the shell of the hull, elliptical bottoms and manhole is steel Fe37B1FN. Material design of connectors - steel 20. The vessel is operated with hydrocarbon gas at an internal overpressure of 0.65 MPa at a temperature of 100 ^{0}C. Container has inner diameter equal to 2400 mm and shell thickness equal to 22 mm. The thicknesses of the left and right elliptical bottoms are 24 mm and 28 mm, respectively. Vessel body has length equal to 10 meters.

Vessel is installed horizontally on two saddle supports, left of which is fixed, and right support is movable. Vessel body consists of 6 separate cylindrical shells of different lengths, which are welded to each other by means of annular seams.

To evaluate the change in stress-deformed state, the license software complex "KOMPAS-3D" was used with the APM FEM system integrated into it. Initially, a 3D capacity model was developed in the COMPAS-3D software complex. The 3D capacity model developed is shown in figure 1 [9-11].

Further, using the strength analysis system "APM FEM," operational loads were applied and the necessary fasteners were installed [12-14].

3. Research results and their discussion

In order to compare the obtained research results, at the first stage, the stress-deformed state of the tank at working pressure and temperature was estimated for the design horizontal position on the foundation without any deviations. The results of the stress-deformed state assessment of the tank at operating pressure and temperature for the design horizontal position on the foundation without any deviations are shown in figure 2.

It can be seen from figure 2 that the value of maximum active stresses is 200 MPa and the zone of maximum stress concentration is the tie-in of the lower nozzle into the cylindrical shell No. 1. Stresses, which in their values are close to maximum stresses, are located in areas of annular welds of the vessel. Moreover, the zone of increased stresses in the zone of welding of the right elliptical bottom to the shell of the body is most clearly indicated due to the different thickness of the jointed elements.

At the second stage, the stress-deformed state of the tank was estimated with deviations from the horizontal design position when mounting 5 mm up and down each of the supports relative to the other support.

The results of stress-deformed state assessment of capacity with maximum stresses are given in figures 3-6.

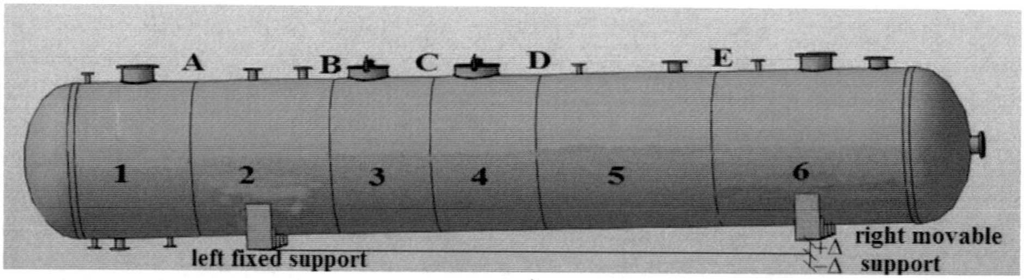

a)

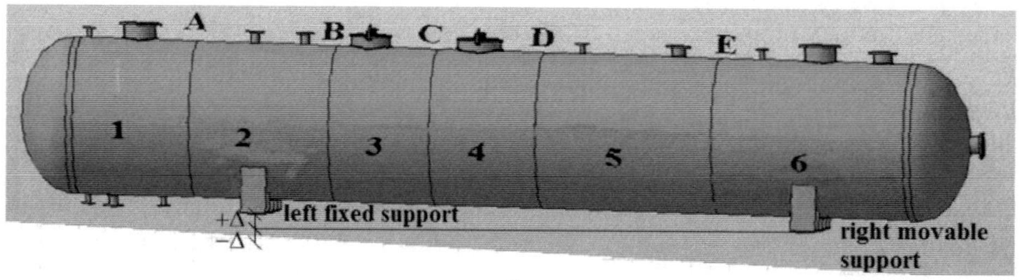

b)

Figure 1. 3D capacity model. A, B, C, D, E – designation of circumferential welds of the container body; 1,2,3,4,5,6 – numbers of welded cylindrical shells of the vessel body. +Δ - an example of a 5 mm vertical displacement upward of fixing one of the tanks supports on the foundation relative to the other support. -Δ - example of 5 mm vertical displacement downward of fixing one of the tanks supports on the foundation relative to the other support.

Figure 2. Results of tank stress-deformed state assessment at operating pressure and temperature without deviations from horizontal design position during installation.

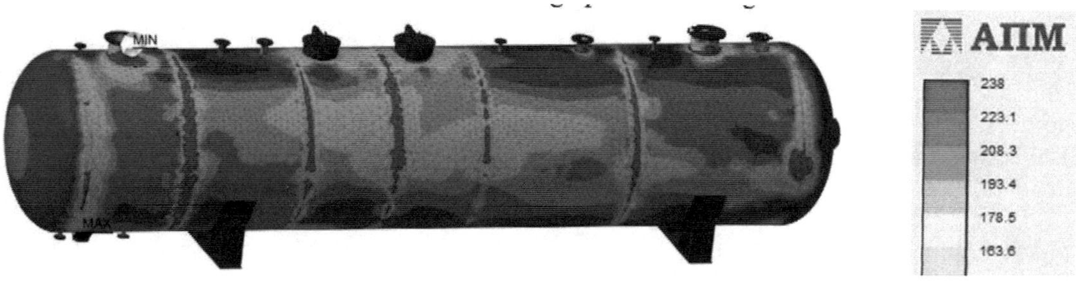

Figure 3. Results of tank stress-deformed state assessment at operating pressure and temperature with deviation from the horizontal design position during installation by 5 mm upwards of the left fixed support relative to the right movable supportresults of calculating the stress- deformed state.

Figure 4 shows that the maximum stresses do not change their location in comparison with figure 2, but are already 238 MPa, which is 38 MPa more than the maximum stresses without deviations from the horizontal. Also, the square of areas of increased stress has decreased.

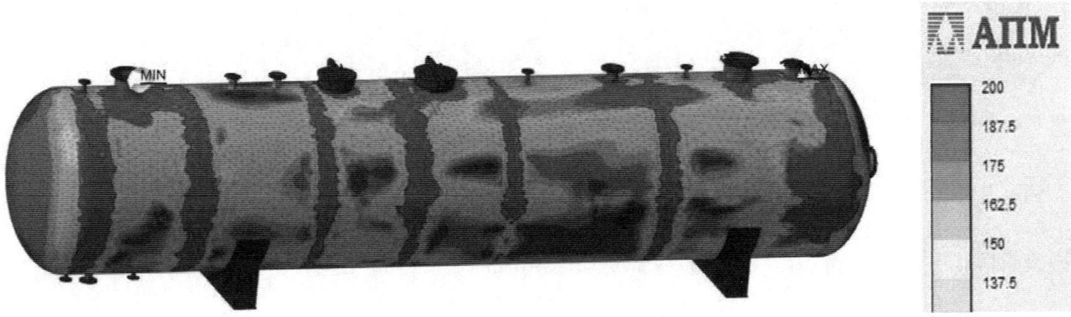

Figure 4. Results of stress-deformed state assessment of the tank at operating pressure and temperature with deviation from horizontal design position during installation 5 mm down of the left fixed support relative to the right movable support.

Analyzing figure 4, a very interesting fact is seen. The values of the maximum effective stresses fully correspond to the values of the maximum effective stresses for the variant without deviations from the horizontal. However, the area of the areas of increased stresses has increased and the zone of concentration of maximum stresses has changed and is concentrated in the rightmost choke of the case.

Figure 5. Results of tank stress-deformed state assessment at operating pressure and temperature with deviation from horizontal design position during installation by 5 mm upwards of RH movable support relative to LH fixed support.

Figure 5 shows that the maximum stresses with a deviation of 5 mm upward of the right movable support relative to the left fixed support are 320 MPa. This value is 82 MPa more than with the same deviation from the horizontal of the left fixed support. The zone of concentration of maximum stresses is concentrated in the extreme right nozzle of the body.

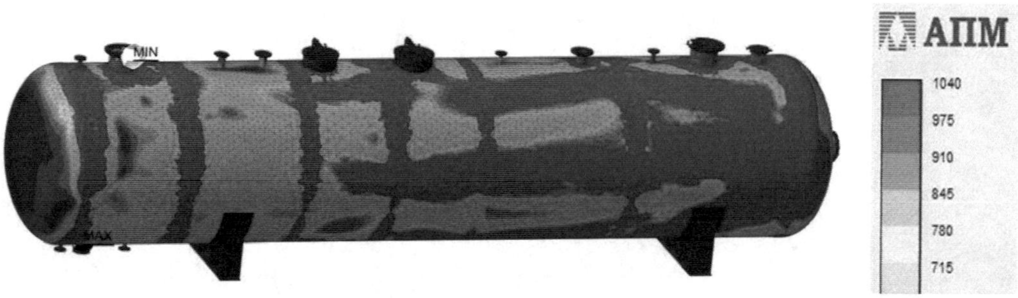

Figure 6. Results of tank stress-deformed state assessment at operating pressure and temperature with deviation from horizontal design position at installation 5 mm downwards of RH movable support relative to LH fixed support.

The results shown in figure 6 show that the maximum stresses are 1040 MPa and have increased by 3.25 times compared to the results in figure 5 and 5.2 times compared to the results in figure 4. The area of increased stresses is predominantly the case.

In order to reduce the values of maximum voltages in the tank for the most unfavorable scenario shown in figure 6 (the presence of a deviation from the horizontal design position during installation by 5 mm down the left movable support relative to the right fixed support; maximum stresses are 1040 MPa), it is proposed to use a stiffening ring. In the 3D model of the container, a ring is installed on the inside to give additional stiffness. The distance from the stiffening ring to the left edge of the cylindrical body of the container is 2400 mm. The profile of the ring is an angle with dimensions of 15 × 15 × 5 mm made of St3 steel. Figure 7 shows the welded stiffness ring on the inside of the cylindrical vessel housing.

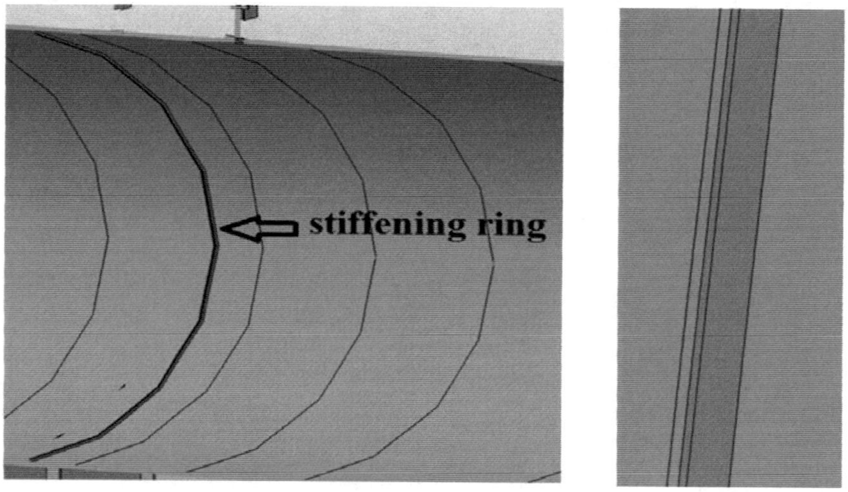

Figure 7. Welded stiffness ring on inner side of cylindrical vessel housing.

The results of the stress-deformed state assessment of the tank with the presence of a stiffness ring on the inside of the housing are shown in figure 8.

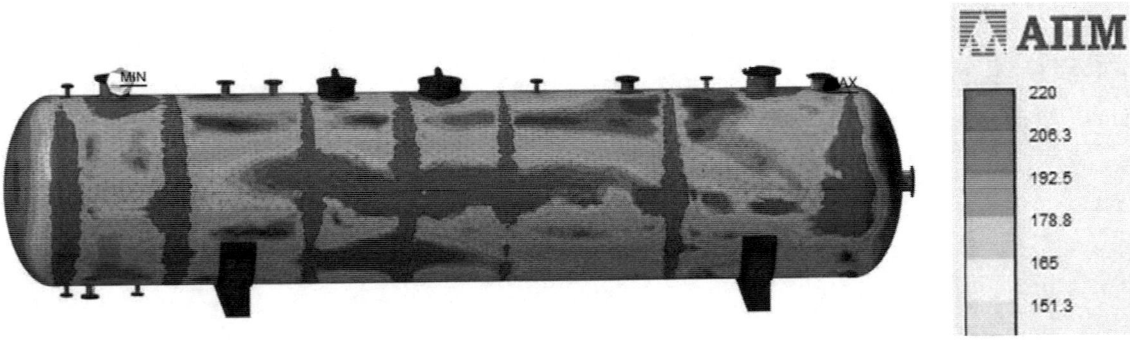

Figure 8. Results of stress-deformed state assessment of the tank with stiffness ring on the inner side of the housing.

As shown in figure 8, the value of the maximum stresses resulting from the installation of the stiffening ring on the inner surface of the container body is only 220 MPa, which is 4.7 times less than the values of the maximum stresses without using the ring.

4. Conclusion

Based on the results of modeling the stress- deformed state of a buffer tank with deviations from the horizontal design position during installation by 5 mm up and down each of the supports relative to the other support, the following conclusions can be drawn:

- it was found that the magnitude and zone of concentration of maximum stresses is influenced by the type of support (movable or fixed) and its location, where there is a deviation from the horizontal during installation, relative to other structural elements of the tank;
- it is shown that the maximum stresses arise when the deviation from the horizontal design position of the right movable support relative to the left fixed support is concentrated downward.
- it has been established that in order to reduce the numerical values of maximum voltages in capacitive equipment in the presence of deviations from horizontality during installation, stiffening rings are used, the dimensions and area of their installation of which should be selected depending on the individual features of the process equipment under consideration.

References

[1] Redutskiy Y, Camitz-Leidland C M, Vysochyna A, Anderson K T and Balycheva M 2021 *Reliability Engineering and System Safety* **210** 107545
[2] De Blasio C, Salierno G and Magnano A 2021 *Energies* **14(10)** 2863
[3] Habib S, Shakoor R A and Kahraman R 2021 *Progress in Organic Coatings* **154** 106218
[4] Mir A W and Ram Kumar K R 2021 *Advances in Intelligent Systems and Computing* **1383** 716-726
[5] Nicoleta T C 2021 *2021 IEEE 14th International Conference on Software Testing, Verification and Validation Workshops, ICSTW* **9440183** 205-212
[6] Mashekov S, Sembayev N, Mashekova A, Bazarbay B and Turgunov Z 2021 *Metalurgija* **60(3-4)** 317-320
[7] Rubtsov A V, Arkhipova K S, Kulakov P A, Zubkova O E and Mukhametzyanov Z R 2021 *Journal of Physics: Conference Series* **1889(4)** 042076
[8] Nikitina O A and Slobodyanik T M 2021 *Lecture Notes in Mechanical Engineering* 999-1005
[9] Tyusenkov A S 2017 *Journal of Chemical Technology and Metallurgy* **4(52)** 766-72
[10] Rubtsov A V, Kulakov P A and Arkhipova K S 2020 *IOP Conference Series: Materials Science and Engineering* **971(5)** 052040
[11] Jie L, Yongfeng W, Jie Y, Yichuan Q and Jiaqi H 2021 *Chinese Journal of Tissue Engineering*

Research **25(33)** 5301-5306
[12] Zhang J, Long X, Han C, Yuan P and Gao J 2021 *Transactions of the Chinese Society of Agricultural Engineering* **37(5)** 20-30
[13] Marruganti M and Frizziero L 2020 *Machines* **8(4),87** 1-34
[14] Song Z, Miao W, Chen G, Zhao G and Ma L 2020 *Chinese Journal of Rare Metals* **44(7)** 706-715

Electromechanical stand for the process of throwing soil with a cutter-thrower

M A Gnusov[1], M V Drapalyuk, D Yu Druchinin and L D Bukhtoyarov

Ministry of Science and Higher Education of the Russian Federation, Voronezh State University of Forestry and Technologies named after G.F. Morozov, 8 Timiryazeva Street, Voronezh 394087, Russian Federation

[1]E-mail: mgnusov@yandex.ru

Abstract. Every year in all countries of the world there are natural disasters, including forest fires. Quality wood is lost. Today the problem is dealt with both by research teams and production companies. The article presents one of the methods for studying the throwing process and adjusting the adequacy of simulation models. The creation of an electromechanical stand for the process of throwing loose elements allows you to monitor the process of separating portions of soil from the main layer, as well as track and digitize movements in molasses with various characteristics of the entire system under study.

1. Introduction

Despite the fact that fires bring diversity to the natural ecosystems of our planet, opening up the possibility of renewal of species biodiversity [1], they are a dangerous natural disaster that requires prompt elimination in the event of its occurrence [2]. The use of soil for churning, filling and stopping the edge of a forest ground fire is carried out manually or using mechanized equipment to extinguish a forest fire with soil [3]. Methods of mechanization and automation of the process of preventing and eliminating forest fires have been actively developing in recent years not only in the territory of the Russian Federation, but also in other countries of the world [4, 5, 6, 7, 8]. Domestic and foreign teams of scientists carry out scientific research aimed at increasing the efficiency of forest fire suppression, including with the help of soil [9, 10, 11]. The complexity of the mechanical system, which requires directing and collecting portions of soil into a general flow to extinguish a fire, requires a step-by-step study and a separate study of each structural element. The formation of a portion of the soil includes raising it to the level of the soil surface (figure 1, a), separation from the blades of the cutter (figure 1, b), the supply of a mechanical impulse (figure 1, c), which is necessary for the passage of the site with the subsequent direction of the soil area to the required quantity at a given distance (figure 1, d). The process proceeds quickly, with the formation of a dust cloud in the area of the blades of the milling body.

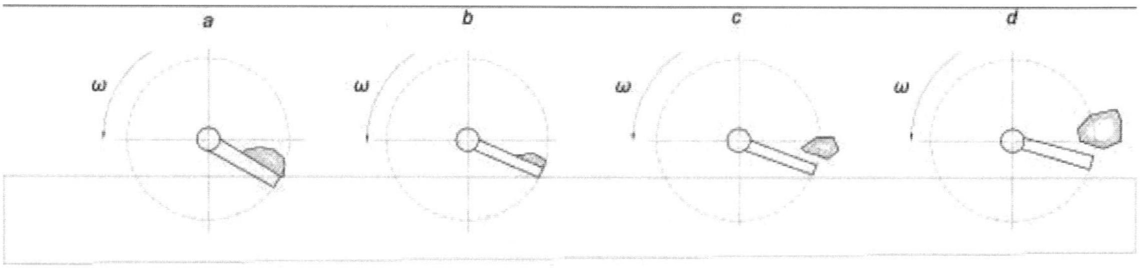

Figure 1. Scheme of the phased formation of a portion of soil.

The development of machines for extinguishing forest fires with soil, the working body of which is made in the form of a cutter with blades throwing soil, and the study of their working processes is carried out on specially created laboratory installations or prototypes.

In 2012, it was created by a team of scientists from Voronezh State Forestry University named after G.F. Morozov under the guidance of professors of the Department of Forestry Mechanization and Mechanical Engineering M.V. Drapalyuk and I.M. Bartenev. A combined machine for extinguishing a forest fire with soil and laying mineralized strips has been developed and manufactured [12]. The team was faced with the task, firstly, to increase the efficiency of preventing and extinguishing ground forest fires by throwing soil, and secondly, to develop a combined ground-throwing machine that combines the process of spreading soil and creating mineralized strips. such as the PF-1 cannon. and type GT-3, respectively, which reduces the greatest costs for design, industrial production, operation and storage of equipment. Laboratory studies were aimed at studying the creation of an optimal flow by carrying out preliminary preparation of a soil shaft using spherical discs (figure 2) [13].

Figure 2. Soil strip-laying machine.

In 2016, these studies were continued in the work of D.S. Stupnikov. under the guidance of Professor M.V. Drapalyuk. The aim of the work was to increase the efficiency and quality of preventive work, as well as to directly extinguish forest fires by justifying the parameters and operating modes of the machine for extinguishing forest fires with soil. The peculiarities of the principle of action of the soil gun are that the soil is delivered to the milling cutter-thrower without additional forced lifting. The separation of soil from the monolith occurs in portions and is directed into an open window in one of the sides of the movement of the unit [15].

Figure 3. Prototype forest fire ground-throwing machine. 1 - hydraulic motor; 2 - pneumohydroaccumulator; 3 - high pressure hoses.

A device for extinguishing a forest fire proposed in [16]. Made on the basis of a closed car. The unit is supposed to be used in the active phase of extinguishing with the possibility of entering the complex directly into the zone of fire development and throwing sand at the edge of the forest fire.

One of the well-known samples of ground-throwing equipment is a series produced by the Spanish company EXiTT [17].

Figure 4. Soil-throwing equipment from the EXiTT company.

The name in Spanish «extinción incendios tierra-tierra» sounds like a ground fire extinguishing unit. The company has developed and created several samples of units capable of extinguishing a forest fire using soil. Soil-throwing machines work mainly on light soils.

2. Materials and methods

The development and creation of a full-fledged unit has both positive and negative sides, which impose their own time and economic constraints. Resource conservation is one of the main challenges facing society. The development of simulation models reduces the number of experiments required. The development of models in CAD programs allows you to optimize the shape of the unit being created, reduce the energy consumption of the structure, and increase the manufacturing accuracy. The results obtained using software products necessarily require confirmation in the course of experimental studies [18, 19].

Modern opportunities offer various solutions to the identified problems of reducing the energy consumption of processes, cost-effective processes, versatility of the created and developed units and stands. Additive technologies allow you to print models to optimize shape and design, complex elements and assemblies. They also allow you to create milling and milling-throwing working bodies. We have developed and created in the program for engineering calculations a stand for studying the process of throwing and flight of soil particles (figure 5).

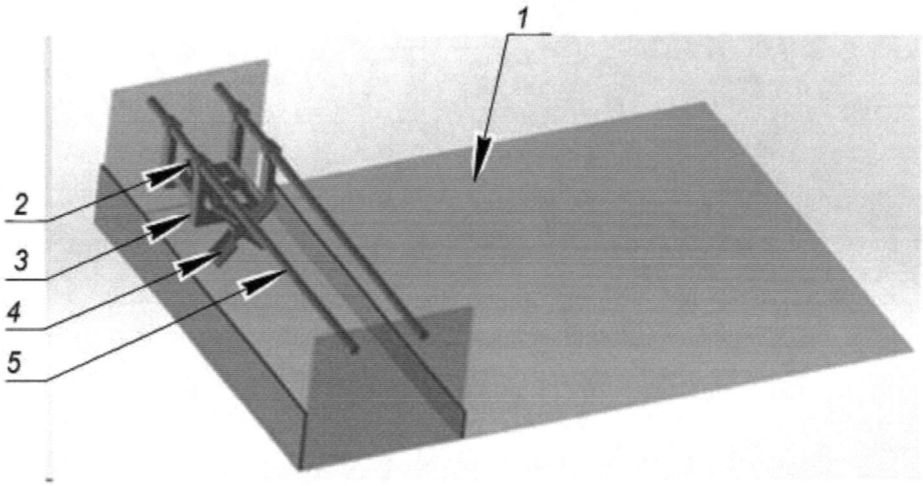

Figure 5. Model of a stand for throwing soil particles with a cutter. 1 - box; 2 - stepper motor; 3 - carriage; 4 - cutter-thrower; 5 – guides.

The cutter is driven by a stepper motor controlled by an Arduino. The wiring diagram is shown in figure 6. The NEMA 14 stepper motor has a 1.7 x 1.7 baseplate and has more torque than comparable smaller stepper motors. This motor has 6 lead wires and operates on a voltage of 12 V. It can also operate on a lower voltage, but its torque will decrease accordingly. With each step, the axis of the NEMA14 motor rotates 1.8 degrees.

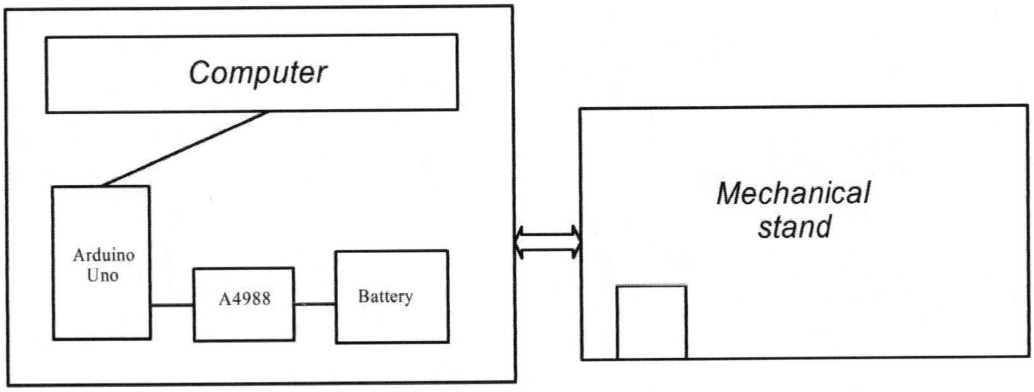

Figure 6. Wiring diagram of the electromechanical stand.

The number of steps per revolution (Steps per Revolution) for each specific stepper motor is calculated using the angle by which the stepper motor is rotated in one step (step angle). The A4988 stepper motor driver module controls the operation of the stepper motor by supplying different phases of power to it at the desired time. The A4988 motor driver can control the stepper motor in 5 different modes: full step, half step, quarter step, 1/8 step and 1/16 step. The stand scheme is launched using the program. For the program to work, you need to download a library for controlling a stepper motor using arduino-libraries / Stepper and add it to the Arduino IDE. After assembling the hardware of the project and uploading the program code to the Arduino board, you will be able to control the direction of rotation of the NEMA 14 stepper motor using a potentiometer.

Parameters of the electromechanical stand are calculated through the coordinates of the approximation from the simulation of the movement of the soil and the setting of the main parameters of the developed and investigated space according to the Pythagorean theorem:

$$r_{ij} = \sqrt{(x_i - x_j)^2 + (y_i - y_j)^2 + (z_i - z_j)^2}.$$

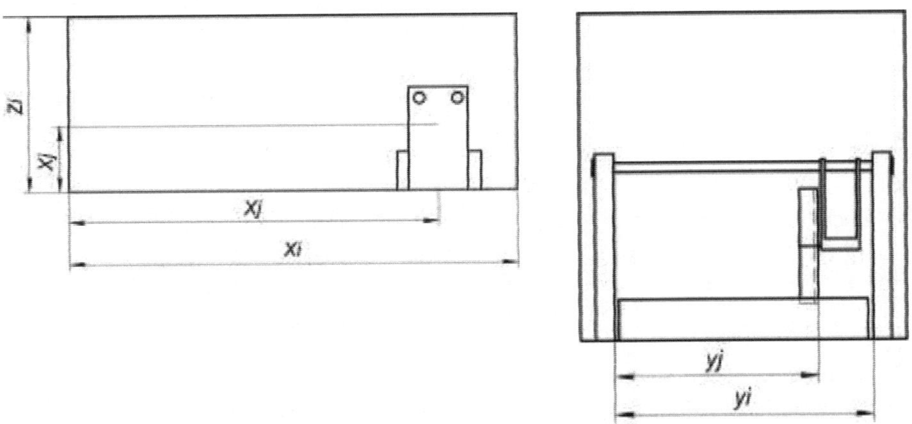

Figure 7. The main parameters of the electromechanical stand.

The stand is made using additive technologies, which made it possible to lighten the main units, reduce the time for manufacturing, and also simplify the operations of collecting elements due to the fact that the fitting was carried out in the SolidWorks software.

Figure 8. Stand for imitating the process of throwing particles with a cutter.

Figure 9. Operation of the thrower cutter with the throwing element.

For laboratory studies of the milling-throwing working body, an electromechanical stand was developed for adjusting the model of throwing soil particles in a given direction, which allows you to get an idea of the system and conduct laboratory studies with various parameters of the working body, rotation frequency, environment and operating conditions of the mill-thrower. On the basis of a series of experiments, in which the parameters of the installation of the geometric shape of the cutter-thrower and the physical and mechanical properties of bulk elements are changed, the nature of the effect of the parameter is established and recommendations are given for choosing the optimal values of the parameter of the working element and the conditions of use.

The design of the electromechanical stand for a complete presentation of the system must correspond to the high accuracy of detailing the main parameters and the properties being laid in order to correctly reproduce the movement of the soil, with a high temporal resolution, to reproduce the interaction of the working planes of the cutter-thrower with the soil (or its imitation in the form of other particles of bulk material) in the process of throwing it and moving it in the air.

The proposed research methodology is focused on using the capabilities of modern technical means and computers. Within the framework of the approximation, the soil is represented as a set of a large number (about 10^3–10^5) of individual elements. The sizes of the elements can be either the same to simplify the study, or different to exclude the adverse effects of the periodicity of the close packing of the elements. Soil elements mechanically interact both with each other and with the working surfaces of the investigated cutter-thrower.

3. Conclusions

The data obtained in the course of a series of experiments are processed by methods of statistical processing of research results or digitized using machine vision. And they can be used to set up simulation models based on the received data about the system. A prerequisite is the analysis of already known data, which are laid down in the form of the parameters of the thrower cutter-thrower.

Acknowledgments

The reported study was funded by RFBR, project number 19-38-60041\20.

References

[1] Pausas J G and Keeley J E 2009 A Burning Story: The Role of Fire in the History of Life *BioScience* **59(7)** 593-601 doi:10.1525/bio.2009.59.7.10
[2] Bartenev I M, Malyukov S V, Gnusov M A and Stupnikov D S 2018 Study of efficiency of soil-thrower and fire-break maker on the basis of mathematic simulation int. *Journal of Mechanical Engineering & Technology* **9(4)** 1008-18
[3] Drapalyuk M V, Popikov P I, Stupnikov D S , Sharov A V and Sherstyukov N A 2019 Increasing

the efficiency of the working process of a forest fire ground-sweeping machine with an energy-saving hydraulic drive of the throwing rotor *Forestry journal* **1(33)** 147-52

[4] Orlovsky S N 2004 Justification of the technology of application and layout of the tractor forest fire ground gun Proceedings of higher educational institutions *Forest Journal* **3(339)**

[5] Drapalyuk M V *et al.* 2019 Forest fires: methods and means for their suppression *IOP Conference Series: Earth and Environmental Science* **226** 012061 http://dx.doi.org/10.1088/1755-1315/226/1/012061

[6] Afzaal H and Zafar N A 2016 Robot-based forest fire detection and extinguishing model *2nd International Conference on Robotics and Artificial Intelligence (ICRAI)* http://dx.doi.org/10.1109/icrai.2016.7791238

[7] Sivaram P, Kumar M and Rajasekaran S 2012 Line based Geometrical Path Planning Algorithm for Extinguishing Forest Fires *International Journal of Computer Applications* **52(8)** 22-6 dx.doi.org/10.5120/8222-1650

[8] Kuznetsov G V, Piskunov M V and Strizhak P A 2017 How to improve efficiency of using water when extinguishing fires through the explosive breakup of drops in a flame: Laboratory and field tests *International Journal of Thermal Sciences* **121** 398-409 dx.doi.org/10.1016/j.ijthermalsci.2017.08.004

[9] Gedjo V M, Dubovyi V K and Leonovich A A 2019 New ways and means of localization and extinguishing surface forest fires *IOP Conference Series: Earth and Environmental Science* **316** 012005 dx.doi.org/10.1088/1755-1315/316/1/012005

[10] Narita D, Gavrilyeva T and Isaev A 2021 Impacts and management of forest fires in the Republic of Sakha, Russia: A local perspective for a global problem *Polar Science* **27** 100573 dx.doi.org/10.1016/j.polar.2020.100573

[11] Rodríguez-Veiga J, Ginzo-Villamayor M and Casas-Méndez B 2018 An Integer Linear Programming Model to Select and Temporally Allocate Resources for Fighting Forest Fires *Forests* **9(10)** 583 dx.doi.org/10.3390/f9100583

[12] Drapalyuk M V, Bartenev I M, Goncharov P E, Bukhtoyarov L D, Popikov P I, Gnusov M A, Druchinin D Yu and Markov O B 2013 *Patent RF*128887

[13] Bartenev I M *et al.* 2012 Combined forest fire-fighting ground gun and recommendations for its use *Polythematic network electronic scientific journal of the Kuban State Agrarian University* **84**

[14] Drapalyuk M V, Goncharov P E, Stupnikov D S and Sharov A V 2016 Forest Fire Ground Thrower *Patent RF*2616021

[15] Drapalyuk M *et al.* 2019 Improving the efficiency of the working process of forest fire soil throwing machine with hydraulic rotor *Forestry Engineering Journal* **9(1)** 151-6 Available at: http://dx.doi.org/10.12737/article_5c9201707790a1.57571839

[16] A fire truck that suppresses fires without water [Electronic resource]. URL: https://wildfiretoday.com/2011/03/27/a-fire-truck-that-suppresses-fires-without-water/ (date of the application: 11.04.2020)

[17] Dirt throwing machine for suppressing wildfire [Electronic resource]. URL: https://wildfiretoday.com/2011/07/04/dirt-throwing-machine/ (date of the application: 11.04.2020)

[18] Lysych M N 2020 Study of driving dynamics of modular forestry tillage machine-tractor units in CAE SolidWorks Motion *IOP Conference Series: Earth and Environmental Science* **595(1)** 012024

[19] Lysych M N 2020 Study driving dynamics of the machine-tractor unit on a virtual stand with obstacles *Journal of Physics: Conference Series* **1515(4)** 042079

Machine-building organization production using quality management methods

E V Priimak and I S Razina

Kazan National Research Technological University, Russia, Kazan

E-mail: ira-a82@mail.ru

Abstract. Mechanical engineering is the leading industry in the world, which has a significant impact on the level of development of a country. It produces various machines, machine tools, mechanisms, devices. The machine-building sector provides equipment to all other sectors of the economy, so the results of the work of related areas depend on its level of development and product quality. Among the main problems of the sector in Russia are high depreciation of fixed assets and low level of innovation. As a result, the low competitiveness of domestic machine-building products in the world market. Competitive advantages can be achieved primarily by improving the quality of manufactured products. Prompt solution of problems related to product quality is achieved by implementing various methods aimed at preventing the appearance of defects. The study presents an algorithm for the development and implementation of the 8D methodology in order to prevent the recurrence of product defects at the enterprise of Aktanyshsky Aggregate Plant. With the help of simple tools that underlie the 8D methodology, the causes of the defect were identified and corrective and preventive measures were proposed.

1. Introduction

Machine-building production is one of the most important industries from the point of view of the Russian economy: it accounts for about 20% of all industrial production in Russia and 1/3 of industrial employment. Machine-building plants in Russia produce a wide range of products, the range of manufactured products is counted in hundreds of positions. However, this sector of the Russian economy has a weak export potential. In the marketable structure of Russian exports, the share of «machinery, equipment» is less than 3%. At the same time, in the structure of imports, the same marketable group accounts for more than 40%. Among the main problems of the sector in the conditions of modern realities in Russia, there is a high depreciation of fixed assets and a low level of innovation. As a result, the low competitiveness of domestic machine-building products in the world market.

First of all, competitive advantages can be achieved by improving the quality of manufactured products. Prompt solution of problems related to product quality is achieved by implementing various methods aimed at preventing the appearance of defects. [1-3]. The list of such methods is quite diverse. In mechanical engineering, and in particular in the automotive industry, the 8D method is often used to solve quality problems [4,5]. The 8D technique means an eight-step approach to problem solving. Actions D1–D4 are aimed at identifying the nature of problems and determining the causes of their occurrence, and actions D5–D8 are aimed at developing and implementing corrective and preventive actions (figure 1).

IOP Conf. Series: Materials Science and Engineering 1181 (2021) 012024 doi:10.1088/1757-899X/1181/1/012024

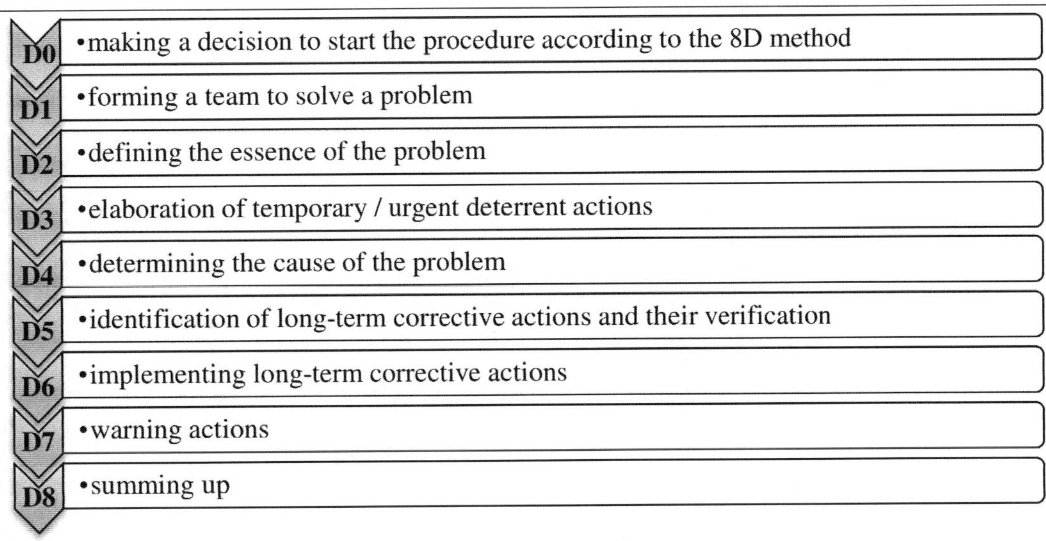

Figure 1. The sequence of the 8D method.

The study presents an algorithm for the development and implementation of the 8D methodology in order to prevent the recurrence of product defects at the enterprise of Aktanyshsky Aggregate Plant. The main activity of Aktanyshsky Aggregate Plant, located in the Republic of Tatarstan, is the production of parts of any complexity according to samples or drawings of the customer, as well as the provision of services for metal machining (in particular, turning, milling). In 2006, the company signed an agreement with KAMAZ for the production of components. As a supplier to the automotive industry Aktanyshsky Aggregate Plant must manage the risks of producing non-conforming products. Preventive and corrective actions should be carried out in accordance with methodology 8D [6].

2. The main part

As the object of research, the balancing axis bracket for the rear suspension of the KAMAZ car was selected (figure 2a). The rear suspension of trucks has a balancing device with a single axle pressed into the bracket. The balance suspension ensures the stability of the car, that is, it prevents strong tilts. During the inspection of the finished product, the quality control department controller (QCD) revealed such a discrepancy as a gas shell in the casting of the bracket, which can cause a decrease in the quality and safety of the product (figure 2 b).

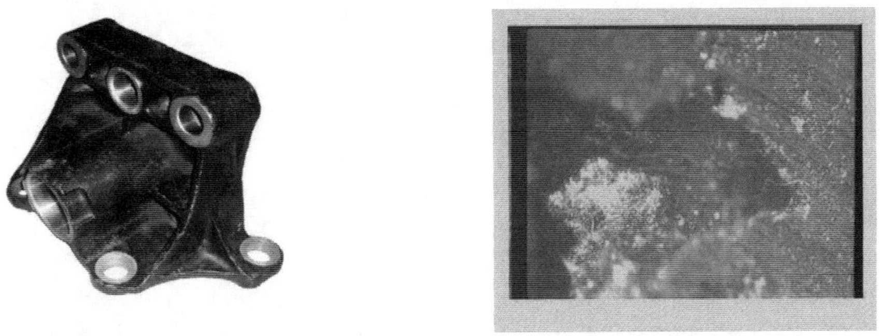

a) b)

Figure 2. a) Balancer axis bracket; b) Enlarged fragment of the casting of the balancing axis bracket with a gas shell.

In steps D1–D4, detailed information about the defect is collected. The input document is the defect controller's report. For this purpose the «5W+1H» method was used, where you can find answers to such questions as «who, what, where, when, why and how». A detailed description of the method is given in the figure 3.

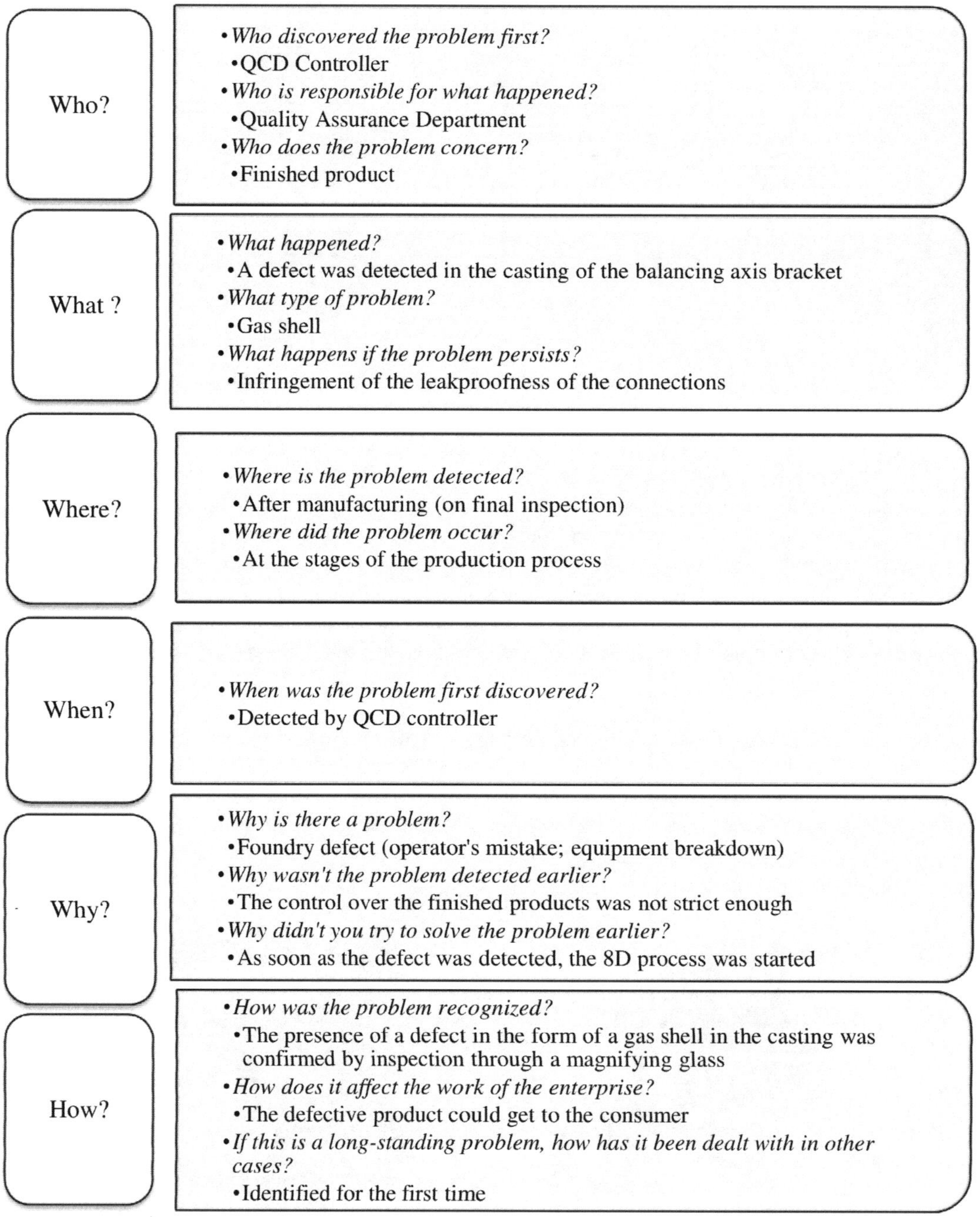

Figure 3. Description of the «5W+1H» method.

As a result, it was found that gas shells with a pore diameter of up to 16 mm were found near the surface of the casting, which can lead to a infringement of the leakproofness of the connections. Therefore, as part of the next stage, it is necessary to introduce short-term measures that immediately protect the consumer from the supply of defective products. This can be a 100% inspection of the finished part in order to detect non-compliance in a timely manner. The control consists of an external inspection and an instrumental measurement. The results are entered in the operational measurement sheet.

Deterrent actions are performed until the root cause of the problem is identified and resolved. Variables such as the human factor, equipment, materials, and process conditions were considered. After analyzing the identified causes, those that may be the main cause were identified (table 1).

Table 1. The main reasons for the discrepancy.

The main reason	Root cause
Methods	Low temperature of the poured metal
Personnel	Non-compliance with modes, instructions, established procedures
Working conditions	Non-compliant workplace
Technology	Infringement of steelmaking technology Incorrect choice of product processing methods
Materials	High gas content in the cast workpiece composition; No ventilation system in the molding mixture; Gas tightness of the molding mixture
Equipment	Damaged appliances; Equipment wear and tear

3. Conclusions

Based on the results of stages D1-D4, the mechanism of the defect occurrence was analyzed and consistently shows how the root cause develops into a specific defect. It was revealed that molds and rods without ventilation channels were used in the manufacture of casting molds (the root cause). Intensive gasification of the combustible components of the mold and rod begins when pouring metal. Excessive gas pressure is created in the mold and rod. It was revealed that the reduction of the gas pressure in the mold was not provided. The total pressure of the gases from the rod and the mold was greater than the pressure of the metal-static head. As a result, a gas shell was formed in the metal (defect). Also, a refrigerator with an oxidized surface was used to cool the casting. Oxides decomposing, form a gas when the melt comes into contact with such a surface. Gas shells appear in the solidifying metal of the casting.

To avoid the appearance of a defect at stages D5-D8, it is necessary to develop long-term corrective measures (for example, 100% control, revision of all relevant documents, equipment audit, additional control of the product after molding and training of production personnel) and, having determined their effectiveness, develop preventive measures.

References

[1] Nikolaeva N G, Gorunova S M, Priimak E V, Garmonov S U and Sopin V F 2009 *Quality management methods* **2** 24-7
[2] Priimak E V, Nikolaeva N G and Gorunova S M 2006 *Competence* **3(32)** 18-21
[3] Priimak E and Razina I 2020 *Materials Science and Engineering: Conf. Ser.* **753** 072002
[4] Saragin B M, Petrenko D S and Letgford L N 2020 *Quality management methods* **9** 38-41
[5] Golub I A and Borisov V V 2020 *Science without Borders* **5(45)** 64-9
[6] GOST R 58139-2018 Quality management systems Requirements for organizations of the automotive industry

Analysis of the stress-deformed state of a tube bundle at different values of partition thicknesses and their deviations from verticality

M M Zakirnichnaya, A V Rubtsov[1], P A Kulakov and N Galiullina

Ufa State Petroleum Technological University, Ufa, 450000, Russia

[1]E-mail: sunset202@mail.ru

Abstract. Shell and tube heat exchangers of hazardous industrial facilities are used to heat or cool working media and represent a rather complex individual design, which can consist of several structural elements, such as a body, bonnets, heat exchange tubes, tube sheets, partitions, etc. The failure of one element contributes to the failure of the entire heat exchanger. In the process of assembling or installing elements of the heat exchanger, the formation of various geometric deviations from design solutions is possible, which may be associated with violations of the technology of the assembly process, installation, etc. The presence of such deviations can negatively affect the reliable and trouble-free operation of the apparatus, serve as a scenario for the formation of places of increased stresses, leading to the initiation of defects in these zones and premature failure. One such deviation may be a change in thickness values and deviations from the vertical design of the transverse baffles of the tube bundle as supports for the heat exchange tubes. In the existing methods, the transverse partitions of the tube bundle do not pay attention and do not take into account their possible impact upon deflection on the stress-deformed state of the apparatus as a whole. In this work, the relationship of the effect of thicknesses, values and directions of deviation from the vertical location according to the design of transverse partitions of the pipe bundle on maximum equivalent stresses is investigated.

1. Introduction

Heat exchangers for oil refining, petrochemical and chemical operations are operated at elevated pressures, temperatures, with toxic and explosive and fire-hazardous environments and, therefore, must meet the highest requirements in the field of reliability and industrial safety.

The operational reliability and industrial safety can be significantly influenced by the quality of assembly or mounting of individual components. When performing the above noted works, the formation of various types of inconsistencies with the requirements of project documentation is possible. This may include deviations from straightness, ovality, offset edges, verticality, etc. All these deviations can further lead to an increase in stresses in certain zones of the heat exchanger, the formation of defects in these zones and the possible creation of an emergency situation with premature shutdown of the equipment. Therefore, for quality control and assessment, a wide arsenal of non-destructive testing methods is used, including one of the most informative methods - visual and measuring control (VMC) [1-5]. VMC allows you to effectively carry out internal and external inspection, identify various deviations, visible defects, size discrepancies with the project and is the primary method in assessing the technical condition. When monitoring and assessing quality, they are guided by the requirements and norms established by the current regulatory and technical documents. However, the non-separable design of the heat exchanger with fixed tube sheets creates an obstacle for a qualitative examination of the inner surface of the apparatus body and the outer surface of the tube bundle. Although, as the analysis of failures of heat exchangers in oil and gas refineries shows, it is precisely the passage of heat exchange tubes that is the most frequent cause of failure of a heat exchanger and accounts for almost 99% of all failures associated with heat exchangers.

The leakage of heat-exchange tubes can occur in places where high stresses are concentrated in their various sections, both in the presence of corrosive and non-corrosive media. The occurrence of stress concentration zones in heat exchange pipes can be facilitated by scenarios such as, for example, changes in wall thicknesses of the transverse partitions of the tube bundle, as well as deviations from the vertical arrangement of the partitions in the case of insufficient fixation of the partitions in the tube bundle. Due to the fact that the baffles in the tube bundle are supports for the heat exchange tubes, such scenarios can have a negative effect on the reliability and safety of the tube bundles as a whole directly on the whole shell-and-tube heat exchanger [6-8].

In this regard, the current task is to simulate the stress-strain state (SDS) of the pipe bundle in order to identify the values of maximum equivalent stresses taking into account the change in wall thickness of the partitions and various values of deviations of the partitions from the vertical design position in different directions.

2. The research method

One-way shell-and-tube heat exchanger of oil refinery with the following technical and operational characteristics was selected for SDS analysis:

- inner diameter of the housing is 400 mm;
- pipe length of tube bundle is 2000 mm;
- total number of heat exchange tubes is 109 pieces;
- the tubes have a smooth surface;
- the size of heat-exchanging tubes is 25×2.5 of mm;
- design temperature of pipe space is 85 °C;
- design temperature of annulus is 50 °C;
- design pressure of pipe space is 3.9 MPa;
- design pressure of annulus is 0.6 MPa;
- material of structural elements: steel 20;
- type of pipe grid - stationary.

Figure 1 shows a tube bundle with fixed tube sheets and numbered transverse partitions.

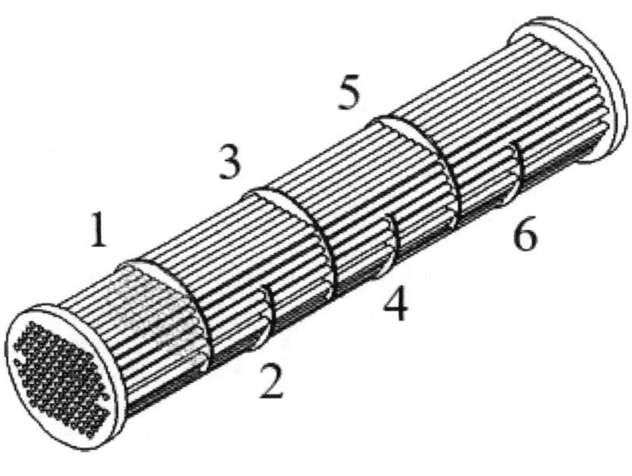

Figure 1. A tube bundle with fixed tube sheets and numbered transverse partitions.

The SolidWorks software package was used to simulate and evaluate the SDS of a tube bundle with fixed tube sheets [9-15].

The SDS of the bundle was estimated for the following cases:

- thickness of transverse partitions is 10 mm, 8 mm, 6 mm and 4 mm without their deviation from the vertical position according to the design;
- transverse partitions No. 1 and No. 6 are deviated from the vertical position according to the project by values from 0 to 15° with a pitch of 3°, in the direction of clockwise direction and similarly counterclockwise direction at thickness of partitions equal to 8 mm;
- transverse partitions No. 1 and 2 with a thickness of 8 mm are mutually deviated from the vertical position along the project in a clockwise direction by values from 3 to 15° with a pitch of 6°;
- transverse partitions No. 1 and 2 with a thickness of 8 mm are mutually deviated from the vertical position according to the project, at that partition No. 1 is deviated in the direction of clockwise direction, and partition No. 2 against the direction of clockwise values from 3 to 15° with a pitch of 6°;
- partitions No. 1 and 3 with a thickness of 8 mm are mutually deviated from the vertical position along the project in a clockwise direction by values from 3 to 15° with a pitch of 6°;
- partitions No. 1 and No. 3 with a thickness of 8 mm are mutually deviated from the vertical position according to the project, at that partition No. 1 is deviated in the direction of clockwise direction, and partition No. 3 is deviated in the direction of clockwise direction from 3 to 15° with a pitch of 6°.

3. Results of the research and the discussion

According to the results of the SDS assessment of the pipe bundle, in the above cases, dependency graphs were obtained, which are shown in figures 2-8.

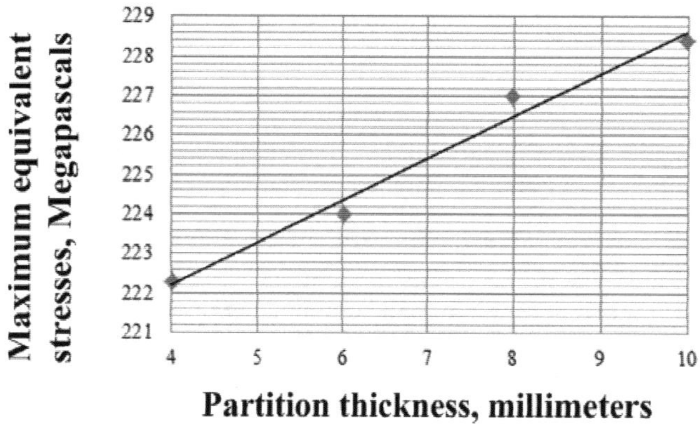

Figure 2. Graph of dependence of maximum equivalent stresses on thickness of transverse partitions.

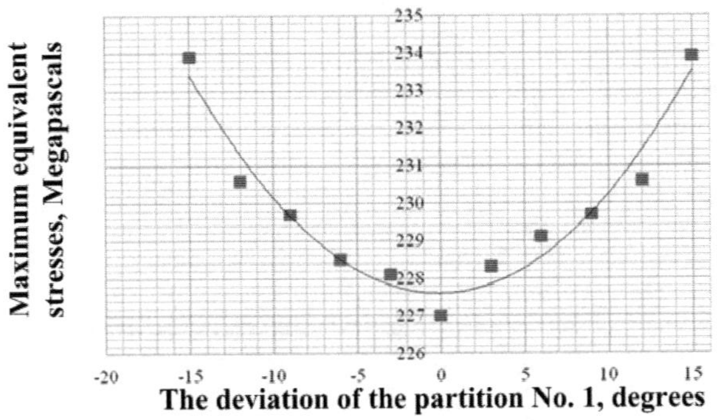

Figure 3. Graph of the dependence of maximum equivalent stresses on the values of the partition No. 1 deviation from the vertical location according to the project: "-" – in the clockwise direction; "+" – counterclockwise.

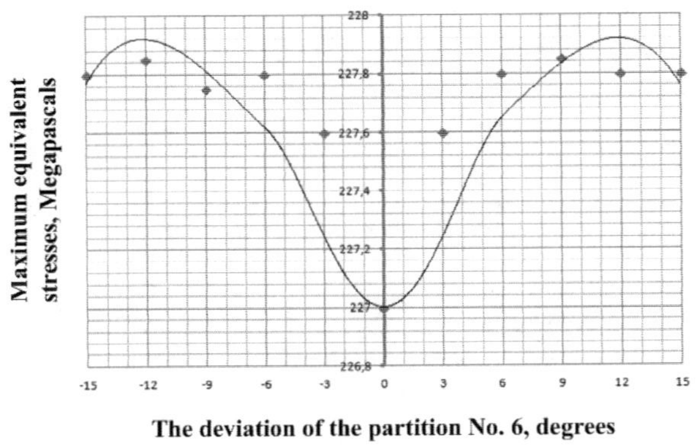

Figure 4. Graph of the dependence of maximum equivalent stresses on the values of the partition No. 6 deviation from the vertical location according to the project: "-" – clockwise direction; "+" – counterclockwise.

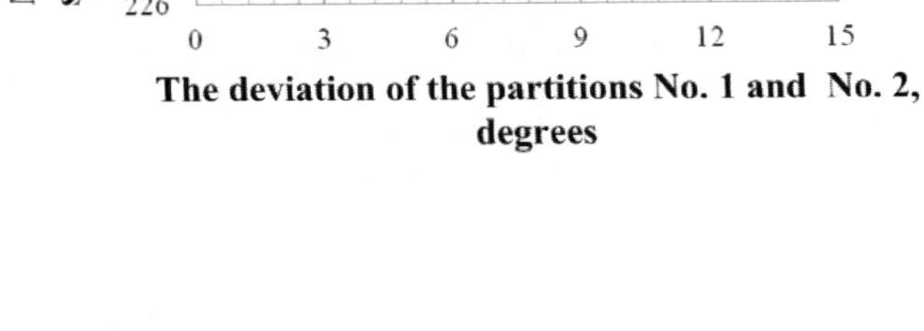

Figure 5. Graph of dependence of maximum equivalent stresses on values of deviations of partitions No.1 and 2 from vertical position in the project in the direction of clockwise direction.

IOP Conf. Series: Materials Science and Engineering 1181 (2021) 012025 doi:10.1088/1757-899X/1181/1/012025

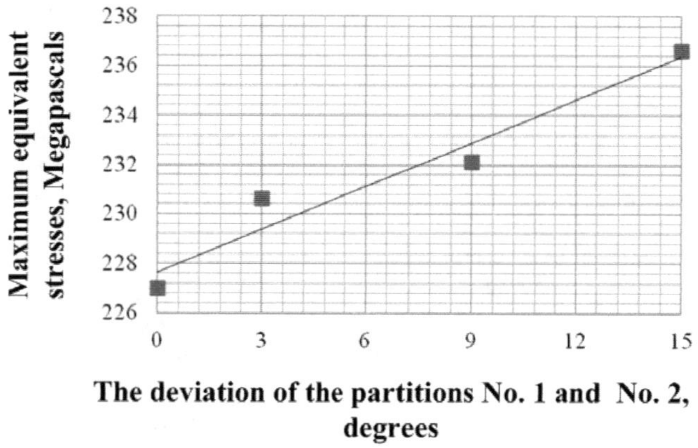

Figure 6. Graph of dependence of maximum equivalent stresses on values of deviations from vertical position according to the design of partition No. 1 in clockwise direction, No. 2 in counterclockwise direction.

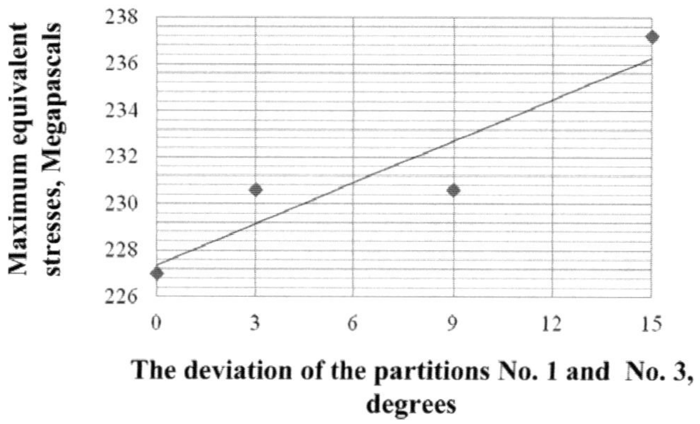

Figure 7. Graph of dependence of maximum equivalent stresses on values of deviations of partitions No. 1 and 3 from vertical position in the project in the direction of clockwise direction.

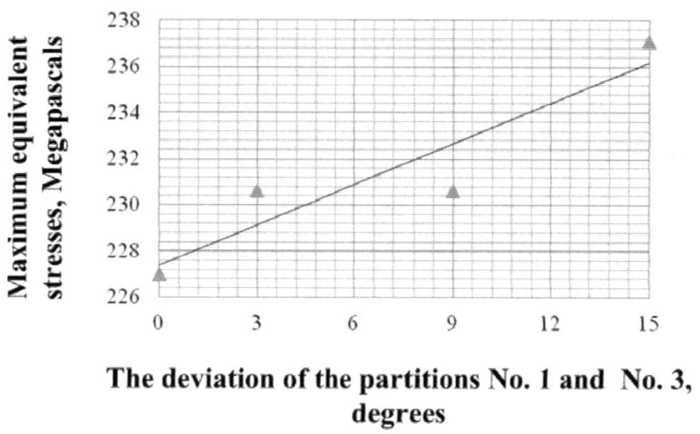

Figure 8. Graph of dependence of maximum equivalent stresses on values of deviation from vertical position according to the design of partition No. 1 in clockwise direction, No. 3 in counterclockwise direction.

4. Conclusion

Based on the results of the SDS simulation of the tube bundle of the shell-and-tube heat exchanger with different thickness values and values of deviation of the partitions from their vertical location according to the project, it is possible to formulate certain conclusions:

- it was found that with an increase in the thickness of the transverse partitions of the tube bundle, an increase in the maximum equivalent stresses in the heat exchange tubes occurs, and the maximum stresses increase by about 2 MPa for every 2 mm of increase in thickness;
- it is shown that the values of the maximum equivalent stresses of the tube bundle depend on the values of the deflection of the partitions. When the values of deviations from the vertical location according to the design of transverse partitions increase, the values of maximum equivalent stresses are observed. This must be taken into account when diagnosing the heat exchanger in order to timely detect the most loaded zones.

References

[1] Valentina D P, Valentina D S, Salvatore M and Stefano R 2021 *Procedia Computer Science* **180** 958-967
[2] Kuusk A and Gao J 2021 *Lecture Notes in Mechanical Engineering* 201-213
[3] The Y-L and Kuusk A G 2021 *Lecture Notes in Mechanical Engineering* 153-163
[4] Freund L and Al-Majeed S 2020 *2020 International Conference on Innovation and Intelligence for Informatics, Computing and Technologies*, 9312015
[5] Kriegel M and Berdermann J 2020 *2020 European Navigation Conference* 9317443
[6] Rubtsov A V, Kulakov P A, Mukhametzyanov Z R, Arkhipova K S and Gimaltdinov I K 2020 *Journal of Physics: Conference Series* **1661(1)** 012079
[7] Galiev I, Ibragimov R, Ashrapov A and Radaykin O 2020 *IOP Conference Series: Materials Science and Engineering* **890(1)** 012140
[8] Mukhametzyanov Z R, Kulakov P A, Rubtsov A V and Churakov Y A 2020 *Journal of Physics: Conference Series* **1582(1)** 012055
[9] Rubtsov A V, Kulakov P A, Mukhametzyanov Z R, Bayazitov M I and Farshatov A R 2020 *Journal of Physics: Conference Series* **1515(5)** 052053
[10] Begalinov A, Almenov T, Zhanakova R and Bektur B 2020 *Mining of Mineral Deposits* **14(3)** 28-36
[11] Nasibullina O A and Tyusenkov A S 2019 *IOP Conference Series: Materials Science and Engineering* **537(2)** 022018
[12] Zheng L Q, Chen X M, Zhang B, Zheng Y Z and Lin Z L 2019 *Chinese Journal of Tissue Engineering Research* **23(36)** 2095-4344
[13] De Miranda D A and Nogueira A L 2019 *Materials Research* **22(2)** e20180564
[14] Erick Fiestas S and Sixto Prado G 2017 *2017 LARS 14th Latin American Robotics Symposium and 2017 5th SBR Brazilian Symposium on Robotics* December 1-6
[15] Qi J, Meng H, Kan Z, Li C and Li Y 2017 *Transactions of the Chinese Society of Agricultural Engineering* **33(24)** 65-71

IOP Conf. Series: Materials Science and Engineering 1181 (2021) 012026 doi:10.1088/1757-899X/1181/1/012026

Thermoelastic non-stationary fields in a rigidly fixed plate

Zh M Kusaeva

Assistant, Samara state technical university, 244, Molodogvardeyskaya St., 443100, Samara, Russian Federation

E-mail: zhkusaeva@mail.ru

Abstract. In this article, a new closed solution of the axisymmetric dynamic problem of the theory of thermoelasticity is constructed for a rigidly fixed circular isotropic plate in the case of temperature changes on its face surfaces. The mathematical formulation of the problem includes linear equations of motion and thermoelasticity in the spatial formulation with respect to the components of the displacement vector, as well as the function of temperature change. The study of non-self-adjoint equations is carried out in an unrelated statement. Initially, we consider the initial boundary value problem of thermoelasticity without taking into account the deformation of the plate, and at the next stage, we study the problem of elasticity theory under the action of a given (defined) temperature change function. To solve the problems, we use a mathematical apparatus for separating variables in the form of finite integral transformations: Fourier, Hankel, and generalized integral transformation (CIP). In this case, at each stage of the study, the procedure is performed to bring the boundary conditions to the form that allows you to apply the appropriate transformation.

1. Introduction

Uneven non-stationary heating of structures for various purposes leads to the thermal deformations and stresses occurence, which must be taken into account in the case of a comprehensive analysis of the strength characteristics of elastic systems of finite dimensions [1, 2]. At present, various theories of thermoelasticity have been developed (CTE, GHI – GHIII, LS) [3, 4], which solves this problem with varying accuracy degrees.

The mathematical formulation of the considered initial-boundary value problems in a linear three-dimensional formulation includes coupled non-self-adjoint differential equations of motion and heat conduction. As a rule, this system of differential equations is considered in an unrelated setting [5-10]. In this case, when an external non-stationary heat load acts on the elastic system, the effect of the rate of change in the volume of the body on the temperature field is not taken into account.

In a related formulation, closed solutions of dynamic problems of thermoelasticity are presented in a few works.

In particular, studies [11-13] were carried out for a finite isotropic cylinder with membrane fixing of its end surfaces. In [11], using the generalized method of finite integral transformations [14], and [12, 13] - the biorthogonal integral transform [15, 16]. Research [17, 18] was carried out using hyperbolic (GHII, GHIII) theories of thermoelasticity and helps to analyze the frequency equations, as well as the forms of harmonic waves in an infinite cylindrical waveguide.

In this work, a rigidly fixed round isotropic plate is investigated. The case of the action on the upper and lower surfaces of an unsteady axisymmetric temperature load (boundary conditions of the

1st kind) is considered. The numerical results of calculating this problem in an unconnected formulation [19] allow us to conclude that the inertial forces of an elastic system affect its stress-strain state only in very thin structures ($h^*/b \le 0.01$, h^*, b – thickness and radius of the plate) under the action of a high-frequency load. Taking into account these results, the inertia forces are not taken into account when solving the system of non-self-adjoint differential equations of the classical (CTE) theory of thermoelasticity, i.e. the constraint is used for the considered constructions $h^*/b \le 0.01$.

The constructed solution of the coupled problem in a three-dimensional formulation makes it possible to take into account the effect of the rate of change in its volume (rate of dilatation) on the nature of the distribution of the temperature field and the stress-strain.

2. Materials and methods

Let a round rigidly fixed plate occupy the region $\Omega: \{0 \le r_* \le b, 0 \le \theta \le 2\pi, 0 \le z_* \le h^*\}$ in the cylindrical coordinate system (r_*, θ, z_*). On the upper and lower surfaces, the temperature is set, the value of which depends on the radial coordinate r_* and time t_*: at $z_* = 0$ $\omega_1^*(r_*, t_*)$, at $z_* = h^*$ $\omega_2^*(r_*, t_*)$ (figure 1).

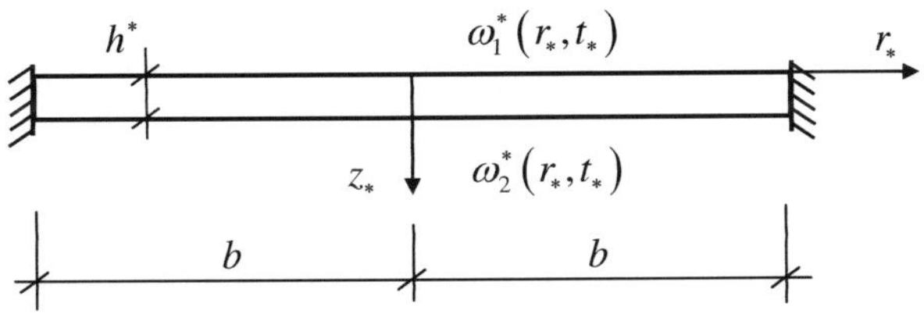

Figure 1. Calculation scheme.

The mathematical formulation of the initial boundary value problem under consideration in a dimensionless form includes a system of linear axisymmetric non-self-adjoint differential equations for the components of the displacement vector $U(r,z,t), W(r,z,t)$ and temperature $\Theta(r,z,t)$:

$$\frac{\partial}{\partial r}\nabla U + a_1\frac{\partial^2 U}{\partial z^2} + a_2\frac{\partial^2 W}{\partial r \partial z} - \frac{\partial \Theta}{\partial r} = 0 \tag{1}$$

$$a_1 \nabla \frac{\partial W}{\partial r} + \frac{\partial^2 W}{\partial z^2} + a_2 \frac{\partial}{\partial z}\nabla U - \frac{\partial \Theta}{\partial z} = 0$$

$$\nabla \frac{\partial \Theta}{\partial r} + \frac{\partial^2 \Theta}{\partial z^2} - \frac{\partial \Theta}{\partial t} - a_3 \frac{\partial}{\partial t}\left(\nabla U + \frac{\partial W}{\partial z}\right) = 0$$

boundary conditions:

$$r = 0,1 \qquad \{U(0,z,t), W(0,z,t), \Theta(0,z,t)\} < \infty, \quad \frac{\partial \Theta}{\partial r}_{|r=1} = 0, \quad \{U(1,z,t), W(1,z,t)\} = 0; \tag{2}$$

$$z = 0, h \qquad \frac{\nu}{1-\nu}\nabla U + \frac{\partial W}{\partial z} = \{\omega_1, \omega_2\}, \quad \frac{\partial W}{\partial r} + \frac{\partial U}{\partial z} = 0, \quad \Theta(r,z,t)_{|z=0,h} = \{\omega_1, \omega_2\};$$

and initial conditions:

$$t = 0 \qquad \Theta(r,z,0) = 0. \tag{3}$$

where $\{U, W, r, z, h\} = \{U^*, W^*, r_*, z_*, h^*\}/b$, $\{\Theta, \omega_1, \omega_2\} = a_4\{\Theta^*, \omega_1^* - T_0, \omega_2^* - T_0\}$, $a_1 = a_2(1-2\nu)$, $a_2 = 0.5(1-\nu)^{-1}$, $a_3 = \frac{\gamma^2(1+\nu)(1-2\nu)}{E(1-\nu)c_\varepsilon}T_0$, $a_4 = \frac{1+\nu}{1-\nu}\alpha_t$, $t = t_*\frac{\Lambda}{b^2 c_\varepsilon}$, $U^*(r,z,t), W^*(r,z,t), \Theta^*(r,z,t)$ –

displacement vector components and temperature increment in dimensional form; $\Theta^* = T - T_0$, Θ^*, T, T_0 – current temperature and temperature of the initial state of the body, in which there are no mechanical stresses; E, ν – elastic modulus and Poisson's ratio of the material; $\alpha_t, c_\varepsilon, \Lambda$ – coefficients of linear thermal expansion, volumetric heat capacity and thermal conductivity of the material.

3. Results and discussion

The initial boundary value problem (1) – (3) is solved by the method of integral transformations, using successively the Hankel transform [20] with finite limits in the variable and the degenerate biorthogonal finite transformation [15] in the coordinate z. At each stage of the solution, the procedure of standardization of the corresponding boundary conditions is carried out [21].

Transformants $R(n,z,t)$, $G(\lambda_{in},n,t)$ and the inversion formulas of the corresponding transformations have the following form:

$$R(n,z,t) = \int_0^1 N(r,z,t) P(n,r) r dr, \quad N(r,z,t) = \sum_{n=0}^{\infty} \Omega_n^{-1} RP; \tag{4}$$

$$G(\lambda_{in},n,t) = \int_0^t R(n,z,t) Y(\lambda_{in},z) dz, \quad R = \sum_{i=1}^{\infty} GH \|K_{in}\|^{-1}; \tag{5}$$

where $N(r,z,t) = [U(r,z,t), W(r,z,t), \Theta(r,z,t)]^T$, $P = [s_{mp}]$ – 3-order diagonal matrix, $(s_{11} = J_1(j_n r), s_{22} = s_{33} = J_0(j_n r))$, $Y(\lambda_{in},z) = [0, 0, K_3(\lambda_{in},z)]^T$, $H(\lambda_{in},z) = [N_1(\lambda_{in},z), N_2(\lambda_{in},z), N_3(\lambda_{in},z)]^T$, K_3, N_1, N_2, N_3 – components of the vector-function of biorthogonal transformations; $\Omega_n, \|K_{in}\|$ – square of the norm of transformation kernels; j_n, λ_{in} – eigenvalues ($n = 0,1,2...,i = 1,2,3...$).

As a result, we obtain an expression for the functions $U(r,z,t), W(r,z,t), \Theta(r,z,t)$ in the form of spectral expansions:

$$N(r,z,t) = \sum_{n=0}^{\infty} \Omega_n^{-1} P(n,r) \left[F_k + \sum_{i=1}^{\infty} G(\lambda_{in},n,t) H(\lambda_{in},z) \|K_{in}\|^{-1} \right], \tag{6}$$

where F_k – matrix is a column of standardizing functions.

The algorithm for solving the initial – boundary value problem of thermoelasticity (1) - (3) is described in detail in [22].

4. Conclusions

As an example, we consider a rigidly fixed round plate($b = 1$m) made of steel, which has the following physical and mechanical characteristics of the material: $E = 2\times10^{11}$ Pa, $\Lambda = 50$, W/(m^{0}K), $\nu = 0.28$, $c_\varepsilon = 3.8\times10^6$ J/(m^3 ^{0}K), $\alpha_t = 1.2\times10^{-5}$ 1/^{0}K.

The case of the action of ($z_* = 0$) a temperature load on the upper front surface in the form of:

$$\omega_1^*(r_*,t_*) = (1-r_*) T_{\max} \left[\sin\left(\frac{\pi}{2t_{\max}} t_* \right) H(t_{\max}^* - t_*) + H(t_* - t_{\max}^*) \right], \omega_2^*(r_*,t_*) = 0, \tag{7}$$

where $H(\tilde{t})$- the single function of Heaviside ($H(\tilde{t})=1$ at $\tilde{t} \geq 0$, $H(\tilde{t})=0$ at $\tilde{t} < 0$), $T_{\max} = T_{\max}^* - T_0$, $T_{\max}^*, t_{\max}^*$ – the maximum value of the external temperature effect and the corresponding time in the dimensional form ($T_{\max}^* = 373\ ^0K$ ($100\ ^0C$), $T_0 = 293\ ^0K$, ($20\ ^0C$)).

Figures 2-4 show graphs of temperature changes $\Theta^*(0, h/2, t)$, components of the vector of displacements $U^*(0.5,z,t), W^*(0,z,t)$ in time and axial coordinate with allowance for (solid line) and also without (dashed line) for the coupling of thermoelastic fields ($t^*_{\max} = 10\,\text{c}$, $t_{\max} = \dfrac{\Lambda}{b^2 c_\varepsilon} t^*_{\max} = 1.3\times10^{-4}$).

The temperature field and stress-strain state are analyzed for plates with a thickness $h = 0.1, 0.2$.

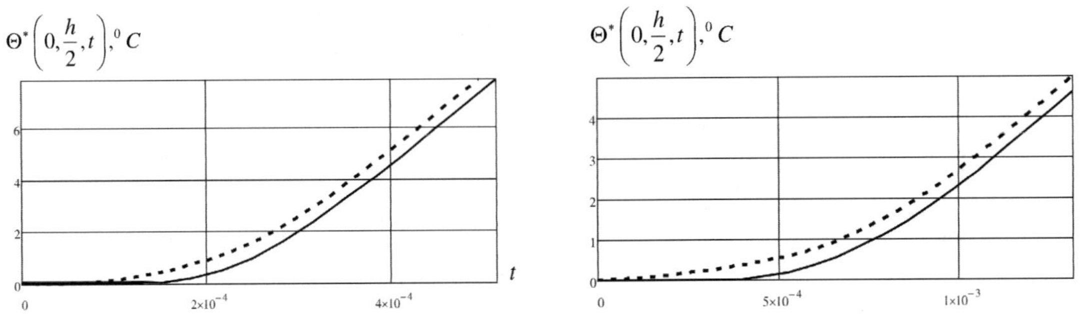

a) $h=0.1$ b) $h=0.2$

Figure 2. Graphs of temperature $\Theta^*\left(0,\frac{h}{2},t\right)$ changes over time t.

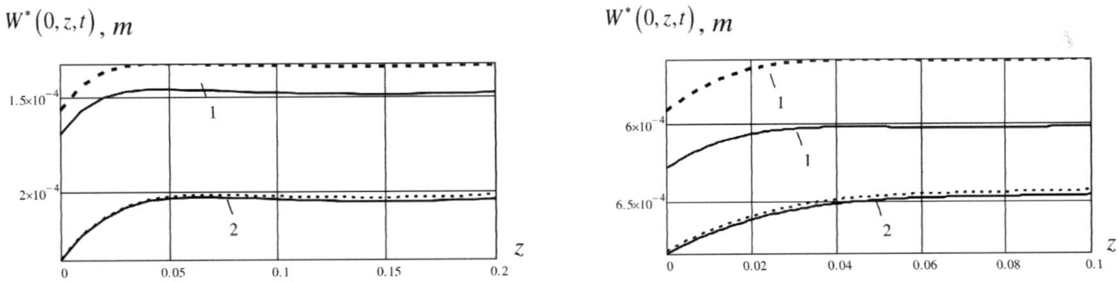

a) $h=0.1$ m b) $h=0.2$ m

Figure 3. Graphs of the $W^*(0,z,t)$ change in the height of the plate (1–$t=t_{\max}$, 2–$t=5t_{\max}$).

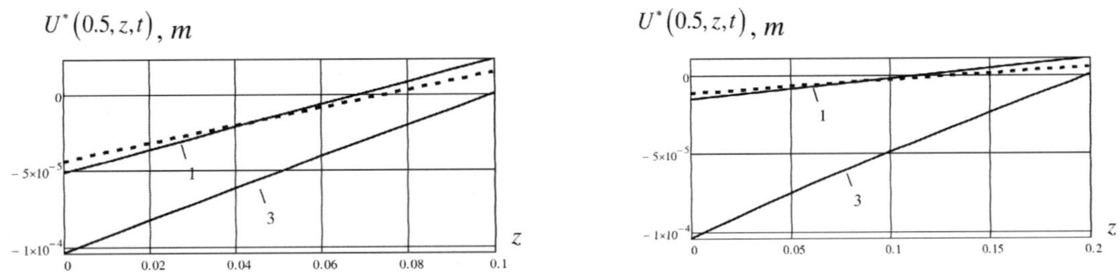

a) $h=0.1$ b) $h=0.2$

Figure 4. Graphs of change $U^*(0.5,z,t)$ along the height of the plate (1–$t=t_{\max}$, 3–$t=100t_{\max}$).

Analysis of the calculation results allows us to draw the following conclusions:

- the connectivity of thermoelastic fields at a given temperature load (7) leads to a slower heating of the plate over time (figure 2). In this case, the rate of change in the volume of the body, which is taken into account in the heat conduction equation (1), has a significant effect at the first stage of the study of the temperature regime, when $t_{\max} < t < 10t_{\max}$. In the future, this effect is not observed;

- a decrease in the rate of temperature change inside the plate due to its dilation gives an increase in the gradients $\partial\Theta/\partial r, \partial\Theta/\partial z$ that are used in the initial differential equations of thermoelasticity (1). As a result, there is an increase in the numerical values of the axial component of the displacement vector (figure 3, graphs 1, 2);
- at a given temperature load, the coupling of thermoelastic fields decreases over time (figure 3). In addition, as a result of warming up the structure, an increase in displacements is observed (figure 3,4), and with a steady temperature regime on the lower surface there are no radial displacements (figure 4, graph 3);
- the linear nature of the change in the radial component of the displacement vector along the height of the plate, allows us to conclude that when solving thermoelasticity problems for homogeneous elastic systems with the help of applied theories, it is possible to use the kinematic hypothesis of plane sections;
- the numerical values of radial displacements $U^*(0.5, z, t)$ at a steady temperature regime do not depend on the thickness of the plate (figure 4, graph 3).

Figure 5 shows graphs of changes in normal mechanical stresses along $\sigma_{rr}(r,0,t)$ the radial coordinate at different points in time taking into account (solid line), as well as disregarding (dashed line) the connectivity of thermoelastic fields.

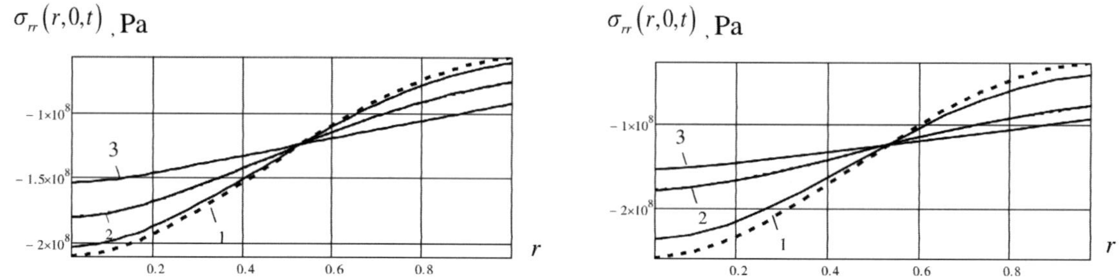

a) $t^*_{max} = 10\ c$ $(t_{max} = 1.3\times10^{-4})$, $1 - t = t_{max}$, $2 - t = 5t_{max}$, $3 - t = 100t_{max}$

b) $t^*_{max} = 1\ c$ $(t_{max} = 1.3\times10^{-5})$, $1 - t = t_{max}$, $2 - t = 50t_{max}$, $3 - t = 500t_{max}$

Figure 5. Graphs of changes $\sigma_{rr}(r,0,t)$ in the radial coordinate $(h = 0.1)$.

Here you can draw a conclusion:

- the greatest influence of the field coupling on the stress tensor component $\sigma_{rr}(r,0,t)$ is observed at $t = t_{max}$. Subsequently, this effect sharply decreases (figure 5);
- in the process of warming up $t > t_{max}$ the structure, when there is a decrease in normal stresses $\sigma_{rr}(r,0,t)$. At $t = t_{max}$ this component of the stress tensor is higher at a faster temperature loading of the plate;
- it should be noted that when the condition is met $\omega_2^*(r_*, t_*) = 0$ the lower face plane $(z = h)$ is a neutral surface $\sigma_{rr}(r,h,t) = 0$, since $\partial W^*/\partial z_{|z=h} = 0$ (figure 3) and (figure 4).

In conclusion, we can conclude that when calculating structures of finite dimensions in the case of a high-speed thermal load, the coupling of temperature and elastic fields has a significant effect on its stress-strain state. Moreover, this feature is more pronounced in thin plates.

Acknowledgments
The reported study was funded by Foundation RFBR and, according to the research project № 20-31-90042.

References

[1] Podstrigach Y S 1984 *Thermal elasticity of bodies of non-uniform structure* (M.: Science)
[2] Novatsky V 1970 *Dynamic problems of thermoelasticity* (M.: World)
[3] Radaev Y N and Taranova M V 2011 Wave numbers of thermoelastic waves in a waveguide with heat exchange on the side wall *Bulletin of the Samara state University Ser. Phys.-Math. Sci.* **2(23)** 53-61
[4] Shashkov A G, Bubnov V A and Yanovsky S Y 2004 *Wave phenomena of thermal conductivity System-structural approach* **2** (M.: Unitorialal URSS)
[5] Kovalenko A D 1965 *Introduction to thermoelasticity* (Kiev: Science Duma)
[6] Sargsyan S H 2013 Mathematical Model of Micropolar Thermo-Elasticity of Thin Shells *J. of Thermal Stresses* **36(11)** 1200-16
[7] Verma K L 2008 Thermoelastic waves in anisotropic plates using normal mode expansion method *World Academy of Science, Engineering and Technology* **(37)** 573-80
[8] Zhornik A I, Zhornik V A and Savochka PA 2012 On a thermoelasticity problem for a solid cylinder *News of the Southern Federal University. Techn. Sci.* **9(1)** 63-9
[9] Makarova I S 2012 Solution of an unrelated thermoelasticity problem with boundary conditions of the first kind *Bulletin of the Samara state technical university. Ser. Phys.-Math. Sci.* **28(3)** 191-5
[10] Harmatij H, Król M and Popovycz V 2013 Quasi-Static Problem of Thermoelasticity for Thermosensitive Infinite Circular Cylinder of Complex Heat Exchange *Advances in Pure Mathematics* **(3)** 430-7
[11] Senitsky Yu E 1982 To the solution of the coupled dynamic thermoelasticity problem for an infinite cylinder and sphere *Applied Mechanics of the AS USSR* **18(6)** 34-41
[12] Lychev S A 2003 The related dynamical problem of thermoelasticity for a finite cylinder *Bulletin of the Samara State University* **4(30)** 112-24
[13] Lychev S A, Manzhirov A V and Uuber S V 2010 Closed solutions of boundary value problems of coupled thermoelasticity *J. of Russian Academy of Sciences* **4** 138-54
[14] Senitsky Yu E 1991 Multicomponent generalized finite integral transformation and its application to nonstationary problems of mechanics *Proceedings of high schools* **4** 57-63
[15] Senitskiy Y E 1996 Biorthogonal multi-component finite integral transform and its application to boundary value problems of mechanics *Proceedings of universities Matematika* **8** 71-81
[16] Lychev S A and Senitsky Yu E 2002 Nonsymmetric integral transformations and their applications to problems of viscoelasticity *Bulletin of Samara un-ta. Natur. Sci. Series* 16-38
[17] Kovalev V A, Radaev Yu N and Semenov D A 2009 Related dynamic problems of hyperbolic thermoelasticity *Bulletin of the Saratov Uni. Ser. Math-s. Mechan. Inform.***4(2)** 94-127
[18] Kovalev V A, Radaev Yu N and Revinsky R A 2011 Passage of a generalized GHIII-thermoelastic wave through a waveguide with a heat-permeable wall *Bulletin of the Saratov University, New.ser. Ser. Math. Mechan. Inform.* **11(1)** 59-70
[19] Shlyakhin D A and Dauletmuratova Zh M 2018 Nonstationary axisymmetric problem of thermo-elasticity for a rigidly fixed circular plate *Eng. J.: Science and Innovation* **5(77)** doi.1018698 / 2308-6033-2018-5
[20] Sneddon I N 1955 *Fourier Transforms* (Moscow: Foreign Literature Publishing House)
[21] Butkovskiy A G 1979 *Characteristics of systems with distributed parameters* (M.: Science)
[22] Shlyakhin D A and Dauletmuratova Zh M 2019 Nonstationary coupled axisymmetric thermoelasticity problem for a rigidly fixed round plate *Bulletin of the PNRPU. Mechan.* **4** 191-200

To the question of determining indicators for evaluating quality compaction and efficiency for pavers

A P Prokopev, Zh I Nabizhanov, R T Emelyanov and V L Sabinin

Siberian Federal University, 79, Svobodny ave., Krasnoyarsk, RU-660041, Russia

E-mail: prok1@yandex.ru

Abstract. The results are considered of a study on the development of indicators of the quality of road materials compaction and efficiency for pavers. On the basis of the dimensionality analysis method, a dimensionless complex for the size of the ramming bar stroke is obtained, for use in the synthesis of systems of continuous non-destructive testing, automatic control and evaluation of the efficiency of stackers. The indicators for evaluating the efficiency of existing and newly created pavers are obtained, which have a close correlation with the compaction coefficient of the asphalt concrete mixture. The results are presented of calculation of specific and generalized performance indicators of various models of pavers.

1. Introduction

The quality of asphalt road surfaces largely depends on the technical operation of compacting the mixture during construction. Under-compaction of layers of road pavement leads to premature numerous damages during operation. According to the code of rules of CR 78.13330.2012 "Highways" (updated version of SNiP 3.06.03-85), at the final stage of construction of road surfaces a set of machines is used – stackers, a link of rollers. The link of rollers consists of light (weighing 5-6 t), medium (8-10 t) and heavy (10-15 t). The paper [1] presents data from studies of the compaction capacity of stackers and road rollers.

Studies [1] have shown "that light rollers provide a compaction coefficient (K_c) of 0.85-0.88, medium – K_c = 0.92-0.94 and heavy – K_c = 0.97-1.0". Taking into account these data and the results of other scientific works [2-5, et al.], we can conclude that a promising direction is to significantly improve the quality of road surfaces, reduce the types and number of road rollers, and increase the economic efficiency of construction, through the use of modern stackers with high-compaction working bodies.

The paper [6] provides data on the high compacting capacity of asphalt pavers with such working bodies: degree of compaction close to the normalized is provided when laying asphalt coarse-grained mixtures (type A and B, with a crushed stone content of at least 40 %), the stacker at a speed of 1.0-1.5 m/min, the frequency of the ramming bar 1400-1650 min^{-1} [6]. In this case, no additional compaction of the mixture is required. High efficiency is confirmed in research publications [3, et al.]. An important factor in the effectiveness of such an application of stackers is the temperature of the mixture in the operating range of 140-110 °C, since because of continuous changes in the temperature and density of the mixture, a constant increase in strength and deformation characteristics occurs. The ultimate strength and the modulus of deformation increase progressively with a decrease in temperature. Moreover, the modulus of deformation, which is a measure of the stiffness of the material, increases faster than the ultimate strength. Thus, when the temperature of the mixture decreases from 140 °C to 90 °C, the modulus of deformation increases sixfold, while the ultimate strength only triples [7]. Therefore, it is very important to compact the asphalt mixture in the optimal mixture's temperature range of the 140 °C - 110 °C.

2. Methods and results

The effectiveness of pavers can be objectively determined by analyzing them under the same operating conditions. The generalized indicator P_{NG}, used at the design stage with known technical characteristics, makes it possible to evaluate pavers belonging mainly to one type-sized group in the range of limited changes in the main parameter. The generalized indicator is the product of the specific energy intensity N_S by the specific metal content G_S [8].

The definition of the values N and G included in the P_{NG}, N_S, G_S expressions is not difficult, as for domestic and foreign stackers they are always given in the prospectuses and technical specifications. The main task is to determine the productiveness of stackers.

The productiveness of stackers can be determined by the following dependency [8]:

$$P = 3600 \cdot B \cdot h \cdot V \cdot \rho, \tag{1}$$

where B – width of the roadbed to be laid, m; h – thickness of the layer to be laid, m; V – working speed of the stacker, m/s; ρ – density of asphalt pavement compacted by the working body of the stacker, t/m^3. However, the direct application of this formula for evaluating the effectiveness of stackers is difficult due to the lack of clear data on the parameters V and ρ, which depend not only on the parameters of the pavers, but also on the type of compacted material.

Let's consider the workflow of the stacker and the parameter included in the productiveness formula. The stacker accepts the asphalt mix, transports it, distributes it, and compacts it. The useful work of the stacker can be attributed to the work on the distribution and compaction of asphalt concrete mixture. This work is carried out by compacting working bodies.

Stackers are used for laying asphalt concrete mixes of different types. When evaluating the effectiveness of stackers, it is possible to exclude parameters that characterize the properties of the compacted material, as it is possible to establish dependencies between the values included in the productiveness formula and the parameters of the stacker.

The density of the asphalt concrete mixture, taking into account the parameters of the compacting working body of the stacker and the road material, can be represented by the following function:

$$\rho = \varphi(e, f_t, V, h),$$

where $e = 2 \cdot r_t$ – bar stroke, m; r_t – eccentricity of drive shaft of the ramming bar, m; f_t – vibration frequency of the ramming bar, Hz.

In paper [6] to determine the parameters of stackers it is proposed to use the parameter n – the number of impacts on a surface element of the material:

$$n = \frac{f_t \cdot l_t}{V}, \tag{2}$$

where l_t – length of the ramming bar, m; V – m/s.

According to the manufacturers the length of the ramming bars of modern stackers is from 0.023 to 0.043 m. There is a need of additional factors, which have a significant impact on the efficiency of the compaction process of road materials, to develop a system for controlling the operating parameters of stackers. Based on the results of experimental studies [2, 11], we can identify a variable corresponding to this task. The stroke of the ramming bars is set when preparing the stackers for work on the object from 0.004 to 0.012 m [12]. The choice of the size of the bars stroke is carried out depending on the thickness of the layer and the type of road material.

Taking into account the functional dependences of the compaction coefficient on the variable parameters of the working process of stackers obtained in scientific works [2, 11, et al.], a dimensionless group for the size of the ramming bar stroke is proposed on the basis of the dimensional analysis method considered in the work of Schenk H. Jr. [13]:

$$n_{t.e} = \frac{f_t \cdot e}{V}. \tag{3}$$

On the basis of the dimensionless group proposed in this article, it is possible to obtain expressions for determining the compaction coefficient of an asphalt concrete mixture:

$$K_c = K \cdot \frac{f_t \cdot e}{V},\ K_c = \frac{\rho}{\rho_{st}},$$

where K – proportionality coefficient; ρ – density of asphalt concrete mixture compacted by the working body of the stacker, t/m^3; ρ_{st} – the density of the same reshaped sample, compacted in the standard way, t/m^3.

Hence, the working speed of the paver is

$$V = K \cdot \frac{f_t \cdot e \cdot \rho_{st}}{\rho}. \tag{4}$$

Substituting expression (5) into formula (1), we get the following dependence:

$$P = 3600 \cdot B \cdot h \cdot f_t \cdot e \cdot \rho_{st} \cdot K. \tag{5}$$

Assuming $h,\ \rho_{st},\ K \approx \text{const}$, the performance can be represented as follows:

$$P = 3600 \cdot B \cdot f_t \cdot e. \tag{6}$$

Considering that the compacting working bodies of stackers can have several ramming bars, the expression (6) can be represented as follows:

$$P = 3600 \cdot B \cdot f_t \cdot \sum_{i=1}^{n} e_i,$$

where n – number of compacting ramming bars.

The dependencies obtained above make it possible to switch to specific indicators. For stackers equipped with working bodies with ramming bars the specific energy intensity, specific metal content and the generalized indicator [8, 10], respectively, are determined by the following dependencies:

$$N_s = \frac{N}{3600 \cdot B \cdot f_t \cdot \sum_{i=1}^{n} e_i}, \tag{7}$$

$$G_s = \frac{G}{3600 \cdot B \cdot f_t \cdot \sum_{i=1}^{n} e_i}, \tag{8}$$

$$P_{NG} = \frac{G \cdot N}{\left(3600 \cdot B \cdot f_t \cdot \sum_{i=1}^{n} e_i\right)^2}. \tag{9}$$

Based on the obtained formulas (7)-(9), the calculation of specific and generalized efficiency indicators was performed to compare different models of pavers. The calculation results are shown in table 1.

Table 1. Performance indicators of pavers.

Paver model	B (m)	G/N (t/kW)	P (m²/h)	G_s/N_s (t h/m²) / (kW h/m²)	P_{NG} (t kW h²/m⁴)
Titan-420	12.5	29.5/167	5513	0.0053/0.030	$1.6 \cdot 10^{-4}$
Titan-410S	12.0	22.3/88	7409	0.0030/0.0119	$3.6 \cdot 10^{-5}$
DR-150S	12.5	23.5/165	5625	0.0042/0.0293	$1.2 \cdot 10^{-4}$
Super-2000	12.5	25.0/150	10773	0.0023/0.0140	$3.2 \cdot 10^{-5}$
Titan-280	7.5	15.1/65	3308	0.0046/0.020	$9.2 \cdot 10^{-5}$
DR-120S	8.5	19.7/82	4131	0.0048/0.020	$9.6 \cdot 10^{-5}$
Super-1700	8.5	18.0/94	6120	0.003/0.015	$4.5 \cdot 10^{-5}$
11011 K	7.0	15.5/54	2520	0.006/0.021	$1.2 \cdot 10^{-4}$
DS-179, two tampers	7.0	18.0/77	5103	0.0035/0.015	$5.2 \cdot 10^{-5}$
DS-179, one tamper	7.0	18.0/77	3402	0.0053/0.023	$1.2 \cdot 10^{-4}$

Thus, based on the results of calculating the performance indicators of pavers according to technical parameters B, e, f_t, it can be concluded that among the considered models the most effective for G_s, N_s, P_{NG} are the models "Super-1700" (Germany), DS-179 with two bars ("Dormashina", plant) (screed width of 7...8.5 m), and "Titan-410S" (Germany), "Super-2000" (Germany) (screed width of 12...12.5 m). Table1 shows that to improve the efficiency of the domestic asphalt paver DS-179, it is necessary to reduce the metal consumption and increase the stroke of the bars.

It is advisable to make unambiguous assessments of the technical level of pavers using the method of calculating the technical level coefficient [10]:

$$K_{t.l} = \frac{P_o}{P_m},$$

where $K_{t.l} < 1$ – average level; $K_{t.l} > 1$ – for the best options of paver models; P_o – basic (standard or average) indicator; P_m – performance indicator.

Then, using the obtained dependencies, you can proceed to specific calculations, for example, on generalized indicators:

$$K_{t.l} = \frac{G_o \cdot N_o \cdot \left(B_m \cdot f_{tm} \cdot \sum_{i=1}^{n} e_{mi} \right)^2}{G_m \cdot N_m \cdot \left(B_o \cdot f_{to} \cdot \sum_{i=1}^{n} e_{oi} \right)^2}.$$

Indicators make it possible to evaluate existing and newly created machines, but they cannot serve as a basis for calculating specific parameters e, f_t, N, G, V, etc. These parameters of asphalt pavers should be calculated according to known dependencies and be within certain restrictions that ensure optimal quality of work (compaction coefficient, evenness, etc.) in the construction of asphalt roads.

3. Conclusion

On the basis of the dimensionality analysis method, we proposed a dimensionless complex for the size of the stroke of the ramming bars of the working body of the pavers $n_{t.e} = f_t \cdot e/V$, which can be recommended for use in the design of systems for continuous non-destructive quality control of compaction and automatic control of compaction of road materials by pavers. The dimensionless complex can be used to create a neural network (intelligent) system for monitoring and controlling the compaction of road materials.

Specific and generalized indicators for evaluating the effectiveness of pavers are obtained as a result of transformations of the dimensionless complex for the size of ramming bars stroke and analytical dependencies.

A formula for evaluating the technical level of pavers by the coefficient of technical level is obtained, which is useful at the stages of creating new models, upgraded versions, as well as for justifying a commercial contract for the purchase of specific models of pavers.

Acknowledgments

The research was performed in Siberian Federal University and supported by Russian Foundation for Basic Research (RFBR grant no. 19-37-90052).

References

[1] Permyakov V B and Zakharenko A V 1989 Substantiation of contact pressure values for compaction of asphalt-concrete mixtures *Construction and Road Building Machinery* **5** 12-3
[2] Böhmer P 1974 Untersuchunger uber die Verdichtung – swirkung von Schwarzdeckenfertigern *Baumaschine und Bautechnik* **7-8** 233-8
[3] Prang R 1989 Hochverdichtungsbonle der ABG *Die Asphaltstrasse* **8** 22-6
[4] Jian Sun and Guiyun Xu 2013 Dynamics Modeling and Analysis of Paver Screed Based on Computer Simulation. *Appl. Sci.* **13** 1059-65
[5] Prokopyev A P, Ivanchura V I, Emelianov R T and Palchikov P A 2018 Implementation of the concept of automation and intellectualization of management of road construction processes *Proc. of the Moscow State University of Civil Eng.* **13 1(112)** 61-70
[6] Maryshev B S, Blind B M, Lieberman M A, etc. 1983 *High-speed construction of asphalt concrete pavements* (Moscow: VPTITRANSSTROY) **1** 39
[7] Boguslavsky A M 1965 *Road asphalt concrete pavement* (Moscow: Higher school publ.)
[8] Balovnev V I, Ermilov A B and Novikov A N 1988 *Road-building machines and complexes* (Moscow: Mashinostroenie Publishers)
[9] Kustarev G V, Pavlov S A and Ushkov A V 2019 *High-performance complexes for high-speed construction of asphalt concrete coverings* (Moscow: MADI publ.)
[10] Balovnev V I 2014 *The similarity in the design and modeling of the system of road-building machines* (Moscow: MADI publ.)
[11] Prokopev A P 1991 *Optimization parameters of the compacting equipment of the paver with swinging tampers* Thesis diss. Ph.D.. (Moscow: MADI publ.)
[12] WIRTGEN GROUP 2019 Concentrating on the essentials: high quality paving *RoadNews* **7** 28-47 Retrieved from: https://media.voegele.info/media/03_voegele/aktuelles_und_presse/roadnews_magazin/roadnews_07/RadNews_07__en.pdf
[13] Schenck Hilbert Jr. 1979 *Theories of engineering experimentation* (McGraw-Hill Book Company, New York)

Dubrovin's method and Ablowitz-Kaup-Newell-Segur hierarchy

A O Smirnov and A S Kolesnikov

Sankt-Petersburg State University of Aerospace Instrumentation, B. Morskaya, 67A, St.-Petersburg, 190000, Russia

E-mail: alsmir@guap.ru

Abstract. Previously, the Dubrovin method was used to construct spectral curves of multiphase solutions of such integrable models of nonlinear optics as vector and derived nonlinear Schrödinger equations. The method of constructing spectral curves of multiphase solutions of the scalar nonlinear Schrödinger equation was based on replacing matrix differential operator from the Lax pair by the scalar one. The present paper shows that Dubrovin's method gives the same results. At the same time, it has the advantage that it consists in a simpler construction of multiphase solutions expressed in terms of elementary functions.

1. Introduction

Dubrovin's method of constructing spectral curves [1] is very useful for studying multicomponent integrable models. With its help, it is not difficult to obtain higher integrable nonlinear equations from the corresponding hierarchies, stationary nonlinear equations that are satisfied by multiphase solutions, ansatzes for equations of spectral curves, as well as the simplest solutions of equations from the hierarchies under study. In particular, in [2], using Dubrovin's method, uniform formulas were obtained for three main types of derivative nonlinear Schrödinger equations: Kaup-Newell equation [3-6], Chen-Lee-Liu equation [5-7] and Gerdjikov-Ivanov equation [5,6, 8, 9].

Since until now a different method has been used to find the equations of spectral curves of multiphase solutions of equations from the Abovitz-Kaup-Newell-Segur hierarchy (see, for example, [10]), the problem arose of comparing the results obtained by different methods. This work is devoted to solving this problem.

2. Monodromy matrix and its properties

Let us consider the equation

$$\Psi_x = U\Psi, \tag{1}$$

where

$$U = \lambda J - Q, \quad J = \begin{pmatrix} -i & 0 \\ 0 & i \end{pmatrix}, \quad Q = \begin{pmatrix} 0 & ip \\ -iq & 0 \end{pmatrix}, \tag{2}$$

λ is a spectral parameter, p and q are functions, $i^2 = -1$.

As was shown by Dubrovin [1], multiphase matrices Q correspond to solutions Ψ of the equation (1) such that matrices $M\Psi$ are also solutions of the equation (1).

$$(M\Psi)_x = U(M\Psi). \tag{3}$$

The matrix M is called a monodromy matrix. From the equation (3), it follows that the matrix M satisfies the equation

$$M_x + MU - UM = 0. \tag{4}$$

Assuming that the Matrix M equals to

$$M = \sum_{k=1}^{n} M_k \lambda^k,$$

we substitute it into the equation (4) and equate to zero the coefficients at all powers of the spectral parameter λ. After performing calculations, we get the following representation for the matrix M:

$$M = a_0 J + a_1 U + \sum_{k=1}^{n-1} a_{k+1} W_k, \tag{5}$$

where a_k are some constants,

$$W_1 = \lambda U - W_1^0, \quad W_{k+1} = \lambda V_k - W_{k+1}^0, \quad W_k^0 = \begin{pmatrix} F_k & H_k \\ G_k & -F_k \end{pmatrix}, \quad k \geq 1.$$

Equation (4) also implies the following relations for the elements of the matrices W_k^0:

$$H_1 = -\frac{1}{2} p_x, \quad G_1 = -\frac{1}{2} q_x,$$
$$(F_k)_x = -iqH_k - ipG_k,$$
$$H_{k+1} = pF_k + \frac{i}{2}(H_k)_x, \quad G_{k+1} = -qF_k - \frac{i}{2}(G_k)_x.$$

In particular,

$$H_2 = -\frac{i}{4}(p_{xx} - 2p^2 q), \quad G_2 = \frac{i}{4}(q_{xx} - 2pq^2), \quad F_2 = \frac{1}{4}(pq_x - qp_x),$$
$$H_3 = \frac{1}{8}(p_{xxx} - 6pqp_x), \quad G_3 = \frac{1}{8}(q_{xxx} - 6pqq_x).$$

3. Multiphase solutions of the Ablowitz-Kaup-Newell-Segur hierarchy equations

Let us consider the following Lax pair

$$\Psi_x = U\Psi,$$
$$\Psi_{t_k} = V_k \Psi,$$

where $V_k = 2^k W_k$. From the equality of derivatives $(\Psi_x)_{t_k} = (\Psi_{t_k})_x$ we get the compatibility condition

$$U_{t_k} - (V_k)_x + UV_k - V_k U = 0.$$

As a result, we obtain the evolutionary nonlinear equations

$$p_{t_k} = -2^{k+1} H_{k+1}, \quad q_{t_k} = -2^{k+1} G_{k+1}. \tag{6}$$

It is easy to see that the first equations defined by formulas (6) are equations from the Ablowitz-Kaup-Newell-Segur hierarchy. In particular, for k=1 the equations (6) are the coupled nonlinear Schrodinger equations [10,11]

$$ip_{t_1} + p_{xx} - 2p^2 q = 0,$$
$$iq_{t_1} - q_{xx} + 2q^2 p = 0.$$

And for k=2, we get the coupled modified Korteweg–de Vries equations [10,11]

$$p_{t_2} + p_{xxx} - 6pqp_x = 0,$$
$$q_{t_2} + q_{xxx} - 6qpq_x = 0.$$

Substituting λ=0 into the equations (4) and (5), we obtain the following equality

$$(M_0)_x - M_0 Q + Q M_0 = 0, \tag{7}$$

where

$$M_0 = a_0 J - a_1 Q - \sum_{k=1}^{n-1} a_{k+1} W_k^0. \tag{8}$$

From the equations (7) and (8), we get the stationary matrix differential equation

$$a_0 (JQ - QJ) + a_1 Q_x + \sum_{k=1}^{n-1} a_{k+1} \left(\left(W_k^0\right)_x - W_k^0 Q + Q W_k^0 \right) = 0,$$

the scalar form of which is

$$\begin{aligned} 2a_0 p + ia_1 p_x - 2i \sum_{k=1}^{n-1} a_{k+1} H_{k+1} = 0, \\ 2a_0 q - ia_1 q_x + 2i \sum_{k=1}^{n-1} a_{k+1} G_{k+1} = 0. \end{aligned} \tag{9}$$

Let us recall that the equation of the spectral curve for the n-phase solution has the form [1]

$$\det(i\mu I - M) = 0,$$

where I is an identity matrix. Therefore, the spectral curve of n-phase solutions has the following form

$$\mu^2 = \det M = a_{n+1}^2 \lambda^{2n+2} + \sum_{k=1}^{2n+2} f_k \lambda^{2n+2-k}, \tag{10}$$

where f_k there are some constants defining the solution. Let us note that, without losing generality, we can put $a_{n+1} = 1$.

4. Spectral curves and the simplest solutions of the nonlinear Schrodinger equation

Using the stationary equations and relations between constants f_k and the functions p and q, we can find the simplest solutions of the Ablowitz-Kaup-Newell-Segur hierarchy equations.

In particular, for n=1 the monodromy matrix is

$$M_1 = a_0 J + a_1 U + W_1.$$

It follows from equations (9) that one-phase solutions for any value of t satisfy the stationary equations

$$\begin{aligned} &2a_0 p + ia_1 p_x + p^2 q - \frac{1}{2} p_{xx} = 0, \\ &2a_0 q - ia_1 q_x + q^2 p - \frac{1}{2} q_{xx} = 0. \end{aligned} \tag{11}$$

For n=1 the equation (10) has the following form

$$\mu^2 = \lambda^4 + 2a_1\lambda^3 + \left(a_1^2 + 2a_0\right)\lambda^2 + f_3\lambda + f_4,$$

where

$$\begin{aligned} &f_3 = 2a_0 a_1 - a_1 pq + \frac{i}{2}\left(pq_x - qp_x\right), \\ &f_4 = a_0^2 + \left(a_0 - a_1^2\right)pq + \frac{i}{2}a_1\left(pq_x - qp_x\right) + \frac{1}{4}\left(p^2q^2 - p_x q_x\right). \end{aligned} \tag{12}$$

Let us express the functions p and q in terms of their Wronskian $w = pq_x - qp_x$ and their product $u = pq$:

$$p = \sqrt{u}\exp\left(-\frac{1}{2}\int \frac{w}{u}dx\right), \quad q = \sqrt{u}\exp\left(\frac{1}{2}\int \frac{w}{u}dx\right). \tag{13}$$

Substituting (13) into (11) and (12), we obtain the following relations

$$\begin{aligned} &w = 2i\left(2a_0a_1 - f_3 - a_1 u\right), \\ &u_x^2 = 4u^3 + 4\left(4a_0 - a_1^2\right)u^2 + 8\left(2a_0^2 - 2a_0a_1^2 + a_1 f_3 - 2f_4\right)u - 4\left(f_3 - 2a_0a_1\right)^2. \end{aligned} \tag{14}$$

It follows from (14) that in the general case, the function $u(x)$ is expressed in terms of elliptic Weierstrass $\wp$-function [12,13]

$$u(x) = \wp(x \mid g_2, g_3) + A,$$

where

$$\begin{aligned} &A = \frac{1}{3}\left(a_1^2 - 4a_0\right), \\ &g_2 = \frac{4}{3}\left(a_1^2 + 2a_0\right)^2 - 8a_1 f_3 + 16 f_4, \\ &g_3 = \frac{8}{27}\left(a_1^2 + 2a_0\right)^3 - \frac{8}{3}\left(a_1^3 + 2a_0a_1\right)f_3 + 4f_3^2 + \frac{16}{3}\left(a_1^2 - 4a_0\right)f_4. \end{aligned}$$

It is well-known that there are values of the constants when the equation (14) has an elementary solution. In particular, let the spectral curve is given by the equation

$$\mu^2 = \left(\left(\lambda + a\right)^2 + b^2\right)^2.$$

Then, expressing the parameters of the solution in terms of the parameters of the spectral curve, we have

$$a_0 = a^2 + b^2, \quad a_1 = 2a, \quad f_3 = 4a\left(a^2 + b^2\right), \quad f_4 = \left(a^2 + b^2\right)^2. \tag{15}$$

Substituting (15) into (14), we get

$$u_x^2 = 4u^3 + 16b^2u^2, \quad w = -4iau. \tag{16}$$

Solving the equations (16), we find

$$u = -4b^2 \operatorname{sech}\left(2bx + \varphi(t)\right)^2, \quad w = 16iab^2 \operatorname{sech}\left(2bx + \varphi(t)\right)^2. \tag{17}$$

Substituting (17) into (13) and simplifying, we obtain

$$p(x,t) = 2ib \operatorname{sech}(2bx + \varphi(t)) e^{2iax + i\theta(t)},$$
$$q(x,t) = 2ib \operatorname{sech}(2bx + \varphi(t)) e^{-2iax - i\theta(t)}.$$

We can find the functions $\varphi(t)$ and $\theta(t)$ from the corresponding evolutionary equations. In particular, in the case the coupled nonlinear Schrödinger equation these functions are equal

$$\varphi(t) = -8abt, \quad \theta(t) = 4\left(b^2 - a^2\right)t.$$

5. Concluding remarks

Dubrovin's method gives classical evolutionary equations from the Ablowitz-Kaup-Newell-Segur hierarchy. Equations of the spectral curves are independent of calculation method. At the same time, it should be noted that the solution expressed in terms of the elliptic Weierstrass $\wp$-function was previously unknown to the authors. We believe that the method considered in this paper is promising for finding multiphase solutions expressed in terms of elementary functions.

Acknowledgments

This work was supported by the Ministry of Science and Higher Education of the Russian Federation (grant agreement № FSRF-2020-0004).

References

[1] Dubrovin B A 1985 Matrix finite-zone operators *J. Soviet Math.* **28(1)** 20
[2] Smirnov A O 2021 Spectral curves for the derivative nonlinear Schrödindger equations *Symmetry* **13** 1203
[3] Kaup D J and Newell A C 1978 An exact solution for a derivative nonlinear Schrödinger equation *J. Math. Phys.* **19** 798
[4] Kamchatnov A M 1997 New approach to periodic solutions of integrable equations and nonlinear theory of modulational instability *Physics Reports* **286** 199
[5] Xu S, He J and Wang L 2011 The Darboux transformation of the derivative nonlinear Schrödinger equation *J. Phys. A* **44(30)** 305203
[6] Zhang Y S, Guo L J, He J S and Zhou Z X 2015 Darboux transformation of the second-type derivative nonlinear Schrödinger equation *Lett. Math. Phys.* **105** 853
[7] Chen H, Lee Y C and Liu C S 1979 Integrability of nonlinear Hamiltonian systems by inverse scattering method. Special issue on solitons in physics. *Phys. Scripta* **20** 490
[8] Gerdjikov V S and Ivanov M I 1983 The quadratic bundle of general form and the nonlinear evolution equations. I. Expansions over the "squared" solutions are generalized Fourier transforms *Bulgarian J. Phys.* **10** 13
[9] Gerdjikov V S and Ivanov M I 1983 A quadratic pencil of general type and nonlinear evolution equations II Hierarchies of Hamiltonian structures *Bulgarian J. Phys.* **10** 130
[10] Matveev V B and Smirnov AO 2018 AKNS hierarchy, MRW solutions, Pn breathers, and beyond

IOP Conf. Series: Materials Science and Engineering 1181 (2021) 012028 doi:10.1088/1757-899X/1181/1/012028

J. Math. Phys. **59** 091419
[11] Ablowitz M J, Kaup D J, Newell A C and Segur H 1974 The inverse scattering transform-Fourier analysis for nonlinear problems *Studies in Appl. Math.* **53(4)** 249
[12] Abramowitz M and Stegun I A 1972 Handbook of mathematical functions with formulas, graphs and mathematical tables *Nat. Bureau Standards Appl. Math. Ser.* **55**
[13] Akhiezer N I 1990 *Elements of the theory of elliptic functions* (American Mathematical Society, Providence, RI)

Model for assessing technogenic risks at the enterprises of the fuel and energy complex of the Krasnoyarsk Territory

Y V Grebnev[1,2] and A K Moskalev[1]

[1] Siberian Federal University, Krasnoyarsk Territory, Krasnoyarsk, Svobodny Ave. 79, Russia
[2] Department of Emergency Situations, Main Directorate of the Ministry of Emergency Situations of Russia for the Krasnoyarsk Territory, Krasnoyarsk, Prospect Mira 68, Russia

E-mail: yaroslav.grebnev@gmail.com, ak_moskalev@mail.ru

Abstract. Every year on the planet there are many natural and man-made accidents that have a significant impact on the infrastructure and normal life of society. In 2020, on the territory of the Krasnoyarsk Territory, there were 2 major technogenic accidents associated with the bottling of oil products and causing significant damage to the environment, which indicates the high importance of the problem of monitoring and forecasting, as well as the need to supervise enterprises that store and refuel equipment with oil products. This issue also acquires great importance in connection with the development of the design resource, tank farm and technological equipment, which are still in operation. The solution to this problem lies in the mainstream of organizing continuous monitoring, increasing supervisory measures, organizing forecasting scenarios for the development of the situation in order to develop action plans for the implementation of one of the calculated scenarios, for a prompt response and reduce the consequences of an emergency. In this work, the problem of developing scenarios for the development of emergency situations at a tank farm and predicting possible consequences in order to reduce the likelihood of these events was solved. To solve this problem, 12 scenarios of the development of emergency situations were simulated and the situation with the ingress of oil products into the Yenisei River was simulated in order to assess the area of bottling of oil products. Modeling of emergency situations was carried out using the software product Toxy + risk and neural network forecasting of the bottling area until localization using the neural network simulator NeuroPro, developed at the Institute of Computational Modeling of the Federal Research Center of the KSC SB RAS. A comparison is made between the analytical method for determining the area of the oil spill in a reservoir and the neural network method.

1. Introduction

Despite the development of green technologies, oil and its derivatives, which also play a major role in the chemical industry, remain the main source of energy for technological machines and equipment. But the active use of petroleum products often leads to local and regional environmental disasters. According to the analysis of accidents that occurred in the Russian Federation, one of the main reasons for the bottling of oil products remains the use of fuel tanks whose service life has expired, but despite this, the operation of this equipment continues. On the territory of Russia, according to the results of the analysis of the literature [1-4], according to experts, about 65% of fuel tanks have exhausted their service life, and this also applies to the equipment installed for servicing these oil depots. An example of a disaster of this kind is the accident that occurred in the city of Norilsk in 2020 at a fuel storage, as a result of which more than 30,000 tons of diesel fuel got into the soil and water bodies.

Oil spills may result in contamination of soil, atmosphere and water bodies. The consequence of this is the disruption of the habitat of plants, animals, birds, fish, etc. Possible destruction of food chains, displacement from habitual habitats, disruption of migration routes, etc. The severity of such consequences largely depends on the season of the spill (period migration, spawning, etc.), as well as hydrometeorological conditions.

In terms of the severity of environmental consequences, two types of mutual location of the accident site with natural objects can be distinguished: accidents with pollution of water bodies and accidents with pollution of the territory.

An urgent scientific task is to model possible accidents and consequences in order to take prompt measures and reduce the consequences of the negative impact of such accidents on the ecosystem.

The importance of emergency forecasting issues was raised in the works of researchers such as G.I. Korovin, A.A. Molchanov, G.A. Dorer, A.A. Volchek, I.A. Chuchuev [5-17].

2. Materials and methods

In this work, we set the goal of developing a model for assessing the risks of an emergency situation caused by the bottling of oil products. As a model object, a tank farm was chosen, located directly near the water protection zone of the Yenisei River, in the area of the Abalakovo settlement of the Krasnoyarsk Territory.

Modeling of scenarios of emergency situations at the oil depot was carried out using the TOXI + Risk software product [18] and the neuroimitator NeuroPro 0.25 developed at the Federal Research Center of the KSC SB RAS [19].

To solve the problem of modeling of oil falling into the Yenisei River, used multilayer neural network with sigmoid activation functions, neural network learning algorithm and parameters for the simulation are presented in [20]

Modeling of scenarios of oil product combustion (spill, fire-flash, explosion) in accordance with the Methodology for determining the calculated values of fire risk at production facilities [21], using the TOXI + Risk software product.

3. Results and discussion

As part of this work, we have analyzed various scenarios for the development of an emergency. Any scenario begins with an initiating event that can occur with a certain frequency. Possible causes and factors contributing to the occurrence and development of accidents at the facility.

When assessing the frequencies of initiating events, a statistical assessment of malfunctions and accidents was carried out by types of equipment, taking into account the possibility of initiating an accident from external causes (lightning strikes, sabotage, terrorism). According to statistical data, the occurrence of sources of initiation of an emergency scenario in petrochemical plants is implemented with the frequencies shown in table 1.

Table 1. Basic frequency indicators of failures of individual items technological equipment.

The source of the accident	Alarm triggering event	Hole diameter expiration, mm	Frequency depressurization, year[1]
Pipelines	Depressurization with subsequent outflow of liquid or two-phase medium	5	$4.3 \cdot 10^{-3}$
		12.5	$6.1 \cdot 10^{-4}$
		25	$5.1 \cdot 10^{-4}$
		50	$2.0 \cdot 10^{-4}$
		Complete destruction	$1.0 \cdot 10^{-4}$
Pumps	Failure of pumping equipment	-	$5 \cdot 10^{-4}$
	Depressurization with	All types	$1.0 \cdot 10^{-4}$

Reservoirs for storage at pressure close to atmospheric	subsequent expiration	-	$5.0 \cdot 10^{-6}$
Reservoirs with stationary roof	liquid in the dump	-	$9.0 \cdot 10^{-5}$
	Quasi-instant	-	$9.0 \cdot 10^{-5}$
Bulk devices (tips, hoses, sleeves)	Fistula	Up to 5	$1.2 \cdot 10^{-3}$
	The gap	-	$4.5 \cdot 10^{-5}$
Tanks (auto and railway) for transportation petroleum products	Depressurization with	-	$1.4 \cdot 10^{-5}$
	Quasi-instant destruction	-	$2.0 \cdot 10^{-6}$
	Depressurization with	5	$4.3 \cdot 10^{-3}$

According to the statistics given in Table 1, the frequency of occurrence of the most dangerous event - depressurization of an oil storage tank - is 5.0×10^{-6} 1/year.

Let us calculate all possible scenarios for the development of an emergency in accordance with the methodology described in Section 1.

3.1. Scenario 1

Emergency depressurization of the ground tank. Complete depressurization of the above ground tank with diesel fuel with a volume of 5000 m^3.

Tank 1 is located in tank farm 1. The height of the earth embankment is 2 m. The cover material inside the embankment is soil. The embankment area of the second group of tanks is 9430 m^2.

If the tank is depressurized, the spill area will be 9430 m^2, the oil product will not go beyond the embankment. The thickness of the spilled oil product layer will be 0.5 m.

3.2. Scenario 2

In the event of a quasi-instantaneous destruction of the above ground tank 1 with diesel fuel, with a volume of 5000 m^3, which is in the embankment, with an area of 9430 m^2, with a height of an earthen embankment of 2 m, as a result of a hydrodynamic shock, the embankment may be destroyed, a part of the spilled oil product will go beyond the embankment and spill over the adjacent territory. The projected spill area is 25,000 m^2. The spill, with an area of 15570 m2, will go beyond the embankment of the tank. The coating material inside and behind the embankment is soil. The layer thickness will be 0.05 m.

3.3. Scenario 3

Emergency depressurization of the above-ground tank 2. Complete depressurization of the above-ground tank with diesel fuel, volume 5000 m^3.

Tank 2 is located in tank farm 1. The height of the concrete embankment is 1.3 m. The coating material inside the embankment is concrete. The embankment area of the third group of tanks is 7310m^2.

If the tank is depressurized, the spill area will be 7310 m^2, the oil product will not go beyond the embankment. The thickness of the spilled diesel fuel layer will be 0.6 m.

3.4. Scenario 4.

In the event of a quasi-instantaneous destruction of a ground tank 2 with diesel fuel with a volume of 5000 m^3, located in a concrete embankment, with an area of 7310 m^2, a height of 1.3 m, as a result of a hydrodynamic shock, the embankment may be destroyed, a part of the spilled oil product will go beyond the embankment and spill over the adjacent territory. The projected spill area is 25,000 m^2. Concrete surface area 7310 m^2. The spill area on the ground surface will be 17690 m^2. The thickness of the spilled oil product layer is 0.2 m.

3.5. Scenario 5

Emergency depressurization of the ground tank 3. Complete depressurization of the ground tank with gasoline, volume 2000 m^3.

Tank 3 is located in tank farm 2. The height of the earth embankment is 1.5 m. The cover material inside the embankment is soil. The area of the park is 9830 m^2.

When the tank is depressurized, the spill area will be 9830 m^2, the oil product will not go beyond the embankment. The thickness of the spilled gasoline layer will be 0.2 m.

3.6. Scenario 6

Emergency depressurization of the ground tank 4. Complete depressurization of the ground tank with fuel oil, volume 400 m^3. Tank 4 is located in a tank farm with an area of 4860 m^2, an earthen embankment height of 1.5 m.

When the tank is depressurized, the spill area will be 2000 m^2, the oil product will not go beyond the embankment. The thickness of the spilled fuel oil layer will be 0.2 m.

3.7. Scenario 7

Quasi-instant destruction of railway tank cars on a two-way railway unloading overpass No. 1 (gasoline A-95, as the most dangerous substance). Depressurization of 50% of the total volume of tanks in a railway train, which is 12 railway tanks, because the simultaneous reception of up to 24 railway tank cars with a volume of 161.6 m^3 each is provided.

Complete depressurization of a train can occur as a result of its derailment, overturning and fracture of the hulls when hitting the ground. This can happen due to improper actions of the locomotive driver, train builder, or a faulty track.

The total volume of 12 railway tanks is 1939.2 m^3.

The spill will occur on a concrete surface with an exit to the ground. Concrete surface area 1158 m^2. The estimated spill area is 9696 m^2. The layer thickness will be 0.05m.

3.8. Scenario 8

Quasi-instant destruction of railway tank cars on railway unloading rack 1 with gasoline. Depressurization of 50% of the total volume of tanks in a railway train, which is 3 railway tanks, because It is provided for the simultaneous reception of up to 6 railway tanks, each with a volume of 161.6 m^3.

The reasons are the same as described in scenario 7.

The total volume of 3 railway tanks is 484.8 m^3.

The spill will occur on a concrete surface with partial exposure to the ground. Concrete surface area 355 m^2. The spill area will be 2424 m^2. The layer thickness will be 0.05 m.

3.9. Scenario 9

Quasi-instant destruction of railway tank cars on the railway unloading rack 2 with gasoline. Depressurization of 50% of the total volume of tanks in a train, which is 1.5 railway tanks, because provided for the simultaneous reception of up to 3 railway tanks, each with a volume of 161.6 m^3.

The reasons are the same as described in scenario 7.

The total volume of 1.5 rail tank cars is 242.4 m^3.

The spill will occur on a concrete surface with partial exposure to the ground. Concrete surface area 116.76 m^2. The spill area will be 1212 m^2. The layer thickness will be 0.05 m.

3.10. Scenario 10

Accidental spill of gasoline (as the most dangerous substance) when a tanker is at the site for loading light oil products. Loading area - 1200 m^2. The site has an asphalt surface. The volume of the tanker is 17 m^3.

The predicted spill area may be 85 m^2, the spill layer thickness is 0.05 m.

The filling is carried out under the supervision of the operator and the driver of the tanker truck. Therefore, the time it takes to detect an emergency is minimal. In the event of a leak of oil products, it is necessary to immediately shut off the supply of oil products, and then take measures to prevent fire.

3.11. Scenario 11

Complete emergency depressurization of the blocking pipeline with gasoline (as the most dangerous substance), with a diameter of 200 mm, the maximum distance between the valves - 1323 m. Transportation of oil products through the pipeline occurs by gravity. The difference in heights is 31.5 m. The possible time for the expiration of oil products from the pipeline is taken as 300 seconds (the time to turn off the emergency unit). Thus, if the blocking pipeline ruptures, the spill volume may reach 111.8 m^3, the spill will occur on the ground surface, the spill area will be up to 1342 m^2, the layer thickness is 0.05 m. The amount of gasoline is 82.7 tons.

3.12. Scenario 12

Complete emergency depressurization of the hose filling device at the berthing facilities, DN = 100 mm, the maximum distance between the valves - 10 m. Oil product - AI-92 gasoline.

Oil products are transported through the pipeline by gravity. The difference in heights is 31.5 m. The possible time for the expiration of oil products from the pipeline is taken as 300 seconds (the time to turn off the emergency unit). If the hose device is destroyed, the volume of the spilled oil product will be 17.64 m^3. The oil product will freely spill over the water surface and move downstream of the Yenisei River. This scenario is the most dangerous from the point of view of damaging the ecosystem of the Yenisei River. We will carry out further modeling in order to predict the possible consequences.

In accordance with the methodology described in Section 2 of this work, using the means of using the TOXI + Risk software product, the following results of modeling the risk of an accident at a tank farm were obtained (table 2).

Table 2. Forecasting the volumes and areas of oil spills in the water area.

Accident	Volume gasoline spill, m^3	Spill area of SGK at water area, m^2	Time after the start of diesel fuel (DF) discharge into the Yenisei River, s.	The distance to which the DF slick will spread downstream of the Yenisei River from the place where the stream flows into, calculated by the analytical method, m.	Distance to which the DF slick will spread downstream of the Yenisei River from the confluence of the brook stream, calculated by the neural network method, m.
Destruction of the onshore process pipeline in the area of possible impact on the Yenisei River	17.64	5880 m^2 – area of pollution of the Yenisei River	600	1223	1352
			1800	2730	2590
			3600	5880	5720

Figure 1 shows the zone of maximum possible contamination by spilled oil products (with full containment in 1 hour).

Figure 1. Simulated oil products bottling zone 1 hour after entering the reservoir (scale 1: 25000).

The most dangerous scenarios associated with environmental risk are accidents with the ingress of oil products into the water area of the Yenisei River with further spread along its course.

As a result of the ingress of oil products into the reservoir, toxic pollution of the environment will occur (pollution of the soil (coastal strip) and the water area of the Yenisei River, atmospheric air with products of evaporation of oil products).

The estimated area of the spill spot will be 5880 m^2. Coastal contamination area - 250 m^2. The average thickness of the spilled oil product layer is 0.003 m. The depth of the coastal strip contamination is 5 cm.

One hour after the accident occurred and oil products got into the water area of the Yenisei River, full containment of oil products will be carried out, and therefore no further increase in scale is forecasted.

4. Conclusion

This paper presents an approach to modeling the scenarios of emergencies at a tank farm using neural network forecasting of the bottling of oil products when they enter the Yenisei River water area. The paper compares the analytical and neural network approaches to predicting the bottling of oil products in the event of an emergency. The neural network was trained on the basis of real data obtained during the bottling of petroleum products at the oil depot in the settlements Tukhard and Norilsk in 2020. The considered scenarios of the development of emergency situations indicate that the most dangerous event for the ecology of the region is the scenario of accident No. 12, which implies a complete emergency depressurization of the hose filling device at the berthing facilities. The simulation results indicate the applicability of this approach to predicting emergency situations at oil depots.

The neural network approach, in comparison with the analytical forecasting method, allows you to quickly perform the necessary calculations by adapting and automating the model for the needs of calculating the risk at a potentially dangerous facility and developing action plans to respond to these emergencies. In the event of an emergency, perform operational modeling and forecasting in order to assess the situation and make a decision on attracting additional forces and means necessary to eliminate the accident.

The results of the work can be applied in the development of plans for the prevention and elimination of emergency spills of oil and oil products.

Acknowledgment

The reported study was funded by RFBR according to the research project№ 20-37-90090.

References

[1] Zykov A V 2010 Possibilities of forecasting the residual life of vertical steel tanks *Problems of defense technology. Series 16: Technical means of countering terrorism* **9(10)** 109-12
[2] Kulneva V V, Zvyagintseva A V, Sazonova S A and Akamsina N V 2020 Models of diffusion of emergency emissions from railway tanks with petroleum products *Journal of Physics: Conference Series* (Krasnoyarsk, Russia) 22077 DOI 10.1088/1742-6596/1679/2/022077
[3] Kondrashov OG and Nazarov MN 2004 causal analysis of accidents vertical steel tanks *Electronic scientific journal Oil and Gas Business* **2** 19
[4] Afanasyev EA 2021 Technologies for preventing the spread of accidents in accidents on tanks with oil products *Problems of technosphere safety: materials of the international scientific-practical conference of young scientists and specialists* **10** 166-71
[5] Moskvichev V , Nicheporchuk V *et al* 2018 Information support for monitoring and risks of socio-natural and man-made systems development *Problems of Risk Analysis* **15(2)** 56-77
[6] Medvedeva O and Solovieva S 2016 Methodology for the valuation of damage caused by atmospheric air pollution *Issues of valuation* **4 (86)** 2–6
[7] Mikheeva A and Ayusheeva S 2014 Comparative analysis of economic damage to the environment in the model areas *Bulletin of East Siberian State Uni of Tech and Man* **3 (48)** 92–96
[8] Korovin GN 1969 Methodology for calculating some parameters of ground forest fires *Collection of scientific research works on forestry Proceedings of LenNIILH* **12** 244-62
[9] Dorrer G A 2008 Dynamics of forest fires (Novosibirsk: Publishing house of the SB RAS) p 404
[10] Grishin A M Mathematical models of forest fires Text (Tomsk: Publishing house of Tomsk University) p 278
[11] Volchek AA 2010 Neural network forecasting tools for flood control Materials of the International Scientific and Practical Conference "Ensuring environmental safety - the path to sustainable development of Kazakhstan" pp 83-16
[12] Chuchueva I A 2010 Forecasting time series using an extrapolation model based on a sample of maximum similarity *Science and modernity: collection of materials of the International Scientific and Practical Conference* (Novosibirsk) pp 187-92
[13] Shoigu S K (ed.) 2010 *Manual on the organization of management and operational (emergency) response in emergency response* (Moscow: EMERCOM of Russia) p 138
[14] Nicheporchuk V V 2018 Modern approaches to forecasting emergencies *Scientific and analytical journal Siberian Fire and Rescue Bulletin* **1** 89-95
[15] Grebnev Y, Moskalev A, Vershkov A and Gazizulina A 2019 The practice of connectionist model for predicting forest fires in the Arctic zones of the Krasnoyarsk Territory *International Journal of System Assurance Engineering and Management Int J Syst Assur Eng Manag.* 1-9
[16] Carlon C, Critto A, Ramieri E and Marcomini A 2007 DESYRE: Decision support system for the rehabilitation of contaminated megasites *Integrated Env Ass and Man* **3(2)** 211-22
[17] Grebnev Y and Moskalev A 2020 Influence of hydroelectric power station on flood situation and modeling of hydrological characteristics *E3S Web Conf. Volume* **221** 1-7
[18] *Automated system of simulation of the situation Toxy + risk* 2021Retrieved from: https://toxi.ru/
[19] *Neuroimmitter NeuroPro* 2021 Retrieved from: http://www.neuropro.ru/
[20] Grebnev Y and Moskalev A 2021 Modelling the accidental oil spills at potentially hazardous facilities in the Arctic zone of Krasnoyarsk Krai *IOP Conference Series: Earth and Environmental Science* **816** 1-9
[21] Ministry of the Russian Federation for Civil Defense, Emergencies and Elimination of Consequences of Natural Disasters 2009 Methodology for determining the estimated values of fire risk at production facilities (Moscow)

Implementation and investigation of a pulsed aluminum-air flame

A G Egorov

Togliatti State University, ulitsa Belorusskaya, 14, Togliatti, 445020, Russia

E-mail: eag@tltsu.ru

Abstract. To implement a pulse combustion process in a flow of aerosolized particles, flow aerodynamics investigation and determination of the time of aluminum particle residence in a sudden expansion combustion chamber was conducted. The identified flow aerodynamics features and obtained values of the local residence time of aluminum particles made it possible to develop a combustion chamber for a pulse power plant. The temperature field of a pulsed aluminum-air flame was obtained and the type of temperature-time relationship along the flame axis was established in the course of plant tests. Integrated radiation flux density and distribution of radiant energy intensity of a pulsed aluminum-air flame was determined in the wavelength spectral range $\lambda = 1.0...4.5$ μm. The investigation results obtained in this work can be applied to create pulse power plants fired with powdered metal combustible.

1. Introduction

The use of powdered metals in propulsion and technological installations [1-4] is due to their high energy characteristics. In installations pulsed combustion may be promising use of gas-dispersion jet streams containing high-powdered metals. Depending on the purpose of the power plant can obtain gas-dispersion pulse streams with preset parameters: speed, dimensions, temperature, frequency, etc.

At present, the technological capabilities of pulsed combustion are far from being fully disclosed. Studies in [5] demonstrated the ability to create rapid combustion based on aluminum powders in effective radiating plasma source with a volume of about 100 m^3 and the energy of the radiation is above 10 MJ.

The use of energy-consuming metals as a powdered metallic fuel (PMF) in pulsed power plants is associated with the organization of a stable combustion process in a non-stationary air suspension flow. Pulse combustion mode aerosuspension flow of the mixture may be arranged by turning on and off supply of fuel to the combustion chamber with a sudden expansion.

In this work, in order to solve the problem of organizing a pulsed combustion regime in an air suspension flow, the aerodynamics of the flow are investigated and the local residence time of aluminum (Al) particles in a chamber with a sudden expansion is determined. Based on the obtained data the installation and the radiating characteristics defined aluminum-air pulsed flame.

2. Experimental apparatuses

Investigation activities were being performed on a test bench built by the Research and Education Center “Energy-Rich Material Combustion Physics” at Togliatti State University. The test bench design and control systems are described in several monographs and our previous publications in periodicals. The schematic of the experimental apparatus for implementation of pulse combustion in a flow of aerosolized aluminum particles is shown in figure 1. The apparatus functions as follows: under the effect of vacuum created by an active air jet streaming out of ejector nozzle 1 powdered Al is fed from hopper 9 through intake tube 8 into ejector inlet chamber 2. From the inlet chamber Al powder passes to ejector mixing chamber 3, where it is mixed with the air streaming out of the ejector nozzle, to form an aerosolized flow which then is conveyed to combustor 6. A pulsed operating mode of the combustion chamber is implemented by turning on and off the feed of powdered aluminum into the combustor using valve. The valve is activated by electronic control unit 7 that allows varying the time of turning on and off the feed of Al powder into the combustor. The combustion process is initiated by ignition plug 5.

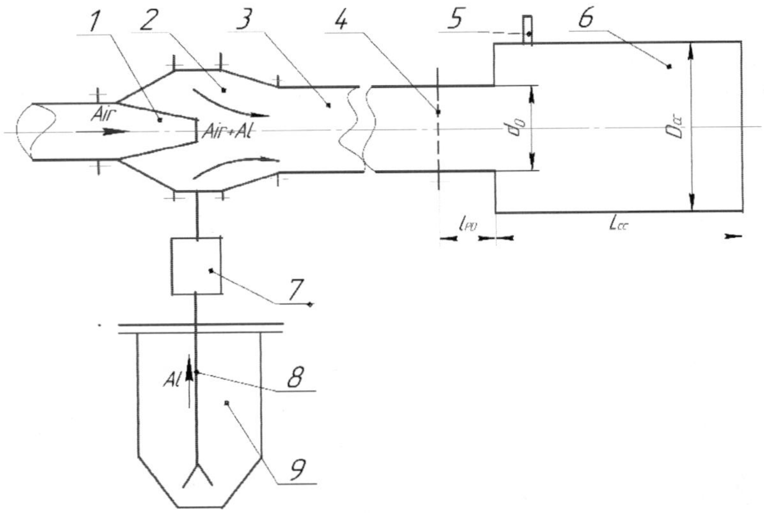

Figure 1. Schematic of the experimental apparatus for implementation of pulse combustion in a flow of aerosolized Al particles. *1* – air supplied from the system; *2* – ejector inlet chamber; *3* – ejector mixing chamber; *4* – perforated disk plate; *5* – ignition plug; *6* – combustor; *7* – electronic control unit; *8* – intake tube; *9* – hopper.

Commercial aluminum powder grades ASD-4, ASD-1, PA-4 and PA-1 were used as PMF. Air at a temperature of 293 K served as oxidizer. The in-flame temperature was measured using an AGA 782 thermographic system including a camera unit, thermal image display and highlighted-line display. The screen of the highlighted-line display was being photographed by an ANF camera with a frequency of 1 frame/s. The thermal image display visualized relative in-flame temperature distribution (a more intense background corresponded to a higher temperature). The highlighted-line display provided visualization of temperature distribution along the jet axis in real time with a frequency of 16 lines/sec. It is well known [6] that the most efficient methods of metal combustion mechanism's investigation include high-speed micro & macro-cinematography and photographing. The high-speed cinematography method was therefore applied to study aerosol flow aerodynamics in the combustor and determine the local residence time of Al particles. It permits measuring the local residence time of Al particles with an accuracy of ~ 5%. Isothermal and short-term tests with combustion were performed in combustion chamber models made of Pyrex heat resistant glass.

In the above experiments, initial turbulence of the incident aerosol flow was taken to be equal to turbulence of the "pure" air flow, i.e. disregarding the presence of Al particles therein.

Pyroelectric sensors were employed to measure thermal radiation energy in different spectral ranges. The relationship between instantaneous power and time $W(\tau)$ in a spectral range λ of 1.0…4.5 μm was determined by differentiating the time dependence of radiant energy $E(\tau) = 4\pi L^2 \cdot F(\tau)$ where F is the energy density measured by the pyroelectric sensors at a distance L from the pulsating dispersed gas flame. Radiation power of the flame as a whole was determined by the relationship $W_0 = I \cdot S$ where I is the energy emitted by a unit of volume per unit of time into a solid angle and S is the lateral area of the flame.

3. Results and discussion

The time of combustible and oxidizer residence in the flow is determined by the structure of its aerodynamics. This parameter is of a critical importance for molecular processes (molecular diffusion and chemical reaction) occurring during, or after, turbulent (convective) diffusion. For that reason, aluminum-air flow aerodynamics determines the local time of Al particle residence in the combustion chamber, on which in turn the work process parameters will depend.

To enable a pulse combustion chamber operation mode at a maximum frequency, the opening time of the combustible feed valve should be minimal, but sufficient to allow filling the recirculation zone with Al particles, their igniting and firing the mainstream flow of aluminum-air mixture. In such a case, a minimum time required for filling the recirculation zone with Al particles and their ignition, as well as firing the mainstream aerosolized flow can be calculated using the expression $\tau_{min} = \tau_{RZ} + \tau_{CC}$. The values of τ_{RZ} and τ_{CC} found for ASD-4 Al powder were $10 \cdot 10^{-3}$ and $4.2 \cdot 10^{-3}$ s. For ASD-1 those values were $15 \cdot 10^{-3}$ and $18.2 \cdot 10^{-3}$ s, respectively. Consequently, τ_{min} for aerosolized Al particles of $d_{32} =$ 7.4 μm will be equal to $14.2 \cdot 10^{-3}$ s, and $33.2 \cdot 10^{-3}$ s for particles of $d_{32} =$ 17.4 μm. With the turning on and off times of PMF feed to the combustor taken as equal, the period duration (pulse + interval) will be $28.4 \cdot 10^{-3}$ s for ASD-4 and $66.4 \cdot 10^{-3}$ s for ASD-1 aerosolized particles. Based on the obtained values of the local time of Al particle residence in the combustor models, a combustion chamber for a pulse power plant has been developed.

Power plant operation in the pulse mode is shown in figure 2.

a

b

Figure 2. Power plant working in the pulse mode. $D_{CC} = 0.07\ m$; a) pulse time: 0.2 s; b) interpulse time: 0.8 s; PMF: ASD-4 + PA-1 (in a ratio of 10 to 4).

According to figure 2a the flame assumes a near-cylindrical shape during the pulse time and a spherical shape during the interpulse time (see figure 2b). Temperature fields of the pulsed aluminum-air flame were recorded at the pulse and interpulse instant employing a thermographic system. Figure 3 shows that the area with a maximum temperature of 2274 K is localized on the flame axis. The temperature during the interpulse time is the same throughout the reacting volume, amounting to 1069 K.

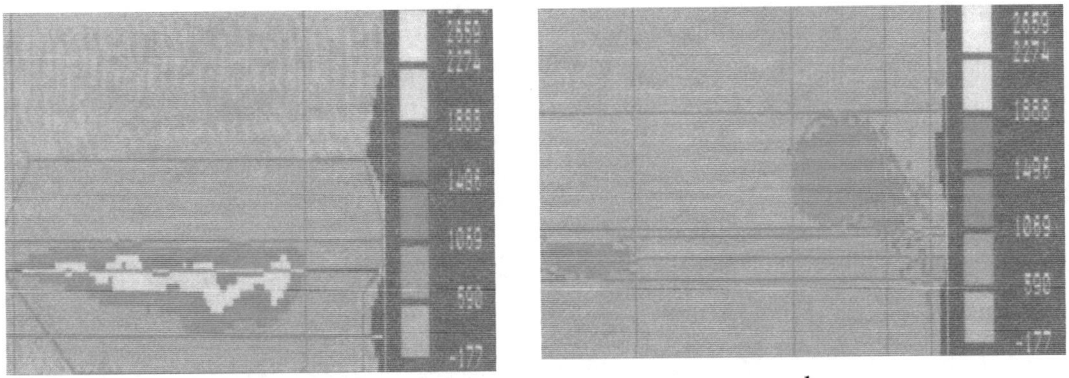

a b

Figure 3. Temperature field of a pulsed flame: a) pulse time: 0.2 s; b) interpulse time: 0.8 s; PMF: ASD-4 + PA-1 (in a ratio of 10 to 4).

Radiative fluxes in a dispersed gas flame are determined by flame macro-parameters (mass concentrations of oxidizer and combustible, metal particle fineness, geometric dimensions of the flame, etc) and optical characteristics of the initial combustible and condensed combustion products, such as absorption and dissipation efficiency, as well as complex refractive index. Also, the above parameters may depend heavily upon the temperature, wavelength and the radiating particle size [7]. Figure 4 shows temperature variation along the axis of a pulsed aluminum-air flame with the time.

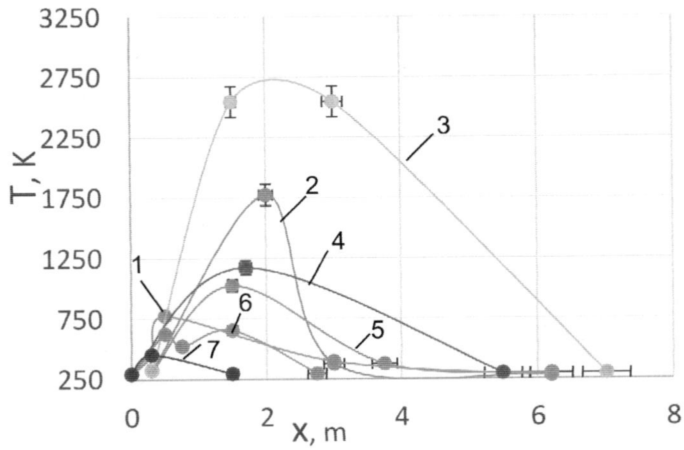

Figure 4. Temperature variation along the axis of a pulsed aluminum-air flame with the time. PMF: ASD-4 + PA-4 (in a ratio of 10 : 2); 1 – $\tau = 0.06\, s$; 2 – $\tau = 0.13\, s$; 3 – $\tau = 0.19\, s$: 4 – $\tau = 0.25\, s$: 5 – $\tau = 0.31\, s$; 6 – $\tau = 0.38$ s; 7 – $\tau = 0.44\, s$; curves 1-3-impulse, 4-7- pause.

The measured maximum temperature value is close to the boiling point of Al $(T_b = 2723K)$ and is significantly lower than the dissociation temperature Al_2O_3 $(T_d \approx 4000K)$, which corresponds to the temperature of the condensed phase in the flame front obtained by the poly-color method [8].

By controlling the operating modes of the apparatus and, respectively, PMF consumption it was established that the pattern of temperature distribution along the axis of the pulsed aluminum-air flame did not change. Such a relationship between the condensed-phase temperature in the combustion zone and the mass concentration of combustible was observed in a laminar aluminum-air flame [7].

Our experimental findings confirm the assumption of the authors of [7] that the condensed-phase temperature is "stabilized" not by phase transitions or Al_2O_3 dissociation, but by radiative heat losses that increase significantly with the rise of temperature due to a heavy dependence of radiative characteristics of Al_2O_3 particles on the temperature. It has been established that the type of the dependence of $T(\tau)$ prevailing during combustion of the aluminum-air flame and of aluminum-iron oxide thermite is the same. The maximum temperature is reached in both cases within the same time of ≈ 0.2 s [9]. The dependence of $T(\tau)$ for a pulsed aluminum-air flame is shown in figure 5.

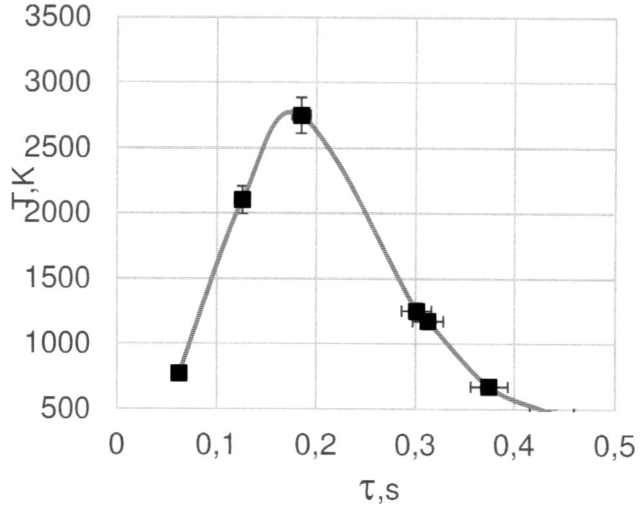

Figure 5. Temperature change during an intense flashю PMF: ASD-4 + PA-4 (in a ratio of 10: 2).

Figure 6 shows a calculated curve of integrated radiation flux density prevailing during an intense aluminum-air flame flash. As can be seen, the shape of the dependence curve $F(\tau)$ is in agreement with the dependence $T(\tau)$ of the aluminum-air flame burning during an intense flash (see figure 5).

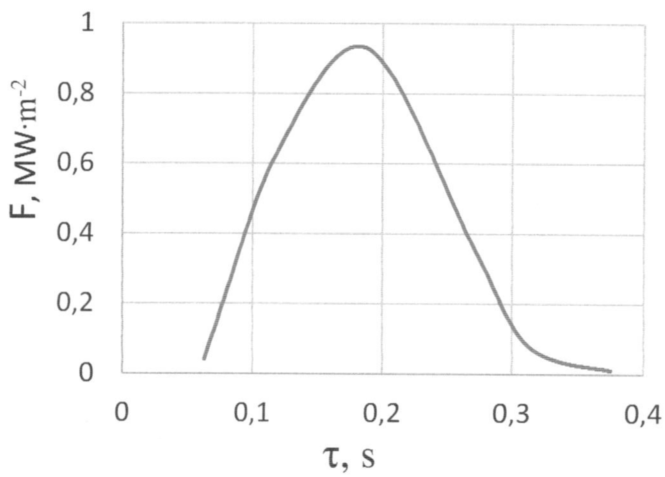

Figure 6. Integral radiation flux density during an intense flash. PMF: ASD-4 + PA-4 (in a ratio of 10:2).

The value of de-excitation time of the bulk of energy of a pulsed aluminum-air flame is consistent with the adjusted de-excitation time of the bulk of energy when burning stationary aerosolized Al particle clouds, which amounts to ~ $0.2 - 0.3\ s/kg^{1/3}$ [10]. It turned out that the dependence of integrated radiant energy flux density F on the temperature is of a linear type both for an aluminum-air flame in the range of 400...1723 K and a flame formed by heterogeneous products of aluminized combustible mixture firing [11].

In order to reveal the relationship between the radiant energy intensity of a pulsed aluminum-air flame and the wavelength, temperature measurements were performed during the pulse and interpulse time. Then the spectral distribution of radiant energy intensity I_λ in the wavelength range $\lambda = 1.0...4.5$ μm was calculated. The spectral distribution of radiant energy intensity for a pulsed aluminum-air flame is shown in figure 7. The nature of the change in spectral distribution of the intensity of radiation energy of a pulsed aluminum-air flame in the investigated wavelength range correlates with the nature of the change I_λ of a stationary aluminum-air flame [7].

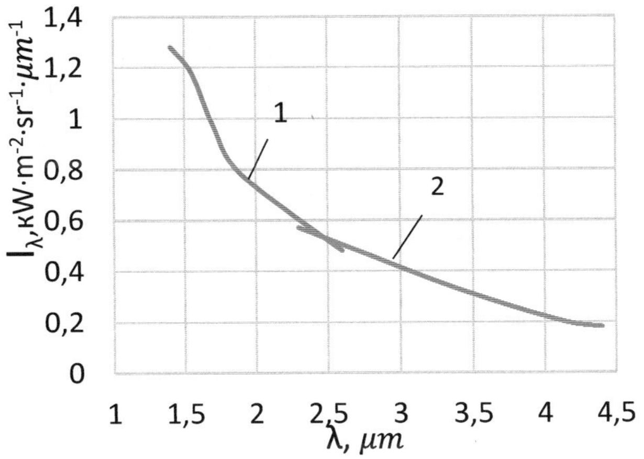

Figure 7. Spectral distribution of radiant energy intensity for a pulsed aluminum-air flame I_λ in the wavelength range $\lambda = 1.0...4.5$ μm:1-pulse;2- pause; PMF: ASD-4 + PA-4 (in a ratio of 10: 2).

By varying the operating mode of the apparatus (changing the pulse and interpulse times) it has been established that the energy emitted by the flame in the investigated spectral range of $\lambda = 1.0...4.5$ μm is proportional to powdered aluminum consumption per pulse. Figure 8 shows the dependence of radiated energy ε on the consumption G_{Al} of powdered Al per pulse.

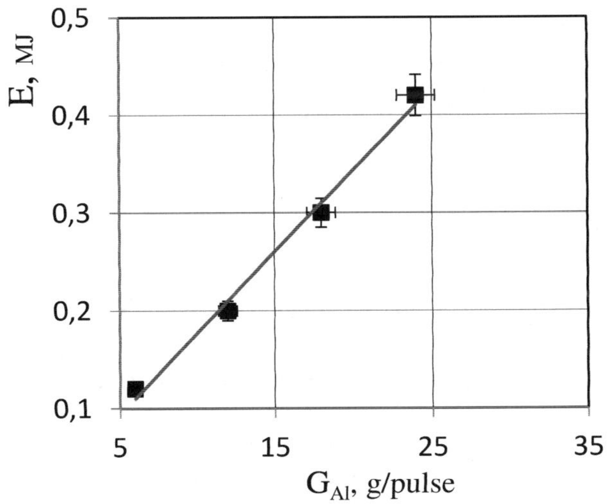

Figure 8. Dependence of the radiated flame energy E on the consumption of aluminum powder per pulse G_{Al} in the wavelength range $\lambda = 1.0...4.5$ μm: PMF: ASD-4 + PA-4 (in a ratio of 10:2).

The established linear relationship between the energy emitted by a pulsed aluminum-air flame and the powder quantity is in good agreement with the data obtained according to [7]. By varying the work mode of the apparatus, two reacting volumes of aerosolized Al particles have been obtained simultaneously: one with a radiation power of 6 kW and the other with 2.5 kW.

4. Conclusion

The results of the above investigations allowed implementing a pulse-combustion mode in an aerosolized aluminum particle flow. Based on the obtained values of the local time of Al particles residence in a sudden-expansion combustion chamber model, a full-sized power plant combustor has been developed. The results of the tests performed confirmed a reliable power plant operation in the pulse mode. Plant work mode variation admits of controlling the energy, power and the dimensions of a pulsed aluminum-air flame.

It has been ascertained that the nature of the relationship between the integrated radiation flux density of a pulsed aluminum-air flame and the time corresponds to that of the temperature change. Spectral distribution of radiant energy intensity of a pulsed aluminum-air flame I_λ in the wavelength range $\lambda = 1.0...4.5$ μm was determined. The dependence of the emitted energy of flame E on the Al powder consumption per pulse G_{Al} in the wavelength range $\lambda = 1.0...4.5$ μm was established.

The investigation results of the present work can be applied to create pulse power plants fired with powdered metal combustible, and also simulate radiative and gasdynamics processes related to the entry into Earth's atmosphere and ground impact of asteroids and meteorites.

References

[1] Chao Li, Chunbo Hu, Xin Xin, Yue Li and Haijun Sun 2016 Experimental study on the operation characteristics of aluminum powder fueled ramjet *Acta Astronautica* **129** 74-81

[2] Krykov A Yu, Malinin V I and Yarushin S G 2005 Opportunity estimate of industrial production of ultrafine powders by a burning of gas suspensions of powder materials *Rocket Motors and Problems of Space Exploration* (Moscow: Torus Press) 433-43

[3] Yue Li, Chunbo Hu, Xiaofei Zhu, Jiaming Hu, Xu Hu, Chao Li and Yupeng Cai 2019 Experimental study on combustion characteristics of powder magnesium and carbon dioxide in rocket engine *Acta Astronautica* **155** 334-49

[4] Egorov A G and Pivneva S V 2006 Organization of the working process in combustion chambers of new generation propulsion and power plants *Bulletin of the Samara State Aerospace University named after S.P. Queen* **2** 381-6

[5] Arkhipov V V, Divnov I I, Zotov N I *et al.* 1966 Research of radiation characteristics of emitters based on combustion of air suspension of aluminum powders *Combustion and explosion physics* **32** 67-71

[6] Pohil P F, Belyaev A F and Frolov Yu V 1972 *Combustion of powdered metals in active media* (The Science)

[7] Poletaev N I and Florko A V 2007 Radiation characteristics of an aluminum dust flame Condensed phase *Combustion and explosion physics* **43** 49-58

[8] Belinsky N I, Kozitsky S V and Florko A V 1985 High-speed spectral device for studying the radiative characteristics of dispersed combustion products *Physics of aerodispersed systems* **28** 38-42

[9] Maltsev V M, Maltsev M I and Kashporov L Ya 1977 *Main characteristics of combustion* (Publishing house: Chemistry)

[10] Anderzhanov E K, Divnov I I, Zotov N I *et al.* 1989 Dependence of foil electric explosion parameters and of a subsequent discharge in the air on the input power *Theoretical physics* **59** 17-23

[11] Kuzmin V A 2004 *Thermal radiation in engines and power plants* (Polex Company Ltd)

Selection of an effective catalyst for the stage isoamyl alcohol dehydration during synthesis isoprene from isopentane

E K Aminova and A A Islamutdinova

Ufa State Petroleum Technological University, Branch in the Sterlitamak, October Pr., 2, Sterlitamak, Republic of Bashkortostan, 453100, Russia

E-mail: k-elmira-k@yandex.ru

Abstract. The annual growth in the consumption of isoprene rubbers dictates manufacturers to searching for the new methods of its production. An intermediate product in the production of isoprene rubbers is the synthesis of a monomer - isoprene. The main industrial methods for its production in Russia are synthesis processes from isobutylene and formaldehyde and two-stage dehydrogenation of isopentane. The article analyzes and selects an effective catalyst for the stage of isoamyl alcohol dehydration in the synthesis of isoprene from isopentane.

Synthetic rubbers are in demand every year all over the world, thanks to the ability to regulate their performance properties.

The cost of obtaining them is approximately 25 to 30% of their cost price, which indicates the feasibility of their multi-tonnage production. Thus, all scientific research is aimed at improving this technology, which consists either in the production of a high-quality product using a low-cost technology, or from cheap and affordable raw materials, or using a waste-free method of production.

Large-tonnage production of 2-methylbutadiene-1,3 began in 1959 the company "Shell" in TORRANCE (USA), which consists in the allocation of isoamylene fraction of oil, followed by catalytic dehydrogenation in his 2-methylbutadiene is 1.3.

Currently, in Russia, is still common processes monomer is a way of getting from 2-methylbutanol and matamala and catalytic dehydrogenation of 2-methylpentane in two stages.

the main industrial methods of its production are the processes of synthesis from isobutylene and formaldehyde and two-stage dehydrogenation of isopentane. Separation from the pyrolysis fraction gives low productivity for the finished product. Therefore, the search for effective methods for obtaining a valuable monomer in the world practice remains an urgent task for any production of synthetic rubbers of various brands.

It is worth noting that the existing methods for producing isoprene have not changed in the last fifty years, these are the production of 2-methylpropene and formic aldehyde, extraction from pyrolysis fractions of oil, catalytic dehydrogenation of isobutylene in one stage, the method of ethine and primary propyl alcohol, and two-stage dehydrogenation of the pentane isomer.

The paper studies the liquid-phase method of synthesis of 2-methylbutaiene-1,3 by oxidative dehydrogenation of an isopentane-isoamylene mixture, as well as the one-stage method, where the raw materials are diluted with steam and hydrogen, the method of condensation of betenes with synthesis gas, accompanied by further decomposition of 2-methylbutanal. For the first time, various brands of catalysts for the dehydration of tertiary alcohols obtained by domestic and imported manufacturers were tested

Despite the advantages described above, the technology using iron-potassium contacts is expensive and energy-intensive. To make the technology less expensive, it is necessary to increase the degree of conversion of methylbutenes and selectivity for isoprene. This technology requires continuous improvement in order to reduce the cost of isoprene in a competitive market. However, the control levers of this technology are almost exhausted, and isoprene synthesized by the two-stage dehydrogenation of isopentane has the highest cost. Therefore, the development of less energy-intensive technologies for the production of isoprene is a key task of Russian researchers. Of particular interest to scientists is the process of epoxidation of isobutylene. The advantage of the method: cost-effective, liquid-phase and implemented at moderate temperatures. The dehydration stage of this technology is intended for the production of isoamylenes from isoamyl alcohol, which is formed in the epoxidation stage as a by-product.Two-stage catalytic dehydrogenation of 2-methylbutane is a domestic method.It widely known and implemented in Russia [1,2]. Process conditionsare: temperature from 570 to 630 ° C, ratio C5-fraction: water vapor = 1: 20, oxide iron-potassium contact, self-regenerating [3,4].

Software materials are widely used for dehydrogenation processes in the modern world, in particular, products of mathematical modeling of dehydrogenation reactions, which make it possible to select the process conditions that ensure high conversions and selectivity of the formation of reaction products [5, 6].

After having studied the regularities of the chemical transformations of the catalytic dehydrogenation process [7, 8], it was identifired that they are of a qualitative nature.

Based on the results of processing the literature data, it can be noted that one of the promising directions for the production of isoprene monomer is the method of eposidation of 2-methylbutene-2 with the production of 2-methyl-2,3-epoxybutane followed by its dehydration into isoprene. in the epoxidation stage, in addition to 2-methyl-2,3-epoxybutane, tertiary amyl alcohol is formed. The qualified method of using a by-product - tertiary amyl alcohol (TAS) by converting it into the main raw material - 2-methylbutene-2. Itmakes this method even more attractive. Therefore, in this work, we study the possibility of carrying out the process of dehydration of tertiary amyl alcohol into 2-methylbutene-2 with the latter returning to the epoxidation stage.

Thus, this scientific work is based on the searching and selection of an effective catalyst for the dehydration of isoamyl alcohol and laboratory tests of selected catalyst samples in the process of TAS dehydration.

We selected inorganic acids and ion-exchange resins as catalysts for this process.

Currently, well-known manufacturers of catalysts based on ion exchange resins are:

- Lewatit (Levatite).
- Amberlite (Amberlite).
- Purolite.
- Dowex (Davex).
- Tulson (Tulsion).
- Relite.
- Resinex (Rezinex).
- With a smaller assortment: Bybron, Diaion, Granion, Jacobi, Polyion, Resindion, Resintech, Trilite.

Domestic producers of cation exchangers are OOO PO Tokem (Kemerovo), catalysts - OOO PKF Polyplast (Ishimbay), ZAO RTN (St. Petersburg).

To study the initial characteristics and tests, cation exchangers with a macroporous structure were selected: "Pyurolyte", T-103, T-103W (MTBE), Resinex, T-8052MP and the domestic catalyst KU-2-8.

The most effective brands were identified by laboratory tests in the dehydration of tertiary amyl alcohol using the catalysts and cation exchangers proposed above based on ion-exchange resins.

Table 1. Characteristics of cation exchangers.

Nameindicators	CCationite K "Pyurolyte"	CCationite TT-103	CCationite TT-103W (MTBE)	CCationite RResinex	CCationite T -8052MP	Catalyst KKU-2-8
Manufacturer	GGreat Britain	PPO «Tokem» LLC	PPO «Tokem» LLC	JJacobi Carbons	TTulsion	RTN CJSC
Appearance	Spherical grains up to 1.3 mm					Granules
Mass fraction of moisture, %	556.3	662.2	555.5	558.8	554.4	54.4
Catalytic activity (TAS conversion), %	994.4	996.0	994.8	882.7	990.5	886.1
Full static exchange loan. (PSOE), -mmol / g;	55.3	55.6	55.6	55.1	55.7	33.9
- mol / l	11.7	00.5	22.0	11.8	22.0	11.1
Mass fraction of free acid in terms of sulfuric acid, %	00.2	00.2	00.9	00.3	00.3	00.5
Specific volume, cm^3 / g	33.1	110.5	22.8	22.8	22.8	33.5
Bulk density, g / cm^3	00.78	00.25	00.81	00.86	00.78	00.62

In terms of activity indicators, the catalysts can be arranged in descending order of T-103, T-103W, Purolyte, T-8052MR, KU-2-8, Resinex.

As you can see in table 1, the T-103 cation exchanger differs from other samples in its high specific volume (highly swellable), which makes it not very attractive for use in an industrial reactor. When loaded into an industrial reactor, the mass of the cation exchanger (in terms of dry weight) will be less.

The highest value of PSOE of the studied cation exchangers in terms of mol / l was shown by the samples T-103W (MTBE) produced by OOO PO Tokem and the cation exchanger T-8052MP produced by Tulsion.

To carry out the process of dehydration of tertiary amyl alcohol into 2-mutilbutene-2, the "Purolyte" catalysts and the domestic cation exchanger of the T brand are effective and technological.

During the experiment, the total static exchange capacity was calculated and measured by titrimetric determination of the concentration of sulfogroups-SO3H in the catalyst.

The essence of this method is to determine the amount of sodium ions absorbed from a solution of sodium hydroxide by a catalyst in contact with a constant volume of alkali.

The amount of sodium ions absorbed by the catalyst is determined by the difference in the concentrations of the alkali solution before and after contact with the catalyst. The concentration of the NaOH solution is determined by volumetric titration with an acid solution in the presence of an indicator.

In addition, the catalytic activity of the catalyst was determined, which was evaluated by the degree of dehydration of tert-amyl alcohol (TAC) under static conditions at the boiling point of the TAC - water azeotrope (at atmospheric pressure), measuring the amount of 2-MB-2 released during dehydration.

Thus, we have tested the selected catalysts and selected the most effective sample. Based on the analysis of scientific literature, as well as experimentally obtained data, the following conclusions can be drawn:

- T-103 cationite differs from other samples by its high specific volume (highly swellable), which makes it unattractive for use in an industrial reactor. When loaded into an industrial reactor, the mass of the cationite (in terms of dry weight) will be less.
- the highest value of the PSOE of the studied cationites in terms of mol/l was shown by the samples T-103W (MTBE) produced by LLC PO "Tokem" and T-8052MR cationite produced by Tulsion.
- -for the process of dehydration of tertiary amyl alcohol in 2-mutilbutene-2, the catalysts "Purolite" and domestic cationite of the T-103W brand (MTBE) produced by LLC PO "Tokem" in Kemerovo are effective and technologically advanced.
- the domestic catalyst in terms of quality is not inferior to the imported analogue of the Purolite brand, for the industrial implementation of the process, the use of a domestic catalyst of the T-103W brand (MTBE) is recommended.

References

[1] Pavlov S Yu and Surovtsev A A1997 Prospects for the development of production of isoprene and polyisoprene rubber *Chemical industry* **7** 219
[2] Ogorodnikov S K and Idris G S 1973 *Production of isoprene* (Leningrad: Chemistry) 296
[3] Gilmutdinov N R, Hafizov A V, Korshunov A I *et al.* 1996 *Oil refining and petrochemistry* **9** 29
[4] Aminova J I and Bazhenov Yu P 1998 *Industry SK* **1** S Z
[5] Slinko M G 1999 Principles and methods of technology of catalytic processes *Teor. fundamentals of Chemical Technology* **ZZ(5)** 528-38
[6] Slinko M G 2000 *Scientific foundations of the theory of catalytic processes and reactors* **41(6)** 933-46
[7] Balandin A A, Bogdanova O K and Shcheglova A P 1965 Obtaining isoprene by catalytic dehydrogenation of isopentanes *AN SSSR* **162** 1320-2
[8] Bogdanova O K, Shcheglova A P and Balandin A A1959 Catalytic dehydrogenation of isopentane-isopentene mixtures *Izvestiya AN SSSR. Otd. Chemical sciences* **2** 350-2

Gas content of a liquid in threaded roughness recesses and its effect on heat and mass exchange

N A Voinov[1], D A Zemtsov[1], A V Bogatkova[1] and N V Deryagina[2]

[1] Reshetnev Siberian State University of Science and Technology, 31, Krasnoyarsky Rabochy Av., Krasnoyarsk 660037, Russian Federation
[2] Siberian Federal University, 79, Svobodny pr., Krasnoyarsk 660041, Russian Federation

E-mail: NGavgolenko@sfu-kras.ru

Abstract. This paper proposes a mechanism of gas bubble formation in a down-flowing liquid film as a result of reduced pressure in circulation zones of threaded roughness recesses. Simulation modelling on the Comsol Multiphysics software was used to calculate circulation loops, liquid velocities, and pressure. Thus, the accepted of gas bubble formation was justifying. We experimentally determined gas content along film thickness and its average value, and proposed an equation for calculating gas content along liquid layer thickness. The effect of gas content on the average thickness of a liquid film, as well as its heat and mass transfer, was confirmed.

1. Introduction

The use of falling film tubular apparatus is widely encountered in the chemical, petrochemical and other industrial process equipment, where heat and mass transfer phenomena take places, such as absorber, column, falling film reactor, and liquid film evaporator [1-6].

The location of large-scale threaded roughness on the inner surface of tubes in the machine with a down-flowing film (figure 1) contributes to an increase in the interphase surface and liquid output [7-10]. Threaded roughness provides the translational and rotational motion of a liquid, which allows maintaining a stable film flow on the film-forming surface. The flow around the roughness projections causes circulation loops [11-13], which intensifies liquid mixing and ensures gas bubble formation in a liquid.

Various devices have been developed on the basis of this phenomenon. For example, gas-liquid bioreactors in which pipes with helical roughness are installed. This made it possible to increase their productivity and reduce operating costs.[5,14]. We tested the design of a scrubber of high gas and liquid throughput with its tubes having threaded roughness [15]. An anti-foaming agent and a flotator [16] with low power consumption were developed. We additionally studied a barometric condenser in which a gas condenses on the surface of a liquid film, which is actively mixed due to the presence of threaded roughness, which allowed intensifying heat transfer during vapour-gas medium condensation.

A design of a film heat exchanger [17] made of profiled plates forming cylindrical channels with a stable flow of a liquid film provided by threaded roughness was developed.

One of the factors limiting the widespread introduction of the studied film machines with large-scale threaded roughness is insufficient knowledge of the hydrodynamic parameters of a down-

IOP Conf. Series: Materials Science and Engineering 1181 (2021) 012032 doi:10.1088/1757-899X/1181/1/012032

flowing liquid film, and the mechanism of gas bubble formation and distribution in roughness recesses, which affects both the flow hydrodynamics and heat and mass exchange parameters.

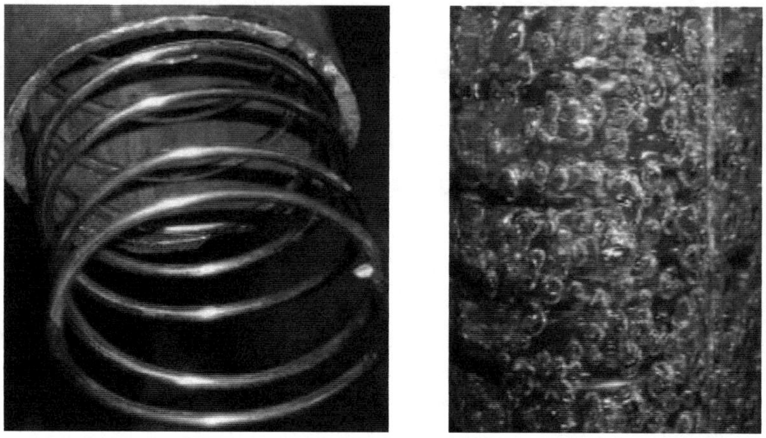

(a) (b)

Figure 1. Tubular packing of a film apparatus with a roughness (a) and a gas-liquid layer on the surface of the packing (b).

We suppose that gas bubbles in a liquid film are formed due to the movement of gas located above the film into the recess of the circulation zones formed behind the roughness projection.

To confirm this hypothesis, we conducted experimental studies and simulated the parameters of a down-flowing liquid film along threaded roughness turns.

2. Results and discussion

The studies were conducted using in tubes with an internal diameter of 0.04 and 0.05 m, up to 2.0 m long. A water film flowed down the inner surface of the tubes and absorbed air from the atmosphere. Threaded roughness was made of wires of diameter h = 1.5-4.0 mm in the form of a spiral of along the tube with parameter s/h = 3-12, where s is the distance between the spiral adjacent turns.

Average flow and angular velocities were determined by the tracer method consisting in injecting dye into a down-flowing film while simultaneously surveying the tracer motion trajectory in slow mode. With the loop closed, the survey was stopped and the time the tracer took following the trajectory determined.

The magnitude of gas content (a gas volume fraction in a liquid) along the height of the liquid layer was determined in the recess using the volumetric method by removing the gas-liquid mixture from a specific distance from the tube wall through a hollow needle equipped with a small transparent pipe of diameter 3.0 mm.

The liquid surface tension was changed by adding ethanol to the tubular nozzle, and ranged from 760 to 550×10^{-4} kg/s^2.

Simulation modeling was carried out using the interactive environment Comsol Multiphysics based on partial differential equations [18]. The calculation was carried out using the "Fluent" package with automatic generation of the mesh and its refining in the area of changing modes and directions of movement.

According to the data obtained, we established water film flow modes on the surface of the threaded roughness tube (figure 2). With small liquid flow rates, there is a jet flow mode where the liquid escapes from the surface of the threaded roughness turns and forms jets, and gas bubbles are locally observed in the roughness recesses (figure 2a). With an increase in liquid flow rates, an annular film flow mode occurs (figure 2b and figure 2c). The roughness recesses have accumulated gas bubbles that are formed at a specific distance (equal to δ_o) from the tube wall. The thickness of the liquid layer in the tube δ_o is influenced by the volumetric flow rate of the liquid and the dynamic viscosity coefficient. When the tube cavity is flooded with a liquid, a flooding mode occurs (figure 2d), and gas bubbles are washed out and taken away with the liquid flow.

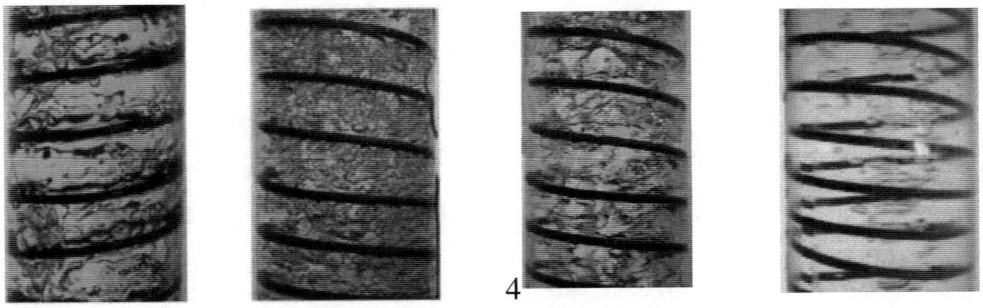

a) b) c) d)

Figure 2. Modes of a falling water film on the inner surface of a pipe with roughness: *d*= 40 mm, *h*= 3 mm, μ= 0.0015 Pa·s, *s/h* = 10.

Based on the hypothesis on the gas bubble in the film are formed in a roughness recess, flow parameters were simulated. Figure 3 shows the research results. According to the data obtained, circulation vortices are observed behind the roughness projection (figure 3a). The formation of these vortices and the occurrence of reverse currents (figure 3b) are influenced by the flow rate, the parameter *s/h*, the increase in the roughness projection and the dynamic viscosity coefficient of the liquid.

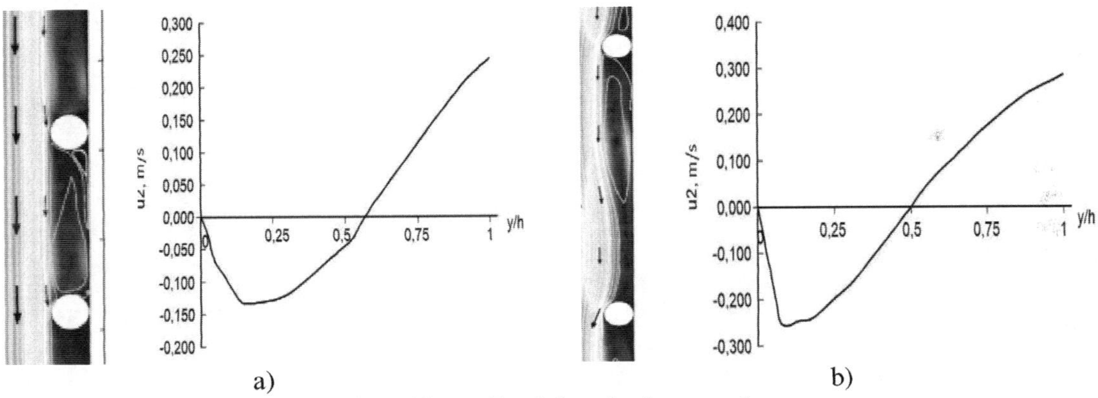

a) b)

D_{in}= 45mm, L= 2.0 m, h=4 mm, t=1 s.

Figure 3. Distribution of fluid velocity in the cavity over the pipe section a) s=15 mm, b) s=38 mm.

It was also established that in the roughness recesses, at a distance of y_l from the wall, local circulation zones are formed with the lowest pressure as compared to the gas pressure above the film surface. With an increase in the roughness projection height, roughness parameter *s/h* and liquid average flow rate, the pressure drop ΔP between the gas above the film surface and the pressure in the circulation zone increases (figure 4a). At that, the distance y_l from the tube wall to the local zone centre increases (figure 4b).

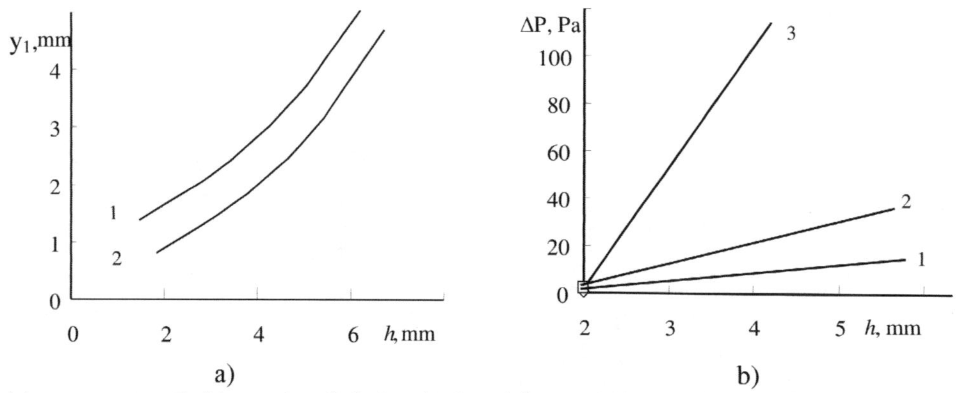

(a) parameter *s/h* (b) at *x/s* = 0,5 a) calculated lines *(1–3)*: *1)* *s/h* =4, u_{av} = 0,8 m/s; *2)* *s/h* =10, u_{av} = 0,8 m/s; *3)* *s/h* =7, u_{av} = 1,0 m/s; б) *h*= 3 mm, u_{av} = 0,8 m/s.

Figure 4. Dependence of the pressure drop on the height of the protrusion of the round profile: (a) and ratio *x/s* (b): a) *s/h* = 6; *x/s*=0,5; calculated lines *(1-2):1) uav*=1,0 m/s;*2)* u_{av}=0,5 m/s; b) *h* = 4 mm; *s/h* = 10; u_{av} = 0,5 m/s.

The fact that the y_1 design value and the δ_o experimental value coincide allows asserting that the local zones with a reduced pressure of 4-100 Pa, Figure 4a are gas bubble formation centres in the roughness recesses.

The experimental change in the local gas content along the thickness of the down-flowing liquid film is shown in figure 5 As the liquid flow rate increases leading to an increase in the liquid film thickness, the gas content decreases.

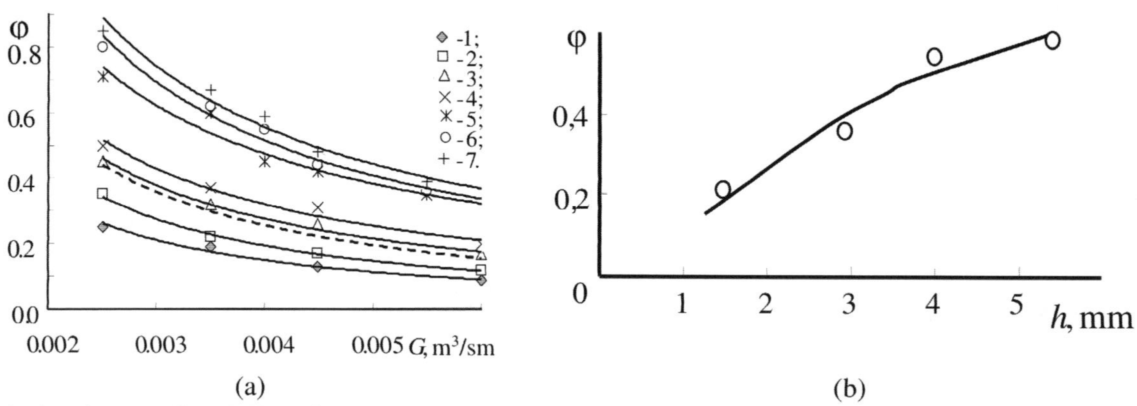

At *h* = 4 mm, *s/h* = 6. Experimental points *(1–4)*: μ = 0.00134 Pa·s;*1)* *y* = 1 mm; *2)* *y* = 2 mm; *3)* *y* = 4 mm; *4)* *y* = 5 mm; *(5–7)*: σ = 550·10⁻⁴ kg/s²; *5)* *y* = 1,5 mm; *6)* *y* = 3,5 mm; *7)* *y* = 4,5 mm.

Figure 5. Dependence the gas content over the thickness of the liquid layer depending on the density of irrigation (a) and the average gas content on the height of the protrusion (b) At Re = 5600, *s/h* = 6, μ= 0,0013 Pa·s (a) and average gas content (b) at *d* = 51 mm; *s/h* = 5; *L*= 2 m; μ= 0,001 Pa·s; *G*= 0,0055 m³/(cm).

Based on experimental data, a dependence was obtained to calculate the gas content along the liquid layer thickness in the recess in the form of

$$\varphi i = A \cdot G^{-1,2} \, (y/h)^{0,4} (\sigma/\sigma_o)^{-1,5}, \quad (1)$$

When A= 4 $\cdot 10^{-4}$, at μ = 0,00035 – 0,0014 Pa·s, σ = (760–500)·10^{-4} kg/s², y/h < 1,25.

It is recommended to calculate the thickness of the wall single-phase layer y1 corresponding to the minimum value y in equation (1) using the numerical modelling method described above.

Figure 6 shows the average velocity measurement results of the gas-liquid medium down-flowing film

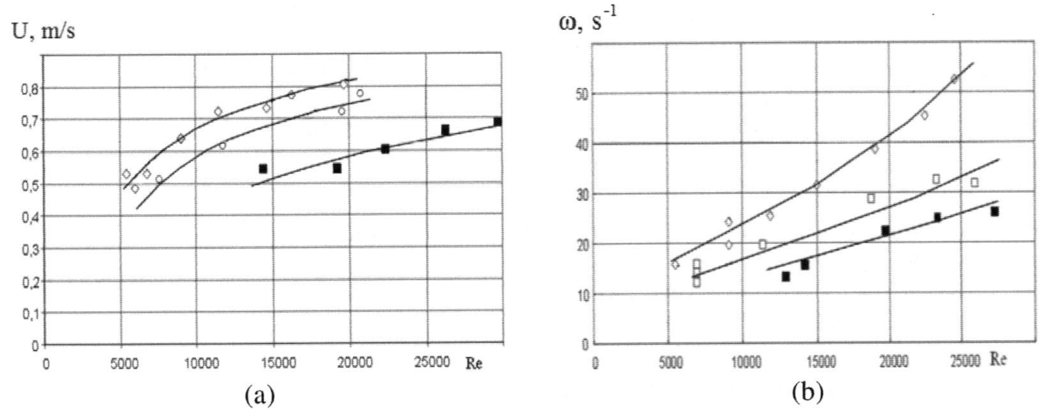

Figure 6. Change in the average speed of the film flow down (a) and the angular speed of rotation (b) from the Reynolds number at d = 46 mm; h = 4mm; s / h = 4-6. Experimental points at water temperature: ◊ – 12 °C; □ – 26 °C; ■ – 60 °C.

According to the data, the average film velocity varies from 0.4 to 0,8 m/s and grows as the Reynolds number of the liquid increases. An increase in the dynamic water viscosity coefficient leads to an increase in the liquid flow rate, which contradicts known information [19-21]. This is apparently due to the presence of gas bubbles in the liquid [22] and their effect on the viscous friction force. The liquid rotation rate increases with an increase in the supplied liquid flow rate and the dynamic viscosity coefficient, and is 10-50 s^{-1} (figure 6b).

As established experimentally, the presence of gas content in the film increases the average film thickness by 1.3 - 1.5 times, which is consistent with the data .

The results of the heat transfer study during heating in the film of liquid [16] flowing down the heat exchange surface with threaded roughness also show a significant effect of the gas content on heat transfer, which is due to the formation of gas bubbles in the liquid and, as a result, a decrease in the thermal conductivity of the gas-liquid medium.

During the absorption in the film of water flowing down the threaded roughness surface [10], in connection with gas bubble formation, gas contact with the film surface and gas bubble interaction in the liquid film are considered mass transfer. Using this approach, the mass transfer coefficient between the gas bubbles and the liquid is (0.8-2.0)10^{-3} m/s, and the magnitude of the mass transfer coefficient in the film is (2 - 8)10^{-3} m/s.

The results obtained are used to calculate and design tubular film machines.

Acknowledgment
Studies were carried out in the laboratory «Deep processing of plant raw materials» of Reshetnev Siberian State University of science and technology.

This work was supported by the Ministry of Science and Higher Education of the Russian Federation within the framework of State Assignment of the «Technology and equipment for the chemical processing of biomass of plant raw materials» project FEFE-2020 0016.

References

[1] Salvagnini W M and Taqueda M E S 2004 A falling-film evaporator with film promoters *Ind. Eng. Chem. Res.* **43** 6832-

[2] Ma X, Chen J, Li S, Sha Q, Liang A, Li W, Zhang J, Zheng G and Feng Z 2003 Application of absorption heat transformer to recover waste heat from a synthetic rubber plant *Appl. Therm. Eng.* **23** 797-806
[3] Leu J S, Jang J Y and Chou Y 2006 Heat and mass transfer for liquid film evaporation along a vertical plate covered with a thin porous layer *Int. J. Heat Mass Transf.* **49** 1937-45
[4] Battisti R, Machado R A F and Marangoni C 2020 A background review on falling film distillation in wetted-wall columns: From fundamentals towards intensified technologies *Chem. Eng. Process. - Process Intensif.* **150** 107873
[5] Voinov N A, Zhukova O P and Nikolaev A N 2013 Intensification of yeast biomass culturing in a film bioreactor *Foods Raw Mater.* **1** 57-66
[6] Guichet V and Jouhara H 2020 Condensation, evaporation and boiling of falling films in wickless heat pipes (two-phase closed thermosyphons): A critical review of correlations *Int. J. Thermofluids* **1-2** 100001
[7] Vojnov N A, Konovalov N M and Nikolaev N A 1993 Features of free flowing of a liquid film along the inner and outer surfaces of pipes with a regular helical roughness *Theor. Found. Chem. Eng.* 638
[8] Kohrt M, Ausner I, Wozny G and Repke J U 2011 Texture influence on liquid-side mass transfer *Chem. Eng. Res. Des.* **89** 1405-13
[9] Vojnov N A, Nikolaev A N and Vojnova O N 2009 Hydrodynamics, warm mass transfer in film bioreactors *Chem. plant raw mater.* **4** 183-93
[10] Voinov N A and Nikolaev N A 2008 *Film tubular gas-liquid reactors* (Moscow: Otechestvo)
[11] Cao B Y, Chen M and Guo Z Y 2006 Effect of surface roughness on gas flow in microchannels by molecular dynamics simulation *Int. J. Eng. Sci.* **44** 927-37
[12] Davies J T and Warner K V. 1969 The effect of large-scale roughness in promoting gas absorption *Chem. Eng. Sci.* **24** 231-40
[13] Brumfield L K, Houze R N and Theofanous T G 1975 Turbulent mass transfer at free, gas-liquid interfaces, with applications to film flows *Int. J. Heat Mass Transf.* **18** 1077-81
[14] Voinov N A, ZHukova O P, Temerov F V and Alashkevich Y D 2015 Film gas-liquid bioreactor *Chem. Ind.* **92** 156-9
[15] Voinov N A, Nikolaev N A and Eremenko N A 2006 Film tubular apparatus with helical roughness *Chem. Ind.* **83** 285
[16] Voinov N A and Nikolaev A N 2011 *Heat Withdrawal in Liquid Film Flow* (Otechestvo)
[17] Vojnov N A and Zemtsov, Denis Andreevich Zhukova, Ol'ga Petrovna Alashkevich J D 2006 RU 2 569 118 1-10
[18] Abd Ali M K 2014 CFD simulation of bubbly flow through a bubble column *Int. J. Sci. Eng. Res* **5** 904-10
[19] Laptev A G, Basharov M M, Lapteva E A and Farakhov T M 2015 *Models of interphase transport and calculation of process efficiency*
[20] Laptev A G and Lapteva E A 2015 The model of heat and mass transfer in rough and irrigated ducts *Thermophys Aeromechanics* **22** 435-40
[21] Picioreanu C, Van Loosdrecht M C M and Heijnen J J 2000 A theoretical study on the effect of surface roughness on mass transport and transformation in biofilms *Biotechnol. Bioeng.* **68** 355-69
[22] Salmi T O, Mikkola J-P and Warna J P 2010 *Chemical reaction engineering and reactor technology* (CRC Press)

AUTHOR INDEX

Institute of Physics
Dirac House, Temple Back
Bristol BS1 6BE UK

ISSN: 1757-8981
ISBN 978-1-7138-4641-3

CURRAN ASSOCIATES INC.
proceedings.com